Acquired Speech and Language Disorders

Acquired Speech and Language Disorders

A neuroanatomical and functional neurological approach

B.E. Murdoch

CHAPMAN AND HALL

London • New York • Tokyo • Melbourne • Madras

UK   Chapman and Hall, 11 New Fetter Lane, London EC4P 4EE

JAPAN   Chapman and Hall Japan, Thomson Publishing Japan,
            Hirakawacho Nemoto Building, 7F, 1-7-11 Hirakawa-cho,
            Chiyoda-ku, Tokyo 102

AUSTRALIA   Chapman and Hall Australia, Thomas Nelson Australia,
            480 La Trobe Street, PO Box 4725, Melbourne 3000

INDIA   Chapman and Hall India, R. Sheshadri, 32 Second Main Road,
            CIT East, Madras 600 035

First edition 1990

© 1990 B.E. Murdoch

Typeset in 10/12 Cheltenham Book by
Leaper & Gard Ltd, Bristol
Printed in Great Britain by
T.J. Press (Padstow) Ltd, Padstow, Cornwall
Distributed exclusively in North America by Paul H. Brookes
Publishing Co. Inc., P.O. Box 10624, Baltimore, MD 21285, USA

ISBN 0 412 33440 2 (PB)

British Library Cataloguing in Publication Data
Murdoch B.E. (Bruce E)
    Acquired speech and language disorders.
    1. Men. Speech disorders. Neurophysical aspects
    I. Title
    616.85507

ISBN 0 412 33440 2

Contents

Preface

1 Neuroanatomical framework of speech and language   1
1.1   The central nervous system   2
1.1.1   The brain   7
1.1.2   The spinal cord   27
1.2   The peripheral nervous system   29
1.2.1   The cranial nerves   31
1.2.2   The spinal nerves   31
1.2.3   The autonomic nervous system   32
1.3   The ventricular system   34
1.4   The meninges   36
1.5   The cerebrospinal fluid   38
1.6   The blood supply to the brain   40
1.6.1   Arterial blood supply   40
1.6.2   Venous blood supply   44
1.6.3   The blood–brain barrier   45
1.7   Speech and language centres of the brain   45
1.8   Neurologically-based communication disorders
            – definitions   48
1.9   Neuropathological substrate of neurogenic speech and
            language disorders   50
1.9.1   Cerebrovascular disorders   51
1.9.2   Neoplasms   54
1.9.3   Head trauma   56
1.9.4   Degenerative disorders   56
1.9.5   Toxic disorders   57
1.9.6   Demyelinating disorders   57
1.9.7   Infectious disorders   58

2 Bostonian and Lurian aphasia syndromes   60
2.1   Models of language – a brief history   60
2.2   Classification of aphasia   67
2.2.1   The Boston Classification System   69

2.2.2	The Lurian Classification System	71
2.3	Bostonian aphasia syndromes	71
2.3.1	Broca's aphasia	71
2.3.2	Wernicke's aphasia	73
2.3.3	Conduction aphasia	76
2.3.4	Global aphasia	78
2.3.5	Transcortical motor aphasia	80
2.3.6	Transcortical sensory aphasia	81
2.3.7	Isolation aphasia (mixed-transcortical aphasia)	82
2.3.8	Anomic aphasia	84
2.3.9	Sub-cortical aphasia syndromes	85
2.4	Lurian aphasia syndromes	85
2.4.1	Efferent (kinetic) motor aphasia	86
2.4.2	Frontal (dynamic) aphasia	87
2.4.3	Pre-motor aphasia	87
2.4.4	Afferent (apraxic) motor aphasia	88
2.4.5	Sensory (acoustic) aphasia	88
2.4.6	Acoustico-mnestic aphasia	89
2.4.7	Semantic aphasia	89
2.5	Methods of lesion localization in aphasia	90
2.5.1	Computed tomography	91
2.5.2	Magnetic resonance imaging	93
2.5.3	Positron emission tomography	95
3 Sub-cortical aphasia syndromes		97
3.1	Neuroanatomy of the striato-capsular region and thalamus	97
3.1.1	Neuroanatomy of the striato-capsular region	98
3.1.2	Neuroanatomy of the thalamus	102
3.2	Blood supply to the striato-capsular region and thalamus	105
3.3	Aphasia associated with thalamic lesions	106
3.4	Aphasias associated with striato-capsular lesions	110
3.5	Role of sub-cortical structures in language	113
4 Speech–language disorders associated with traumatic head injury		120
4.1	Open versus closed injury	120
4.2	Complications of head injury	122
4.2.1	Skull fractures	122
4.2.2	Concussion, contusions and lacerations	123
4.2.3	Increased intra-cranial pressure	124
4.2.4	Vascular lesions	125

4.2.5	Cranial nerve lesions	128
4.2.6	Rhinorrhoea and otorrhoea	128
4.2.7	Post-traumatic epilepsy and post-traumatic vertigo	128
4.3	Mechanisms of head injury	129
4.4	Speech and language disturbances following head injury	130
4.4.1	Nature of the language disturbance following head injury	130
4.4.2	Language impairment and neurological status following head injury	135
4.4.3	Prognostic indicators for language function following head injury	136
4.4.4	Speech disorders following head injury	137
4.4.5	Mechanisms of recovery in head injury	138
4.5	Other neuropsychological sequelae of head injury	141

5 Language disorders subsequent to right-hemisphere lesions — 142

5.1	Lateralization of language function	142
5.2	Linguistic functions of the right hemisphere	143
5.2.1	Language symptoms of right-hemisphere damage	144
5.2.2	Hemispherectomy	152
5.2.3	Commissurotomy	153
5.2.4	Anatomical differences between the left and right hemispheres	156
5.3	Other neuropsychological sequelae of right-hemisphere damage	157
5.4	The role of the right hemisphere in recovery from aphasia	158

6 Language disturbances in dementia — 163

6.1	Types of dementia	164
6.1.1	Characteristics of Alzheimer's disease	166
6.1.2	Characteristics of Pick's disease	169
6.1.3	Characteristics of multi-infarct dementia	170
6.1.4	Characteristics of dementia in extra-pyramidal syndromes	171
6.1.5	Characteristics of Korsakoff's syndrome	172
6.2	Language disorders in cortical dementias	173
6.2.1	Language disorders in Alzheimer's disease	173
6.2.2	Language disorders in Pick's disease	179
6.3	Relationship between the language of dementia and aphasia	180

7 Neurological disturbances associated with aphasia — 184

7.1 Apraxia — 184
7.1.1 Ideomotor apraxia — 185
7.1.2 Ideational apraxia — 188
7.1.3 Limb-kinetic apraxia — 189
7.1.4 Constructional apraxia — 189
7.1.5 Dressing apraxia — 190
7.2 Apraxia of speech — 191
7.2.1 Clinical characteristics of apraxia of speech — 191
7.2.2 Differentiation of apraxia of speech from aphasia — 193
7.2.3 Differentiation of apraxia of speech from dysarthria — 194
7.3 Alexia and agraphia — 194
7.3.1 Alexia without agraphia — 195
7.3.2 Alexia with agraphia — 197
7.3.3 Frontal alexia — 198
7.3.4 Deep alexia — 199
7.3.5 Agraphia — 200
7.4 Agnosia — 201
7.4.1 Visual agnosia — 201
7.4.2 Auditory agnosia — 202
7.4.3 Tactile agnosia — 203
7.4.4 Special forms of agnosia — 203
7.5 Gerstmann syndrome — 203

8 Dysarthria associated with upper and lower motor neurone lesions — 205

8.1 Flaccid dysarthria (lower motor neurone dysarthria) — 207
8.1.1 Innervation of the speech mechanism — 208
8.1.2 Neurological disorders associated with lower motor neurone lesions — 215
8.1.3 Clinical characteristics of flaccid dysarthria — 219
8.1.4 Speech disorders in myasthenia gravis — 224
8.2 Spastic dysarthria (upper motor neurone dysarthria) — 224
8.2.1 Pyramidal and extra-pyramidal systems — 225
8.2.2 Neurological disorders associated with upper motor neurone lesions — 229
8.2.3 Clinical characteristics of spastic dysarthria — 230

9 Dysarthrias associated with extra-pyramidal syndromes — 234

9.1 Hypokinetic dysarthria — 235
9.1.1 Neurological disorders associated with hypokinetic dysarthria — 235
9.1.2 Clinical characteristics of hypokinetic dysarthria — 238

9.2   Hyperkinetic dysarthria   244
9.2.1   Quick hyperkinesias   245
9.2.2   Slow hyperkinesias   249
9.2.3   Essential tremor (organic voice tremor)   254

**10 Dysarthrias associated with lesions in other motor
      systems**   255
10.1   Ataxic dysarthria   255
10.1.1   Neuroanatomy of the cerebellum   255
10.1.2   Function of the cerebellum in voluntary motor
          activities   263
10.1.3   Clinical signs of damage to the cerebellum   266
10.1.4   Diseases of the cerebellum associated with ataxic
          dysarthria   268
10.1.5   Clinical characteristics of ataxic dysarthria   268
10.2   Mixed dysarthria   274
10.2.1   Amyotrophic lateral sclerosis   275
10.2.2   Multiple sclerosis (disseminated sclerosis)   277
10.2.3   Wilson's disease (hepato-lenticular degeneration)   280

11 Acquired aphasia in childhood   282
11.1   Acquired childhood aphasia   283
11.1.1   Clinical features of acquired childhood aphasia   283
11.1.2   Recovery from acquired childhood aphasia   291
11.2   Acquired childhood aphasia of different aetiologies   296
11.2.1   Acquired aphasia following vascular disorders   297
11.2.2   Acquired aphasia following head trauma   298
11.2.3   Acquired aphasia associated with tumours   299
11.2.4   Acquired aphasia following infection   301
11.2.5   Acquired aphasia associated with convulsive
          disorder   301
11.3   Summary   304

References   305

Index   342

Preface

The stimulus for writing this book arose from the author's perception of a lack of available texts which adequately integrate the subjects of neuroanatomy and functional neurology with the practice of speech–language pathology. This perception was gained from almost two decades of teaching in the areas of neuroanatomy and acquired neurological speech–language disorders to speech pathology students initially at the South Australian College of Advanced Education and, for the past five years, at the University of Queensland. Although a plethora of excellent texts devoted specifically to each of the subjects of neuroanatomy, neurology and aphasiology have been published, few have attemped to integrate these individual subject areas in such a way as to provide a more clear understanding of the neurological bases of clinically recognized forms of aphasia and motor speech disorders.

In writing this text, I have attempted to provide a better balance between neuroanatomy–neurology and speech–language pathology. Relevant areas of neuroanatomy and neurology are introduced and discussed in the context of specific speech and language disorders. In this way, I have aimed at providing a better link between the relevant neuroanatomical and neurological knowledge on the one hand, and specific neurologically based communication disorders on the other, in order to enhance the reader's understanding of the origins, course and prognosis of these disorders.

Of course the writing of any book requires the support and encouragement of other people. This text was no exception. I owe much to my colleagues at the Department of Speech and Hearing, The University of Queensland, for their constant encouragement. In particular, Helen Chenery and Anne Ozanne provided me with valuable discussion, helpful suggestions and moral support during the preparation of the manuscript. I am also indebted to Meredith Kennedy whose words of encouragement and constructive criticism were of considerable assistance. Without the competent secretarial support provided by Dorothy Trezise and Lindy Warner, this book could not have been completed. Finally, I must apologize to my wife and children who lacked a husband and father while I laboured to complete this volume.

CHAPTER ONE

Neuroanatomical framework of speech and language

The materials presented in this chapter are intended to provide the reader with an introductory knowledge of the anatomy of the human nervous system. Such a knowledge is necessary prior to discussion of the signs, symptoms and neurological mechanisms underlying the various neurogenic speech–language disorders in later chapters. Where necessary, more detailed information regarding the anatomy of specific brain structures important for speech–language function is provided in subsequent relevant chapters.

Human communication in the form of speech and language behaviour is dependent upon processes which occur in the nervous system. Consequently, a knowledge of the basic structure and function of the human nervous system is an essential prerequisite for the speech–language pathologist to understand the anatomical, physiological and pathological basis of human communication disorders.

The nervous system is an extremely complex organization of structures which serves as the main regulative and integrative system of the body. It receives stimuli from the individual's internal and external environments, interprets and integrates that information and selects and initiates appropriate responses to it. Consider this process in the context of a spoken conversation between two persons. The words spoken by one partner in the conversation, in the form of sound waves, are detected by receptors in the inner ear of the second partner and conveyed to the cerebral cortex of the brain via the auditory pathways where they are perceived and interpreted. Following integration with other sensory information a response to the verbal input is formulated in the language centres of the brain and then passed to the motor areas of the brain (i.e. areas that control muscular movement) for execution. Nerve impulses from the motor areas then pass to the muscles of the speech mechanism leading to the production of a verbal response by the second person.

Speech is produced by the contraction of the muscles of the speech mechanism which include the muscles of the lips, jaws, tongue, palate, pharynx and larynx as well as the muscles of respiration. These muscle contractions, in turn, are controlled by nerve impulses which descend from the motor areas of the brain to the level of the brainstem and spinal cord and then pass out to the muscles of the speech mechanism via the various nerves which arise from either the base of the brain (cranial nerves) or spinal cord (spinal nerves). Likewise, language is also dependent on processes which occur in the brain, particularly in the cerebral cortex.

For the purposes of description, the nervous system can be arbitrarily divided into two large divisions: the central nervous system and the peripheral nervous system. The central nervous system is comprised of the brain and spinal cord while the peripheral nervous system consists of the end organs, nerves and ganglia which connect the central nervous system to other parts of the body. The major components of the peripheral nervous system are the nerves which arise from the base of the brain and spinal cord. These include 12 pairs of cranial nerves and 31 pairs of spinal nerves, respectively. The peripheral nervous system is often further subdivided into the somatic and autonomic nervous systems, the somatic nervous system including those nerves involved in the control of skeletal muscles (e.g. the muscles of the speech mechanism) and the autonomic nervous system including those nerves involved in the regulation of involuntary structures such as the heart, the smooth muscles of the gastrointestinal tract and exocrine glands (e.g. sweat glands). Although the autonomic nervous system is described as part of the peripheral nervous system, it is really part of both the central and peripheral nervous systems. It must be remembered, however, that these divisions are arbitrary and artificial and that the nervous system functions as an entity, not in parts. The basic organization of the nervous system is summarized in Figure 1.1.

1.1 THE CENTRAL NERVOUS SYSTEM

The nervous system is comprised of many millions of nerve cells or neurones, which are held together and supported by specialized nonconducting cells known as neuroglia. The major types of neuroglia include astrocytes, oligodendrocytes and microglia. It is the neurones that are responsible for conduction of nerve impulses from one part of

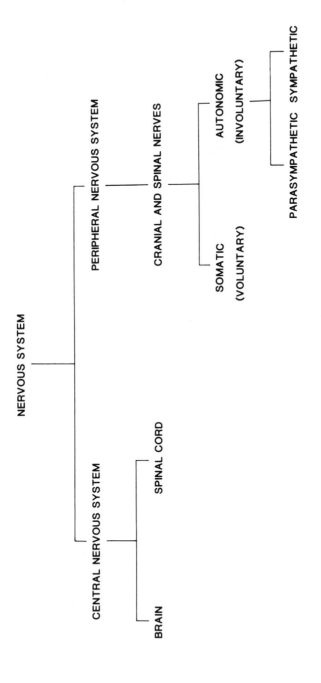

Figure 1.1 Basic organization of the nervous system.

the body to another, such as from the central nervous system to the muscles of the speech mechanism to produce the movement of the lips, tongue, etc. for speech production. Although there are a number of different types of neurones, most consist of three basic parts: a cell body (also known as a soma or perikaryon) which houses the nucleus of the cell; a variable number of short processes (generally no more than a few millimetres in length) called dendrites (meaning 'tree-like') which receive stimuli and conduct nerve impulses; and a single, usually elongated process called an axon, which in the majority of neurones is surrounded by a segmented fatty insulating sheath called the myelin sheath. A schematic representation of a neurone is shown in Figure 1.2.

The cytoplasm of a neurone contains the usual cell organelles (e.g. mitochondria) with the exception of the centrosome. Mature neurones cannot divide or replace themselves because of the lack of a centrosome. In addition to the usual organelles, however, the cytoplasm of nerve cells also contain two organelles unique to neurones – Nissl substance (chromidial substance) and neurofibrils. Seen with the light microscope, Nissl substance (bodies) appears as rather large granules widely scattered throughout the cytoplasm of the nerve cell body. Nissl bodies specialize in protein synthesis, thereby providing the protein needs for maintaining and regenerating neurone processes and for renewing chemicals involved in the transmission of nerve impulses from one neurone to another. Seen with the light microscope neurofibrils are tiny tubular structures running through the cell body, axon and dendrites. Although the function of the neurofibrils is uncertain, it has been suggested that they may facilitate the transport of intra-

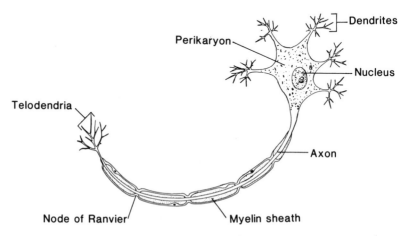

Figure 1.2 Structure of a typical motor neurone.

cellular materials within the neurone. In Alzheimer's disease, the neurofibrils become abnormally twisted, a feature used in the diagnosis of this condition (see Chapter 6).

The area where two neurones communicate with one another is called a synapse. It represents a region of functional but not anatomical continuity between the axon terminal of one neurone (the pre-synaptic neurone) and the dendrites, cell body or axon of another neurone (the post-synaptic neurone). The synapse is an area where a great degree of control can be exerted over nerve impulses. At the synapse, nerve impulses can be either blocked (inhibited) or facilitated. There may be thousands of synapses on the surface of a single neurone. When one considers that there are billions of neurones, the complexity of the circuitry of the nervous system is staggering.

Structurally, each synapse is made up as follows. As the terminal part of an axon approaches another neurone, it decreases in diameter, loses its myelin (if a myelinated fibre) and divides repeatedly forming small branches, termed telodendria. At the end of each telodendron is a small swelling called a bouton terminal or synaptic knob. The structure of the bouton terminal has been elucidated by electron microscopy. It contains a number of structures, in particular mitochondria and synaptic vesicles. The synaptic vesicles contain neurotransmitter substance which is released when a nerve impulse arrives at the bouton. There are many kinds of neurotransmitter substance, some of which facilitate (excitatory transmitters) nerve impulse conduction in the post-synaptic neurone while others inhibit (inhibitory transmitters) nerve impulse conduction in the post-synaptic neurone. Some of the more common neurotransmitter substances include acetylcholine, norepinephrine (noradrenaline), serotonin, dopamine and γ-aminobutyric acid (GABA).

When released from the synaptic knob the chemical transmitter diffuses across a gap called the synaptic cleft between the bouton and the membrane of the post-synaptic neurone to either excite or inhibit the post-synaptic neurone. As neurotransmitter substance is only located on the pre-synaptic side, a synapse can transmit in only one direction. In addition to the chemical synapses just described, in certain parts of the nervous system electrical synapses or gap junctions are present. In this type of synapse the membranes of the pre- and post-synaptic neurones lie in close proximity to one another and comprise a pathway of low resistance which allows current flow from the pre-synaptic neurone to act upon the post-synaptic neurone.

Neuroeffector junctions are functional contacts between axon terminals and effector cells. Structurally, neuroeffector junctions are similar to synapses with the exception that the post-synaptic structure is not a nerve cell but rather a muscle or gland. We will not concern ourselves greatly with junctions with smooth or cardiac muscles or glands, but will rather concentrate on junctions with skeletal muscles as this is the type of muscle tissue that comprises the muscles of the speech mechanism.

In the case of skeletal muscles, the neuroeffector junction is termed a motor end-plate. The structure of a typical motor end plate is shown in Figure 1.3. Each motor nerve fibre branches at its end to form a complex of branching nerve terminals, each terminal innervating a separate skeletal muscle fibre. A single axon of a motor neurone therefore innervates more than one skeletal muscle fibre; the motor neurone plus the muscle fibres it innervates constitute a motor unit. The bouton of each terminal contains synaptic vesicles that contain neurotransmitter substance. The motor neurones running to skeletal muscles use acetylcholine as their transmitter substance. The arrival

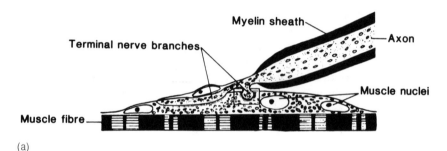

(a)

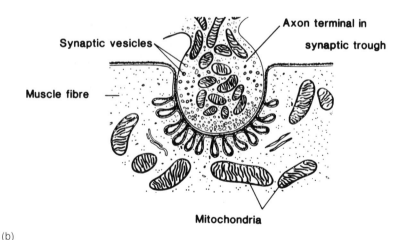

(b)

Figure 1.3 (a) Motor end-plate and (b) close-up of a motor-end plate showing the relationship between structures in nerve and muscle.

of a nerve impulse at the bouton causes release of acetylcholine from the vesicles in a similar manner to that for transmission at the synapse, only in this case the transmitter diffuses across the neuro-muscular junction to cause contraction of the muscle fibre. In the condition called myasthenia gravis there is a failure in transmission at the neuromuscular junction, possibly as a consequence of antibodies that interfere with the transmission of the acetylcholine. The result is that the muscles of the body, including the muscles of the speech mechanism, fatigue very easily when active. Where the muscles of the speech mechanism are involved, this leads to a characteristic speech disorder which is described in Chapter 8.

Both parts of the central nervous system are composed of two types of tissue – grey matter and white matter. The grey matter is made up mainly of neurone cell bodies and their closely related processes, the dendrites. White matter is comprised primarily of bundles of long processes of neurones (mainly axons), the whitish appearance resulting from the lipid insulating material (myelin). Cell bodies are lacking in the white matter. Both the grey and white matter, however, contain large numbers of neuroglial cells and a network of blood capillaries. Within the white matter of the central nervous system, nerve fibres serving similar or comparable functions are often collected into bundles called tracts. Tracts are usually named according to their origin and destination (e.g. cortico-spinal tracts). By contrast, the nerve cell processes that leave the central nervous system are collected into bundles that form the various nerves.

In the brain, most of the grey matter forms an outer layer surround-ing the cerebral hemispheres. This layer, which varies from around 1.5 to 4mm thick is referred to as the cerebral cortex (cortex meaning 'rind' or 'bark'). Within the spinal cord the distribution of grey and white matter is largely the reverse to that seen in the brain, the grey matter forming the central core of the spinal cord which is surrounded by white matter. In some parts of the central nervous system, notably the brainstem (see below) there are regions that contain both nerve cell bodies and numerous myelinated fibres. These regions are therefore comprised of diffuse mixtures of grey and white matter.

1.1.1 The brain

The brain is that part of the central nervous system contained within the skull. It is the largest and most complex mass of nerve tissue in the body and in the average human weighs approximately 1400g. The brain is surrounded by three fibrous membranes called the meninges and is suspended in fluid called cerebrospinal fluid. Within the brain are a series of fluid filled cavities called the ventricles.

The nervous system begins development in the embryo as the neural tube. At the rostral end of the neural tube develop three swellings called the primary brain vesicles. These vesicles are the prosencephalon, mesencephalon and rhombencephalon which eventually become the fore-brain, mid-brain and hind-brain respectively. Shortly after the appearance of the three primary brain vesicles, the prosencephalon divides into the telencephalon (which becomes the cerebral hemispheres) and the diencephalon (which gives rise to the thalamus and hypothalamus). In addition the rhombencephalon is divided by a fold into a rostral part called the metencephalon (which becomes the pons and cerebellum) and the myelencephalon (which forms the medulla oblongata). The mesencephalon remains undivided and becomes the mid-brain of the adult brain. The adult brain, or encephalon, can be divided into three major parts: the cerebrum, the brainstem and the cerebellum (see Figure 1.4).

(a) The cerebrum

The cerebrum is the largest portion of the brain, representing approximately seven-eighths of its total weight. Centres which govern all sensory and motor activities (including speech production) are

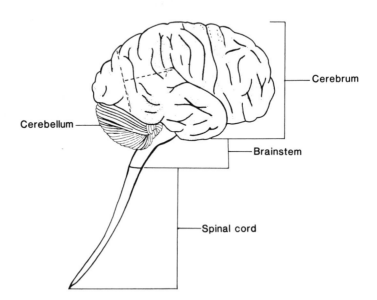

Figure 1.4 Major components of the central nervous system.

located in the cerebrum. In addition, areas which determine reason, memory and intelligence, as well as the primary language centres, are also located in this region of the brain.

The surface of the cerebrum is highly folded or convoluted. The convolutions are called gyri (*sing.* gyrus) while the shallow depressions or intervals between the gyri are referred to as sulci (*sing.* sulcus). If the depressions between the gyri are deep, they are then called fissures. A very prominent fissure, called the longitudinal fissure, is located in the mid-sagittal plane and almost completely divides the cerebrum into separate halves or hemispheres, called the right and left cerebral hemispheres. The longitudinal fissure can be viewed from a superior view of the brain as shown in Figure 1.5.

The cerebral cortex is the convoluted layer of grey matter covering the cerebral hemispheres. The cerebral cortex comprises about 40% of the brain by weight and it has been estimated that it contains in the region of 15 billion neurones. The cellular structure of the cerebral cortex itself is not uniform over the entire cerebrum and many researchers in the past have suggested that areas of the cortex with different cell structures also serve different functional roles. Inference concerning structure and function have been largely drawn from observations on animals, especially monkeys and chimpanzees, as well as from studies of humans undergoing brain surgery. Such studies have shown that some specific functions are localized to certain general areas of the cerebral cortex. These functional areas of the cortex have been mapped out as a result of direct electrical stimulation of the cortex or from neurological examination after portions of

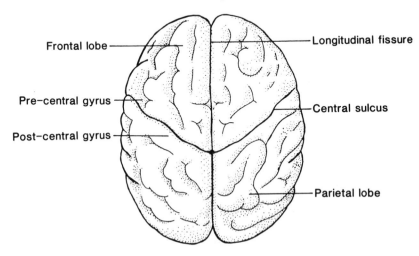

Figure 1.5 Superior view of the cerebral hemispheres.

the cortex have been removed (ablated). Although several systems for mapping the various areas of the cerebral cortex have been developed, the number system developed by Brodmann in the early 1900s has been the most widely used. The Brodmann number system is shown in Figure 1.6. In compiling this system, Brodmann attempted to correlate the structure and function of the cerebral cortex and arrived at a numerical designation of regions showing differential morphology. As a result, the cerebral cortex can be divided into motor, sensory and association areas (see Figure 1.6). The motor areas control voluntary muscular activities while the sensory areas are involved with the perception of sensory impulses (e.g. vision and audition). Three primary sensory areas have been identified in each hemisphere – one for vision, one for hearing and one for general senses (e.g. touch). The association cortex (also called the 'uncommitted cortex' because it is not obviously devoted to some primary sensory function such as vision, hearing, touch, smell, etc. or motor function) occupies approximately 75% of the cerebral cortex. It used to be believed that the association areas received information from the primary sensory areas to be integrated and analysed in the association cortex and then fed to the motor areas. It has been established, however, that they receive multiple inputs and outputs, many of them independent of the primary sensory and motor areas. Three main association areas are recognized: pre-frontal, anterior temporal and parietal–temporal–occipital area. Overall they are involved in a variety of intellectual and cognitive functions.

Beneath the cerebral cortex, each cerebral hemisphere consists of

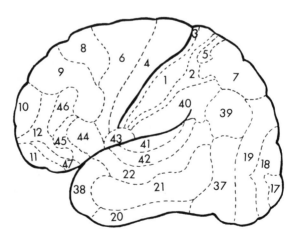

Figure 1.6 Brodmann cytoarchitectural map.

white matter within which there is located a number of isolated patches of grey matter. These isolated patches of grey matter are referred to as the basal nuclei (a nucleus is a mass of grey matter in the central nervous system) or often as the basal ganglia (strictly speaking, however, a ganglion is a group of nerve cells located outside the central nervous system). The basal nuclei or ganglia serve important motor functions and when damaged are associated with a range of neurological disorders including Parkinson's disease, chorea, athetosis and dyskinesia (see Chapter 9), all of which may have associated speech deficits. The anatomy of the basal ganglia is described and their possible role in language discussed in Chapter 3.

The white matter underlying the cerebral cortex consists of myelinated nerve fibres arranged in three principal directions. Firstly there are association fibres. These transmit nerve impulses from one part of the cerebral cortex to another part in the same cerebral hemisphere. One bundle of association fibres that is important for language function is the arcuate fasciculus (a fasciculus is a bundle of nerve fibres in the central nervous system). The arcuate fasciculus connects a language area in the temporal lobe with a language region in the frontal lobe and when damaged is thought to cause a language disorder called conduction aphasia (see Chapter 2). The second fibre group are known as commissural fibres. These transmit nerve impulses from one cerebral hemisphere to the other. The third group of fibres which make up the sub-cortical white matter are projection fibres. These form the ascending and descending pathways that connect the cerebral cortex to the lower central nervous system structures such as the brainstem and spinal cord.

In overall appearance, each cerebral hemisphere is a 'mirror-twin' of the other and each contains a full set of centres for governing the sensory and motor activities of the body. Each hemisphere is also largely associated with activities occurring on the opposite (contralateral) side of the body. For instance, the left cerebral hemisphere is largely concerned with motor and sensory activities occurring in the right side of the body. Although each hemisphere has a complete set of structures for governing the motor and sensory activities of the body, each hemisphere tends to specialize in different functions. For example, speech and language in most people is largely controlled by the left cerebral hemisphere. The left hemisphere also specializes in hand control and analytical processes. The right hemisphere specializes in such functions as stereognosis (the sense by which the form of objects is perceived, e.g. if a familiar object such as a coin or key is placed in the hand it can be recognized without looking at it) and the perception of space. The cerebral hemisphere which controls speech

12 Neuroanatomical framework of speech and language

and language is referred to as the dominant hemisphere. The concept of cerebral dominance is discussed further in Chapter 5.

Although almost completely separated by the longitudinal fissure, the two cerebral hemispheres are connected internally by a number of commissures. By far the largest commissure is the corpus callosum, a mass of white matter which serves as the major pathway for the transfer of information from one hemisphere to the other. The anterior portion of the corpus callosum is called the genu while the posterior part is referred to as the splenium. Between the genu and the splenium is located the body of the corpus callosum. In addition to the corpus callosum three lesser commissures also connect the two hemispheres. These include the fornix, the anterior commissure and the posterior commissure. The location of these various commissures can be seen from a mid-sagittal section of the brain as shown in Figure 1.7.

Each cerebral hemisphere can be divided into six lobes. These include the frontal, parietal, occipital, temporal, central (also called the insula or Island of Reil) and limbic lobes. The six lobes are delineated from each other by several major sulci and fissures, including the lateral fissure (Fissure of Sylvius), central sulcus (Fissure of Rolando), cingulate sulcus and the parieto-occipital sulcus. A superior view of the brain reveals two lobes, the frontal and parietal, separated by the central sulcus (see Figure 1.5).

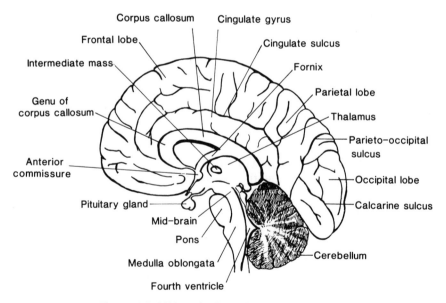

Figure 1.7 Mid-sagittal section of the brain.

Four lobes, namely the frontal, parietal, temporal and occipital lobes, can be seen from a lateral view of the cerebrum (see Figure 1.8). The boundaries of the lobes on the lateral cerebral surface are as follows: the frontal lobe is located anterior to the central sulcus and above the lateral fissure; the parietal lobe is located posterior to the central sulcus, anterior to an imaginary parieto-occipital line (this runs parallel to the parieto-occipital sulcus which is found on the medial surface of the hemisphere in the longitudinal fissure – see Figure 1.7) and above the lateral fissure and its imaginary posterior continuation toward the occipital pole; the temporal lobe is located below the lateral fissure and anterior to the imaginary parieto-occipital line.

The central lobe or insula is not visible from an external view of the brain. It is hidden deep within the lateral fissure. To view the central lobe the lateral fissure must be held apart. Those parts of the frontal, parietal and temporal lobes which cover the external surface of the insula are called the frontal operculum, parietal operculum and temporal operculum, respectively.

The limbic lobe is a ring of gyri located on the medial aspect of each cerebral hemisphere. The largest components of this limbic lobe include the hippocampus, the para-hippocampal gyrus and the cingulate gyrus, some of which can be examined from a mid-sagittal view of the brain.

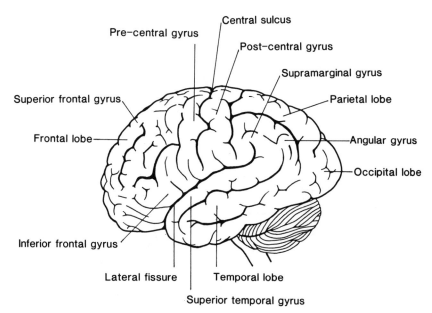

Figure 1.8 Lateral view of the brain.

The boundaries of the lobes on the medial cerebral surface are as follows: the frontal lobe is located anterior to the central sulcus and above the line formed by the cingulate sulcus; the parietal lobe is bounded by the central sulcus, cingulate sulcus and parieto-occipital sulcus; the temporal lobe is located lateral to the para-hippocampal gyrus; the occipital lobe lies posterior to the parietal-occipital sulcus; the limbic lobe is comprised of the gyri bordered by the curved line formed by the cingulate sulcus and the collateral sulcus.

Although there is considerable overlap in the functions of adjacent cerebral lobes, each lobe does appear to have its own speciality. For instance, located in the frontal lobes are the centres for the control of voluntary movement, the so-called motor areas of the cerebrum. Immediately anterior to the central sulcus is a long gyrus called the pre-central gyrus (see Figure 1.8). This gyrus (Brodmann area 4), also known as the primary motor area or motor strip represents the point of origin for those nerve fibres which carry voluntary nerve impulses from the cerebral cortex to the brainstem and spinal cord. In other words the nerve cells in this area are responsible for the voluntary control of skeletal muscles on the opposite side of the body. Electrical stimulation of the primary motor area causes the contraction of muscles primarily on the opposite or contralateral side of the body. The nerve fibres which leave the primary motor area and pass to

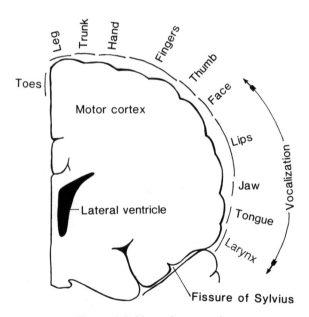

Figure 1.9 Motor homunculus.

either the brainstem or spinal cord form what are known as the pyramidal pathways. (These pathways are discussed in more detail in Chapter 8.)

All parts of the body responsive to voluntary muscular control are represented along the pro-central gyrus in something of a sequential array. A map showing the points in the primary motor cortex that cause muscle contractions in different parts of the body when electrically stimulated is shown in Figure 1.9. These points have been determined by electrical stimulation of the human brain in patients having brain operations under local anaesthesia. The map as shown is referred to as the motor homunculus. It will be noted that the areas of the body are represented in an almost inverted fashion, the motor impulses to the head region originating from that part of the pre-central gyrus closest to the lateral sulcus, while impulses passing to the feet are initiated from an area located within the longitudinal fissure. The size of the area of pre-central gyrus devoted to a particular part of the body is not strongly related to the size of that body part. Rather, larger areas of the motor strip are devoted to those parts of the body which have a capacity for finer and more highly controlled movement. Consequently the area devoted to the hand is larger that that given to the leg and foot. Likewise, because the muscles of the larynx are capable of very discrete and precise movements, the area of pre-central gyrus devoted to their control is as large or larger than the area given to some of the big leg muscles which are capable of only more gross movements.

In addition to the primary motor area, several other motor areas have been located in the frontal lobes by stimulation studies. These latter areas include the pre-motor area (Brodmann area 6), the supplementary motor area, the secondary motor area and the frontal eye field (Brodmann area 8). The pre-motor area lies immediately anterior to the pre-central sulcus. Not only does it contribute fibres to the descending motor pathways, including the pyramidal pathways, it also influences the activity of the primary motor area. Electrical stimulation of the pre-motor area elicits complex contractions of groups of muscles. Occasionally vocalization occurs, or rhythmic movements such as alternate thrusting of a leg forward or backward, turning of the head, chewing, swallowing or contortion of parts of the body into different postural positions. It is believed that the pre-motor area programs skilled motor activity and thereby directs the primary motor area in its execution of voluntary muscular activity. Therefore, whereas the primary motor area controls the contraction of individual muscles and acts as the primary output source from the cerebral cortex for voluntary motor activities, the pre-motor area functions in the control of co-ordinated, skilled movements involving the contraction of many muscles simultaneously.

The secondary motor area is located in the dorsal wall of the lateral fissure immediately below the pre-central gyrus. Its functional significance is unknown. The supplementary motor area is an extension of Brodmann area 6 and is located within the longitudinal fissure on the medial aspect of the hemisphere immediately anterior to the leg portion of the primary motor area. Some researchers consider it to be a second speech area. The frontal eye field lies anterior to the pre-motor area (see Figure 1.6). It controls volitional eye movements. Stimulation of the frontal eye field results in conjugate (joined together) movements of the eyes to the opposite sides.

Another important area of the frontal lobe is Broca's area (Brodmann areas 44 and 45). Also known as the motor speech area, Broca's area is one of two major cortical areas that have been identified as having specialized language functions. Broca's area is located in the inferior (third) frontal gyrus of the frontal lobe (see Figure 1.8) and appears to be necessary for the production of fluent, well-articulated speech. The importance of Broca's area to language production is outlined in more detail later in this chapter and the relationship between lesions of this region and the occurrence of specific speech and language disturbances is discussed in Chapter 2.

The parietal lobe is involved in a wide variety of general sensory functions. The sensations of heat, cold, pain, touch, pressure and position of the body in space and possibly some taste sensation all reach the level of consciousness here. The primary sensory area for general senses (also called the somesthetic area or sensory strip) occupies the post-central gyrus (areas 3, 1 and 2 of the Brodmann cytoarchitectural map) (see Figure 1.6). Each sensory strip receives sensory signals almost exclusively from the opposite side of the body (a small amount of sensory (touch) information comes from the same or ipsilateral side of the face). As in the case of the motor strip, the various parts of the body can be mapped along the post-central gyrus to indicate the area devoted to their sensory control. This map is referred to as the sensory homunculus and is shown in Figure 1.10. It can be seen that some areas of the body are represented by large areas in the post-central gyrus. The size of the area devoted to a particular part of the body is directly proportional to the number of specialized sensory receptors contained in that part of the body. In other words, the proportion of the sensory strip allocated to a particular body part is determined by the sensitivity of that part. Consequently, a large area of the post-central gyrus is assigned to highly sensitive areas such as the lips and hand (particularly the thumb and

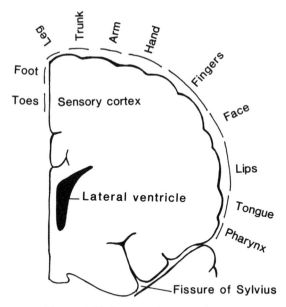

Figure 1.10 Sensory homunculus.

index finger) and a smaller area assigned to less sensitive areas such as the trunk and legs.

In addition to the post-central gyrus, two other gyri in the parietal lobe are also of importance to speech–language pathologists. These are the supramarginal gyrus and the angular gyrus (see Figure 1.8). The supramarginal gyrus wraps around the posterior end of the lateral fissure while the angular gyrus lies immediately posterior to the supramarginal gyrus and curves around the end of the superior temporal gyrus. In the dominant hemisphere (usually the left), these two gyri form part of the posterior language centre, an area involved in the perception and interpretation of spoken and written language. The relationship between damage to these two gyri and the occurrence of specific language deficits is discussed in Chapters 2 and 7.

The temporal lobe is concerned with the special sense of hearing (audition) and at least some of the neurones concerned with speech and language are located here. The primary auditory area is not visible from a lateral view of the brain because it is concealed within the lateral fissure. The floor of the lateral fissure is formed by the upper surface of the superior temporal gyrus. This surface is marked by transverse temporal gyri. The two most anterior of these gyri, called the anterior temporal gyri or Heschl's convolutions, represent the primary auditory area (Brodmann areas 41 and 42). The posterior part of the superior temporal gyrus (Brodmann area 22) which is evident on the lateral surface of the temporal lobe together with that part of the floor of the lateral fissure that lies immediately behind the primary auditory area (an area called the planum temporal) constitute the auditory association area. In the dominant hemisphere the auditory association area is also known as Wernicke's area, another important component of the posterior language centre. The pathological effects on language of lesions in Wernicke's area are discussed in Chapter 2.

The occipital lobe is primarily concerned with vision. The primary visual area (Brodmann area 17) surrounds the calcarine sulcus which is located in the longitudinal fissure on the medial surface of the occipital lobe (see Figure 1.7).

The limbic lobe, also known as the rhinencephalon (smell brain), is associated with olfaction, autonomic functions and certain aspects of emotion, behaviour and memory. Although the functions of the central lobe are uncertain, it is believed that is also operates in association with autonomic functions.

(b) The brainstem

If both the cerebral hemispheres and the cerebellum are removed from the brain, a stalk-like mass of central nervous system tissue remains, the brainstem. The brainstem is comprised of four major parts. From rostral (head) to caudal (tail) these include the diencephalon, mid-brain (mesencephalon), pons (metencephalon) and medulla oblongata (myelencephalon). The relationship of these components to one another can be seen in Figure 1.11. (Note: in some classification systems the diencephalon is included as part of the cerebrum.)

THE DIENCEPHALON

The diencephalon (or 'tweenbrain) lies between the cerebral hemispheres and the mid-brain. It consists of two major components, the thalamus and hypothalamus. The thalamus is a large rounded mass of grey matter measuring about 3 cm antero-posteriorly and 1.5 cm in the two other directions. Located above the mid-brain, it is not visible in surface views of the brain. It can be seen, however, from a mid-sagittal section of the brain (see Figure 1.7). The thalamus is almost completely divided into right and left thalami by the third ventricle. In most people, however, the two large ovoid (egg-shaped) thalami of both sides are connected to one another by a band of grey matter called the inter-thalamic adhesion (intermediate mass). Each thalamic

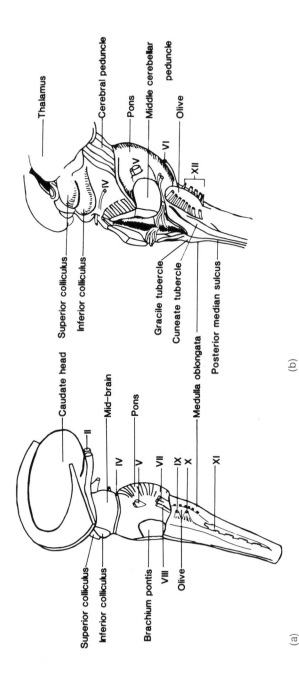

Figure 1.11 (a) Lateral view of the brainstem and (b) dorso-lateral view of the brainstem.

mass contains over 30 nuclei which enable it to perform important sensory and motor functions. In particular the thalamus is one of the major sensory integrating centres of the brain and is sometimes referred to as the gateway to the cerebral cortex. All of the major sensory pathways with the exception of the olfactory pathways pass through the thalamus on their way to the cerebral cortex. The thalamus, therefore, receives sensory information via the sensory pathways, integrates that information, and then sends it on to the cerebral cortex for further analysis and interpretation.

In addition to its sensory activities, the thalamus is functionally interrelated with the major motor centres of the cerebral cortex and can facilitate or inhibit motor impulses originating from the cerebral cortex. In recent years, a number of researchers have also documented the occurrence of language disorders following thalamic lesions, thereby suggesting that the thalamus may play a role in language function. A more complete description of the neuroanatomy of the thalamus together with a discussion of thalamic aphasia is presented in Chapter 3.

The hypothalamus lies below the thalamus (see Figure 1.7) and forms the floor and the inferior part of the lateral walls of the third ventricle. When examined from an inferior view of the brain (see Figure 1.12) the hypothalamus can be seen to be made up of the tuber cinereum, the optic chiasma, the two mammillary bodies and the infundibulum. The tuber cinereum is the name given to the region

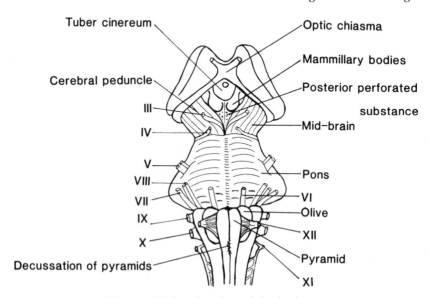

Figure 1.12 Inferior view of the brainstem.

bounded by the mammillary bodies, optic chiasma and beginning of the optic tracts. The infundibulum, to which is attached the posterior lobe of the pituitary gland is a stalk-like structure which arises from a raised portion of the tuber cinereum called the median eminence. The median eminence, the infundibulum and the posterior lobe of the pituitary gland together form the neurohypophysis (posterior pituitary gland). The mammillary bodies are two small hemispherical projections placed side by side immediately posterior to the tuber cinereum. They contain nuclei important for hypothalamic function. The optic chiasma is a cross-like structure formed by the partial crossing over of the nerve fibres of the optic nerves. Within the optic chiasma the nerve fibres originating from the nasal half of each retina cross the mid-line to enter the optic tract on the opposite side.

Although the hypothalamus is only a small part of the brain, it controls a large number of important body functions. The hypothalamus controls and integrates the autonomic nervous system which stimulates smooth muscle, regulates the rate of contraction of cardiac muscle and controls the secretions of many of the body's glands. Through the autonomic nervous system, the hypothalamus is the chief regulator of visceral activities (e.g. it controls the heart rate, the movement of food through the digestive system and contraction of the urinary bladder). The hypothalamus is also an important link between the nervous and endocrine systems and regulates the secretion of hormones from the anterior pituitary gland and actually produces the hormones released from the posterior pituitary.

The hypothalamus is the centre for 'mind-over-body' phenomena. When the cerebral cortex interprets strong emotions, it often sends impulses over tracts that connect the cortex with the hypothalamus. The hypothalamus responds either by sending impulses to the autonomic nervous system or by releasing chemicals that stimulate the anterior pituitary gland. The result can be a wide range in changes of body activity. The hypothalamus controls other aspects of emotional behaviour such as rage and aggression. It also controls body temperature and regulates water and food intake and is one of the centres that maintains the waking state. The hpothalamus also has a role in the control of sexual behaviour.

THE MID-BRAIN

The mid-brain is the smallest portion of the brainstem and lies between the pons and diencephalon. The mid-brain is traversed internally by a narrow canal called the cerebral aqueduct (Aqueduct of Sylvius) which connects the third and fourth ventricles and divides the mid-brain into a dorsal and ventral portion. A prominent elevation lies

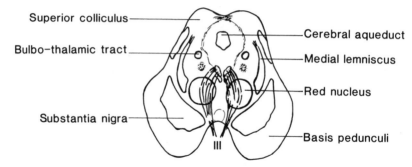

Figure 1.13 Transverse section through the mid-brain at the level of the superior colliculus.

on either side of the ventral surface of the mid-brain. (see Figure 1.12). These two elevations are known as the cerebral peduncles (basis pedunculi) and consist of large bundles of descending nerve fibres. The region between the two cerebral peduncles is the inter-peduncular fossa. Cranial nerve III* (the oculomotor nerve) arises from the side of this fossa. The floor of the fossa is known as the posterior perforated substance due to the many perforations produced by blood vessels that penetrate the mid-brain.

The dorsal portion of the mid-brain contains four rounded eminences, the paired superior and inferior colliculi (collectively known as the corpora quadrigemina) (see Figure 1.11). The four colliculi comprise the roof or tectum of the mid-brain. The superior colliculi are larger than the inferior colliculi, and are associated with the optic system. In particular they are involved with the voluntary control of ocular movements and optic reflexes such as controlling movement of the eyes in response to changes in the position of the head in response to visual and other stimuli. The major role of the inferior colliculi, on the other hand, is as relay nuclei on the auditory pathways to the thalamus. Cranial nerve IV (the trochlear nerve) emerges from the brainstem immediately caudal to the inferior colli-culus and then bends around the lateral surface of the brainstem on its way to the orbit (see Figure 1.11).

The internal structure of the mid-brain as seen in a transverse section at the level of the superior colliculus is shown in Figure 1.13. Each cerebral peduncle is divided internally into an anterior part, the crus cerebri, and a posterior part, the tegmentum, by a pigmented band of grey matter called the substantia nigra. The crus cerebri consists of fibres of the pyramidal motor system (see Chapter 8)

*By convention, cranial nerves are given Roman numerals, see p. 30.

(including cortico-spinal, cortico-bulbar and cortico-mesencephalic fibres) as well as fibres which connect the cerebral cortex to the pons (cortico-pontine fibres). The substantia nigra is the largest single nucleus in the mid-brain. It is a motor nucleus concerned with muscle tone and has connections to the cerebral cortex, hypothalamus, spinal cord and basal ganglia. Another important motor nucleus found in the tegmentum of the mid-brain is the red nucleus, so called because of its pinkish colour in fresh specimens. The red nucleus is located between the cerebral aqueduct and the substantia nigra. Large bundles of sensory fibres such as the medial lemniscus also pass through the tegmentum of the cerebral peduncles on their way to the thalamus from the spinal cord. In addition, the nuclei of cranial nerves III and IV are also located in the tegmentum of the mid-brain.

THE PONS

The pons lies between the mid-brain and medulla oblongata and anterior to the cerebellum, being separated from the latter by the fourth ventricle. The term pons means 'bridge' – the pons takes its name from the appearance of its ventral surface, which is essentially that of a bridge connecting the two cerebellar hemispheres.

As in the case of the mid-brain, the pons may also be divided into a dorsal and ventral portion. The dorsal portion is continuous with the tegmentum of the mid-brain and is also called the tegmentum. The ventral portion of the pons is the basilar pons. The basilar pons is a distinctive brainstem structure, presenting as a rounded bulbous structure (see Figure 1.11). It contains mainly thick, heavily myelinated fibres running in a transverse plane. These fibres connect the halves of the cerebellum and run into the cerebellum as the brachium pontis or middle cerebellar peduncle. Cranial nerve V (the trigeminal nerve) emerges from the lateral aspect of the pons. Each trigeminal nerve consists of a smaller motor root and a larger sensory root. In the groove between the pons and medulla oblongata (the ponto-medullary sulcus) there emerge from medial to lateral, cranial nerves VI (the abducens nerve), VII (the facial nerve) and VIII (the vesti-bulocochlear or auditory nerve). As in the case of the trigeminal nerve, the facial nerve emerges from the brainstem in the form of two distinct bundles of fibres of unequal size. The larger motor root is the motor facial nerve proper. The smaller bundle contains autonomic fibres and is known as the nervus intermedius.

The internal structure of the pons as seen in a transverse section through the mid-pons is shown in Figure 1.14. The dorsal and ventral portions of the pons are separated by the trapezoid body which is comprised of transverse auditory fibres. Although the pons consists

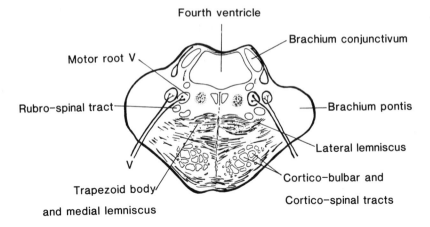

Figure 1.14 Transverse section through the pons at the level of the middle cerebellar penduncle (brachium pontis).

mainly of white matter, it does contain a number of nuclei. Nuclei located in the tegmentum include the motor and sensory nuclei of the trigeminal nerve, the facial nucleus and abducens nucleus. A nucleus involved in the control of respiration, the pneumotaxic centre, is also located in the pons. Major sensory fibres also ascend through the tegmentum of the pons via the medial and lateral lemniscus. The basilar pons near the mid-line contains small masses of nerve cells called the pontine nuclei. The cortico-pontine fibres of the crus cerebri of the mid-brain terminate in the pontine nuclei. The axons of the nerve cells in the pontine nuclei in turn give origin to the transverse fibres of the pons, which cross the mid-line and intersect the cortico-spinal and cortico-bulbar tracts (both components of the pyramidal motor system – see Chapter 8), breaking them up into smaller bundles. Overall, the basal portion of the pons acts as a synaptic or relay station for motor fibres conveying impulses from the motor areas of the cerebral cortex to the cerebellum. These pathways are described more fully in Chapter 10.

THE MEDULLA OBLONGATA

The medulla oblongata is continuous with the upper portion of the spinal cord and forms the most caudal portion of the brainstem. It lies above the level of the foramen magnum and extends upwards to the lower portion of the pons. The medulla is composed mainly of white fibre tracts. Among these tracts are scattered nuclei that either serve as controlling centres for various activities or contain the cell bodies of some cranial nerve fibres.

On the ventral surface of the medulla in the mid-line is the anterior median fissure. This fissure is bordered by two ridges, the pyramids (see Figure 1.12). The pyramids are composed of the largest motor tracts that run from the cerebral cortex to the spinal cord, the so-called cortico-spinal tracts (pyramidal tracts proper). Near the junction of the medulla with the spinal cord, most of the fibres of the left pyramid cross to the right side and most of the fibres in the right pyramid cross to the left side. The crossing is referred to as the decussation of the pyramids and largely accounts for the left cerebral hemisphere controlling the voluntary motor activities of the right side of the body and the right cerebral hemisphere the voluntary motor activities on the left side of the body.

Dorsally, the posterior median sulcus and two dorso-lateral sulci can be identified on the medulla (see Figure 1.11). On either side of the posterior median sulcus is a swelling, the gracile tubercle, and just lateral to this is a second swelling, the cuneate tubercle. Both of these swellings contain important sensory nuclei, the gracile nucleus and cuneate nucleus, respectively. These nuclei mark the point of termination of major sensory pathways called the fasciculus gracilis and fasciculus cuneatus which ascend in the dorsal region of the spinal cord.

The ventro-lateral sulcus can be identified on the lateral aspect of the medulla. Between this sulcus and the dorso-lateral sulcus at the rostral end of the medulla, is an oval swelling called the olive which contains the inferior olivary nucleus. Posterior to the olives are the inferior cerebellar peduncles which connect the medulla to the cerebellum. In the groove between the olive and the inferior cerebellar peduncle emerge the roots of the IXth (glossopharyngeal nerve) and Xth (vagus) nerves and the cranial roots of the XIth (accessory) nerve. The XIIth (hypoglossal) nerve arises as a series of roots in the groove between the pyramid and olive.

The internal structure of the medulla oblongata as seen from a transverse section is shown in Figure 1.15. The medulla contains a number of important cranial nerve nuclei including the nucleus ambiguus (which gives rise to the motor fibres which are distributed to voluntary skeletal muscles via the IXth, Xth and cranial portion of the XIth nerves) and the hypoglossal nucleus (which gives rise to the motor fibres which pass via the XIIth nerve to the muscles of the tongue). As well as containing the nuclei for various cranial nerves, the medulla also contains a number of nuclei that initiate and regulate a number of vital activities such as breathing, swallowing, regulation of heart rate and the calibre of smaller blood vessels.

Located in the central region or core of the brainstem, stretching

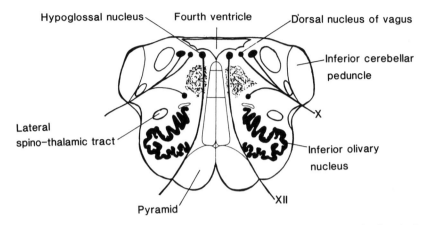

Figure 1.15 Transverse section through the medulla oblongata at the level of the inferior olivary nucleus.

R.F.

through the medulla, pons and mid-brain to the lower border of the thalamus is a diverse collection of neurones collectively known as the reticular formation. The reticular formation receives fibres from the motor regions of the brain and most of the sensory systems of the body. Its outgoing fibres pass primarily to the thalamus and from there to the cerebral cortex. Some outgoing fibres pass to the spinal cord. Stimulation of most parts of the reticular formation results in an immediate and marked activation of the cerebral cortex leading to a state of alertness and attention. If the individual is sleeping, stimulation of the reticular formation causes immediate waking. The upper portion of the reticular formation plus its pathways to the thalamus and cerebral cortex have been designated the reticular activating system because of its importance in maintaining the waking state. Damage to the brainstem reticular activating system, as might occur as a result of head injury, leads to coma, a state of unconsciousness from which even the strongest stimuli cannot arouse the subject.

(c) The cerebellum

The cerebellum (small brain) lies behind the pons and medulla and below the occipital lobes of the cerebrum (see Figure 1.7). Grossly, it may be seen to be composed of two hemispheres, the cerebellar hemispheres, which are connected by a median portion called the vermis. The cerebellum is attached to the brainstem on each side by three bundles of nerve fibres called the cerebral peduncles.

In general terms the cerebellum refines or makes muscle movements smoother and more co-ordinated. Although it does not in itself initiate any muscle movements, the cerebellum continually monitors and adjusts motor activities which originate from the motor area of the brain or peripheral receptors. It is particularly important for co-ordinating rapid and precise movements such as those required for the production of speech.

The anatomy of the cerebellum together with the effects of cerebellar lesions on speech production are described and discussed in more detail in Chapter 10.

1.1.2 The spinal cord

The spinal cord is that part of the central nervous system that lies below the level of the foramen magnum. Protected by the vertebral column, the spinal cord lies in the spinal or vertebral canal and, like the brain, is surrounded by three fibrous membranes, the meninges. It is cushioned by cerebrospinal fluid and held in place by the denticulate ligaments. It is comprised of well-demarcated columns of motor and sensory cells (the grey matter) surrounded by the ascending and descending tracts which connect the spinal cord with the brain (the white matter). A transverse section of the spinal cord shows that the grey matter is arranged in the shape of the letter 'H', with anterior and posterior horns and a connecting bar of grey matter (see Figure 1.16). A lateral horn of grey matter is also present in the thoracic part of the cord. A narrow cavity called the central canal is located in the connecting bar of grey matter.

The spinal cord is divided into five regions, each of which takes its name from the corresponding segment of the vertebral column. These

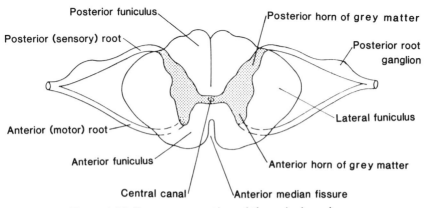

Figure 1.16 Transverse section of the spinal cord.

regions include (from top to bottom) the cervical, thoracic, lumbar, sacral and coccygeal regions. There are 31 pairs of spinal nerves arising from the spinal cord: 8 of these nerve arise from the cervical region, 12 from the thoracic, 5 each from the lumbar and sacral regions and 1 from the coccygeal region. Each spinal nerve is formed by the union of a series of dorsal and ventral roots, the dorsal roots carrying only sensory fibres which convey information from peripheral receptors into the spinal cord, and the ventral roots containing only motor fibres which act as a final pathway for all motor impulses leaving the spinal cord.

The segments of the spinal cord in the adult are shorter than the corresponding vertebrae. Consequently, the spinal cord in the adult does not extend down the full length of the vertebral canal. Rather the spinal cord extends only from the foramen magnum to the level of the first or second lumbar vertebra. The lower-most segments of the cord are compressed into the last 2–3cm of the cord, a region known as the conus medullaris. Due to the relative shortness of the spinal cord compared with the vertebral column, the nerve roots arising from the lower segments of the cord have a marked downward direction in the lower part of the vertebral canal forming a leash of nerves known as the cauda equina (horse's tail).

The white matter of the spinal cord is arranged into funiculi (funiculus meaning 'cord-like') (see Figure 1.16). A posterior median septum divides the white matter into two (right and left) posterior funiculi in the dorsal portion of the spinal cord. The white matter between the

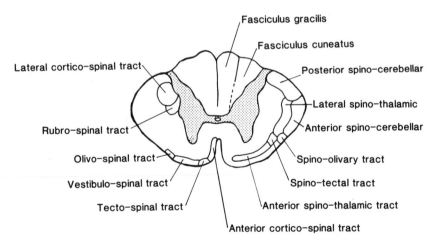

Figure 1.17 Transverse section of the spinal cord showing the general arrangement of the major ascending and descending tracts.

dorsal and ventral nerve roots on each side is called the lateral funiculus. The ventral portion of the spinal cord is divided by the anterior median fissure into two anterior funiculi. Each funiculus contains tracts of ascending and descending fibres. The approximate positions of the various tracts are shown in Figure 1.17.

1.2 THE PERIPHERAL NERVOUS SYSTEM

Nerve impulses are conveyed to and from the central nervous system by the various parts of the peripheral nervous system. Afferent or sensory nerve fibres carry nerve impulses arising from the stimulation of sensory receptors (e.g. touch receptors) towards the central nervous system. Those nerve fibres that carry impulses from the central nervous system to the effector organs (e.g. muscles and glands) are called efferent or motor fibres. The terms afferent and efferent are also used to describe fibres in the central nervous system as well as in the peripheral nervous system. When applied to central nervous system fibres, however, the term afferent describes fibres taking nerve impulses to a particular structure (e.g. afferent supply of cerebellum), while the term efferent refers to fibres taking impulses away from a particular structure, (e.g. efferent supply of cerebellum).

Some nerve fibres are associated with the structures of the body wall or extremities, such as skeletal muscles, skin, bones and joints. These fibres are called somatic fibres and may of course be either sensory or motor. Other nerve fibres, which may also be either sensory or motor, are more closely associated with the internal organs such as the smooth muscles found in the gastrointestinal tract and blood vessels, etc. These fibres are referred to as visceral fibres.

The nerves of the peripheral nervous system are made up of bundles of individual nerve fibres. In most cases these nerves contain all of the types of nerve fibres described above (i.e. somatic afferent and somatic efferent, visceral afferent and visceral efferent). Consequently, although it may be correct to speak of sensory or motor nerve fibres, it is rarely correct to speak of sensory or motor nerves. Only in the case of some cranial nerves is it possible to speak of sensory or motor nerves *per se.* For example, cranial nerve II (the optic nerve) is entirely a sensory nerve. On the other hand, cranial nerve XII (the hypoglossal nerve) is often regarded as a motor nerve.

The three principal components of the peripheral nervous system are the cranial nerves, the spinal nerves and the peripheral portions of the autonomic nervous system. These three morphologic subdivisions are not independent functionally, but combine and communicate with

Table 1.1 Summary of the cranial nerves

Nerve		Function
I	Olfactory	Smell
II	Optic	Vision
III	Oculomotor	Four extrinsic eye muscles (medial, inferior and superior recti, inferior oblique) and levator palpebrae. Parasympathetic to iris diaphragm of eye (constriction) and ciliary muscles of eye (lens accommodation)
IV	Trochlear	One extrinsic eye muscle (superior oblique)
V	Trigeminal	
	Motor root	Muscles of mastication and tensor typani
	Sensory root	Cranial–facial sensation
VI	Abducens	One extrinsic eye muscle (lateral rectus)
VII	Facial	
	Motor root	Muscles of facial expression and stapedius.
	Intermediate root (nervus intermedius)	Parasympathetic innervation of sub-mandibular and sub-lingual salivary glands. Taste from anterior two-thirds of tongue
VIII	Vestibulocochlear nerve	
	Vestibular nerves	Balance
	Cochlear nerve	Hearing
IX	Glossopharyngeal	Stylopharyngeus muscle. Parasympathetic innervation to parotid salivary gland. Sensation from pharynx and taste from posterior one-third of tongue
X	Vagus	Pharyngeal and laryngeal muscles and levator veli palatini. Parasympathetic innervation of thoracic and upper abdominal viscera
XI	Accessory	
	Cranial portion	Joins the vagus to supply the muscles of the larynx and pharynx
	Spinal portion	Sternocleidomastoid and trapezius muscles
XII	Hypoglossal	All intrinsic and most extrinsic tongue muscles

each other to supply both the somatic and visceral parts of the body with both afferent and efferent fibres.

1.2.1 The cranial nerves

Twelve pairs of cranial nerves arise from the base of the brain. With only one exception, the olfactory nerves which terminate within the olfactory bulbs, all cranial nerves either originate from or terminate within the brainstem. The cranial nerves are numbered by Roman numerals according to their position on the brain from anterior to posterior. The names given to the cranial nerves indicate either their function or destination. Some cranial nerves are both sensory and motor. Others, however, are either sensory or motor only. Table 1.1 summarizes the principal features of the twelve cranial nerves including their names and peripheral connections.

Cranial nerves are important to the speech–language pathologist in that they are responsible for the control of the majority of muscles comprising the speech mechanism. In particular, cranial nerves V, VII, IX, X, XI and XII are vital for normal speech production, and for this reason the anatomy of these nerves is described in more detail in Chapter 8.

1.2.2 The spinal nerves

As mentioned previously, each of the 31 pairs of spinal nerves is formed by the union of the dorsal and ventral nerve roots which emerge from each segment of the spinal cord. Once formed in this manner, each spinal nerve leaves the vertebral canal through its intra-vertebral foramen (opening) and ends soon after by dividing into a dorsal ramus (branch) and ventral ramus. The dorsal rami of the spinal nerves segmentally supply the deep back muscles and the skin of the posterior aspect of the head, neck and trunk. The ventral rami are larger than the dorsal rami and behave quite differently. Whereas the dorsal rami show a segmental arrangement, the ventral rami in the cervical, lumbar and sacral regions form four extensive, intermingled networks of nerves called plexuses. Consequently most nerves arising from these plexuses carry nerve fibres of neurones from more than one segment of the spinal cord. The ventral rami in the thoracic region course in the intercostal spaces to supply primarily the intercostal muscles and the skin overlying them. The four plexuses together with the major nerves arising from each are listed in Table 1.2.

Table 1.2 Spinal plexuses

Plexus	Origin	Peripheral nerves arising from plexus
Cervical	Ventral rami of 1st and 4th cervical nerves and part of the 5th cervical nerve	Phrenic nerve*
Brachial	Ventral rami of 5th and 8th cervical nerves and first thoracic nerve	Axillary, median, radial and ulnar nerves
Lumbar	Ventral rami of 1st, 2nd, 3rd and greater part of 4th lumbar nerves	Femoral nerve
Sacral	Ventral rami of 4th and 5th lumbar nerves and first four sacral nerves	Sciatic nerve

*This nerve is important for speech production in that it supplies the respiratory diaphragm.

1.2.3 The autonomic nervous system

The autonomic nervous system regulates the activity of cardiac muscle, smooth muscle and the glands of the body (particularly the exocrine glands). In this way the autonomic nervous system controls the activity of the visceral organs and, among other things, helps to regulate arterial pressure, gastrointestinal motility and secretion, urinary output, sweating, body temperature and various other functions. Normally the autonomic nervous system is an involuntary system that functions below the conscious level.

Although the visceral organs are supplied with both afferent and efferent nerve fibres, most authors when describing the components of the autonomic nervous system only include the efferent (motor) fibres that connect the central nervous system to effector organs such as smooth muscles, glands, etc. in their description. One reason for this is that the sensory fibres coming from the visceral organs are similar to those of the somatic nervous system (i.e. similar to those that come from the skin and voluntary muscles). In both, the sensory fibres run all the way from the receptor to the central nervous system without synapsing. On the other hand the efferent pathways that supply smooth muscles, etc. are arranged very differently from those that supply the skeletal muscles. Anatomically, the efferent pathways of the autonomic nervous system are unique in the following ways: whereas a skeletal muscle fibre is innervated by a neurone with its cell body in the central nervous system and its axon extending without interruption to the muscles, smooth muscle, cardiac muscle and glands are innervated by a two-neurone chain comprised of a pre-ganglionic neurone and a post-ganglionic neurone which synpase in a ganglion outside the central nervous system.

Anatomically and functionally the autonomic nervous system can be divided into two major divisions, the sympathetic nervous system and the parasympathetic nervous system. Most visceral organs are inner-vated by both divisions, each of which has an opposite effect on the organ involved. The sympathetic or thora-columbar division of the autonomic nervous system arises from all of the thoracic spinal nerves and the first two or three lumbar spinal nerves. The parasympathetic or cranio-sacral division of the autonomic nervous system is located within cranial nerves III, VII, IX and X and sacral spinal nerves 2, 3 and 4. All of these nerves also carry somatic motor fibres.

The parasympathetic fibres in the oculomotor nerve (III) supply the ciliary muscles of the lens of the eye and the sphincter of the pupil. Parasympathetic fibres distributed via the facial nerve (VII) regulate the secretion of saliva from the sub-mandibular and sub-lingual glands while secretion from the parotid salivary gland is controlled by the parasympathetic fibres of the glossopharyngeal nerve (IX). The vagus nerve provides parasympathetic innervation for most visceral organs contained in the thorax and abdomen and is the single most important nerve of the parasympathetic division. Vagal activity maintains the normal heart rate and a reduction in vagal tone causes the heart to beat more rapidly. The vagal fibres cause constriction of the bronchi and air passages of the lungs and also supply the digestive tract as far as the transverse colon. The sacral parasympathetic outflow supplies the lower part of the digestive tract not supplied by the vagus as well as the bladder musculature and erectile tissue of the external genitalia.

Under normal conditions, both divisions of the autonomic nervous system work together to maintain homeostasis. In times of stress, however, the sympathetic nervous system accelerates various body activities and prepares the body for 'flight or fight'. Some of the body changes that occur as a result of the actions of the sympathetic nervous system are shown in Table 1.3. Following a period of stress the parasympathetic nervous system tends to slow down body acti-vities and bring the body back to its normal state. Parasympathetic

Table 1.3 Actions of the sympathetic nervous system

1. Dilates the pupils of the eyes to allow more light to enter
2. Increases the heart rate and the force of contraction of the heart muscle
3. Increases the respiratory rate
4. Dilates the airways into the lungs
5. Elevates the blood pressure through increased vasoconstriction and an increase in heart rate
6. Inhibits digestion — the motility of the gastrointestinal system is decreased and blood is diverted from the gut to the skeletal muscles
7. Stimulates the sweat glands to produce more sweat

action stimulates those functions of the body that are most appropriate to times of relaxation (e.g. digestion, bladder and bowel emptying and sexual function).

The autonomic nervous system is closely integrated with the body's metabolism and with the endocrine system. Although it is influenced by the individual's emotional state, it operates without voluntary control. Many parts of the autonomic nervous system are able to function on a spinal basis. However, the activity of the autonomic nervous system is normally under the control of centres located in the medulla oblongata and hypothalamus.

1.3 THE VENTRICULAR SYSTEM

The ventricular system is a series of cavities within the brain which contain a fluid known as cerebrospinal fluid. These cavities develop from the canal within the cranial portion of the neural tube as the latter structure develops into the brain. The system includes two lateral ventricles, the third ventricle, the cerebral aqueduct (Aqueduct of Sylvius) and the fourth ventricle. The shapes and locations of the various brain ventricles are shown in Figure 1.18. One lateral ventricle extends into each of the cerebral hemispheres. They lie below the corpus callosum, each extending in a large 'C' shape from the frontal lobe to the temporal lobe, though with a small spur (posterior horn) extending into the occipital lobe. The lateral ventricles communicate with one another and with the third ventricle through a pair of foramina known as the Foramina of Munro (interventricular foramina). The lateral ventricles are separated medially by a membranous partition known as the septum pellucidum.

The third ventricle is a small slit-like cavity in the centre of the diencephalon. The lateral walls of this cavity are formed mainly by the

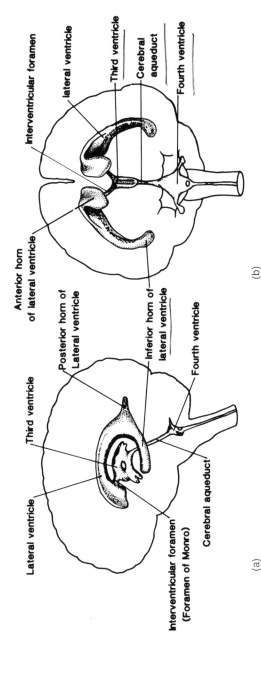

Figure 1.18 (a) Lateral view of the ventricles and (b) frontal view of the ventricles.

(a)

(b)

thalamus and to a lesser extent by the hypothalamus. It is connected posteriorly to the fourth ventricle by the cerebral aqueduct. The cerebral aqueduct is a narrow channel running within the mid-brain between the corpora quadrigemina and the cerebral peduncles. The fourth ventricle is a cavity which lies between the pons and medulla on one side, and the cerebellum on the other. It continues below into a narrow channel, the central canal, which is present in the lower medulla oblongata and throughout the length of the spinal cord. Cerebrospinal fluid escapes from the ventricular system through three foramina that are present in the roof and walls of the fourth ventricle. There are two lateral openings known as the Foramina of Luschka and a medial opening called the Foramen of Magendie.

The ventricles and the central canal are lined by ependymal cells. In each of the four ventricles there are complex tufts of small blood vessels and modified ependymal cells which form what are known as choroid plexuses. These plexuses are concerned with the formation of cerebrospinal fluid.

1.4 THE MENINGES

Three membranes, collectively known as the meninges, surround and protect the brain and spinal cord. From the outside these are the dura mater, arachnoid and pia mater. All three envelope the brain and spinal cord.

The dura mater is a tough, inelastic outer membrane, made of strong white fibrous connective tissue. In the head it is composed of two layers. The outer layer lines and adheres to the skull and is actually the periosteum of the cranial bones. The inner layer of the dura mater covers the brain and in certain locations extends down into the major fissures of the brain, where in doing so, it forms three major folds which divide the skull cavity into adjoining compartments. First, it extends down into the longitudinal fissure and then is reflected back on itself, forming a membranous septum between the two cerebral hemispheres known as the falx cerebri. This septum is actually a double thickness or fold of the inner layer of the dura. A similar, but smaller fold of the inner dura, called the tentorium cerebelli, extends between the occipital lobes of the cerebral hemispheres and the cerebellum in such a way as to form a roof or tent over the cerebellum. Finally, another fold of the inner dura extends between and separates the two cerebellar hemispheres. This latter fold is known as the falx cerebelli.

In certain areas within the skull, the two layers of the dura mater are separated from one another, forming spaces called cranial venous sinuses. These sinuses are filled with blood that flows from the brain to the heart. As we will see later, these sinuses are important in the absorption of cerebrospinal fluid into the blood stream.

In the vertebral canal, the dura mater is separated from bone (i.e. the vertebrae) by an interval, the epidural space, which contains fat and many small veins. It should be noted that the vertebrae have their own periosteal lining and thus the dura mater in the spinal canal is only a single layer. A comparable space to the epidural space is not found in the cranial cavity, except when artificially produced (e.g. by bleeding between the skull and dura mater following head trauma – extra-dural haemorrhage). The main blood supply to the dura mater is the middle meningeal artery, which is a branch of the external carotid artery. Extra-dural haemorrhage classically follows traumatic rupture of middle meningeal artery.

Immediately deep to the dura mater is the second or middle meninge called the arachnoid. The arachnoid is a thin, avascular, delicate, transparent, cobwebby layer. It does not follow each indentation of the brain but rather skips from gyrus to gyrus. The small space between the arachnoid and the dura mater is known as the sub-dural space. This space is ordinarily filled with small amounts of lymph-like material. The arachnoid is loosely attached to the inner meninge (the pia mater) by a fine network of connective-tissue fibres (trabeculae), so that a space is created between the arachnoid and the pia mater. This space is called the sub-arachnoid space. Cerebrospinal fluid circulates through the sub-arachnoid space. Several large spaces called cisterns represent enlargements in the sub-arachnoid space. The cisterna magna (cerebral medullaris) is located dorsal to the medulla and inferior to the cerebellum. The pontine and interpeduncular cisterns are located to the anterior brainstem, and the superior cistern is located posterior to the mid-brain.

The fourth ventricle of the brain communicates with the sub-arachnoid space via the Foramina of Luschka and Foramen of Magendie mentioned above. Cerebrospinal fluid circulates through the ventricles, enters the sub-arachnoid space via these foramina and is eventually absorbed into the venous system. We will look more at this circulation shortly. The arachnoid sends tuft-like extensions up through the inner layer of the dura mater into the venous sinuses. These extensions are called arachnoid villi, and they aid in the return of cerebrospinal fluid to the blood.

The pia mater is the innermost meninge and is intimately attached to the brain and spinal cord. It is composed of delicate connective tissue and contains the blood vessels that nourish the neural tissue of the brain and spinal cord. The cerebral blood vessels are adherent to the external surface of the pia mater. Unlike the other two membranes, it dips down into the invaginations of all the sulci of the brain and closely follows the convolutions of the gyri. The pia mater together with the arachnoid are known as the leptomeninges. Inflammation of the meninges is called meningitis which most often involves the leptomeninges.

In head injuries (see Chapter 4) bleeding may occur into the sub-arachnoid space (sub-arachnoid haemorrhage), into the sub-dural space (sub-dural haemorrhage) and between the outer dura mater and the skull (extra-dural haemorrhage). An extra-dural haemorrhage may result from bleeding meningeal vessels after a fracture of the skull, caused by a blow to the head. A sub-dural haemorrhage can be caused by the tearing of veins crossing the sub-dural space, which may follow after the sudden movement of the cerebral hemispheres relative to the dura and skull (e.g. as caused by head striking an immovable object such as a wall). A sub-arachnoid haemorrhage may result from the rupture of an aneurysm in a branch of the internal carotid or vertebral arteries. The presence of blood-stained cerebrospinal fluid obtained from a lumbar puncture is confirmatory of sub-arachnoid haemorrhage.

1.5 THE CEREBROSPINAL FLUID

Cerebrospinal fluid is a clear, colourless fluid, which is found in the ventricular system and the sub-arachnoid space. The brain and spinal cord actually float in the medium. Most of the cerebrospinal fluid is produced by the choroid plexuses of the ventricles of the brain. The volume of cerebrospinal fluid in the ventricles and sub-arachnoid space is about 120–140 ml, with approximately 23 ml in the ventricular system and 117 ml in the sub-arachnoid space. It has been estimated that cerebrospinal fluid is replaced about once every 6 hours. To maintain a constant volume, therefore, cerebrospinal fluid has to constantly move into the venous sinuses, and hence into the blood stream, via the arachnoid villi.

Cerebrospinal fluid produced in each of the lateral ventricles flows through the inter-ventricular foramen (Foramen of Munro) into the third ventricle. More fluid is produced in the third ventricle and all of it flows through the cerebral aqueduct (Aqueduct of Sylvius) to the fourth ventricle where more fluid is added. From the fourth ventricle, the fluid escapes into the sub-arachnoid space through one of the three foramina mentioned above. It then circulates around the brain and spinal cord and eventually reaches the arachnoid villi, where, by a process of osmosis, it is emptied into the great venous dural sinuses, particularly the superior sagittal sinus.

An obstruction to the passage of cerebrospinal fluid results in a back-up of cerebrospinal fluid and an increase in intra-cranial pressure. This condition, in which there is an accumulation of cerebrospinal fluid in either the ventricular system or sub-arachnoid space, is called hydrocephalus ('water on the brain'). If the obstruction to the flow is within the ventricular system itself (e.g. mid-brain tumours often cause constriction of the cerebral aqueduct leading to accumulation of cerebrospinal fluid in the lateral and third ventricles) the condition is called obstructive or non-communicating hydrocephalus. If the cerebrospinal fluid can get out of the ventricular system, but owing to a blockage in the sub-arachnoid space cannot then circulate properly to reach the arachnoid villi, the condition is called 'communicating hydrocephalus'. This latter form of hydrocephalus can occur if there are adhesions in the sub-arachnoid space due to past inflammation (e.g. meningitis) or may be due to haemorrhage into the sub-arachnoid space.

Hydrocephalus can occur in either adults or children but is most commonly associated with infants who have a congenital abnormality that blocks the flow of cerebrospinal fluid. The cerebral aqueduct and foramina of the fourth ventricle are common sites of obstruction. The flexibility of the infant skull causes the head to enlarge in response to the increased intra-cranial pressure. Initially, therefore in infant cases the compression of neural tissue is moderate. Surgical intervention is usually required whereby a tube (shunt) is placed in a ventricle above the blockage and the excess fluid shunted into one of several areas distal to the block including the cisterna magna, jugular vein or atrium of the heart. Hydrocephalus can also occur in adults as a result of tumours, meningitis and traumatic haemorrhage. As the skull is inflexible in adults, brain tissue can be rapidly compressed and immediate surgical procedures may be necessary to save the patient's life.

The normal functions of the cerebrospinal fluid are still uncertain. The fluid undoubtedly cushions the brain and spinal cord and minimizes damage that might otherwise result from sudden movements or from blows to the head and spine. The fluid plays a role in the diffusion of materials into and away from the brain, and it might well transport specific substances such as neuro-hormones from one part of the central nervous system to another.

Cerebrospinal fluid can be sampled by a procedure known as lumbar puncture and a variety of tests carried out to aid the medical diagnosis of a number of neurological disorders. The same procedure can be used to inject drugs to combat infections.

1.6 THE BLOOD SUPPLY TO THE BRAIN

1.6.1 Arterial blood supply

The arterial blood supply of the contents of the cranial cavity is derived from the paired internal carotid and vertebral arteries. The internal carotid arteries supply blood to the greater part of the cerebral hemispheres. However, the occipital lobes get their chief supply via the vertebral arteries, which also feed the brainstem, and cerebellum. The internal carotid artery gives rise to the ophthalmic, anterior cerebral, anterior choroidal, middle cerebral and posterior communicating arteries. The vertebral artery gives rise to the posterior inferior cerebellar artery, the anterior and posterior spinal arteries and the basilar artery. The basilar artery in turn gives rise to the anterior inferior cerebellar artery, the superior cerebellar artery and the posterior cerebral arteries.

The common carotid arteries ascend in the neck. At the level of the thyroid cartilage each divides into an external and an internal carotid artery. Each internal carotid artery enters the cranial cavity through a canal (the carotid canal) in the base of the skull, emerges alongside the optic chiasma and divides into an anterior and middle cerebral artery. The two anterior cerebral arteries are united by a small communicating branch called the anterior communicating artery. Prior to dividing into the anterior and middle cerebral arteries, the internal carotid gives rise to the opthalmic, posterior communicating and anterior choroidal arteries. The opthalmic artery is the first branch of the internal carotid. It enters the orbit to supply the optic nerve and eye. The posterior communicating artery connects the internal carotid with the posterior cerebral artery and has branches which help supply parts of the hypothalamus, sub-thalamus, thalamus, internal capsule and mid-brain. The anterior choroidal artery usually arises from the internal carotid distal to the posterior communicating artery. It helps supply the choroid plexuses of the lateral ventricles, optic tract, mid-brain, globus pallidus, posterior limb of the internal capsule and thalamus (for details see Chapter 3).

The vertebral arteries ascend in foramina (openings) in the transverse processes of the cervical vertebrae and enter the cranial cavity through the foramen magnum. On the ventral surface of the brainstem they join to form a single arterial stem, the basilar artery. This artery ascends in front of the brainstem and ends by dividing into two

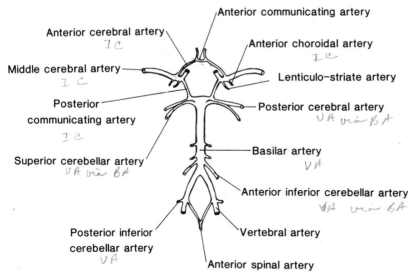

Figure 1.19 The Circle of Willis.

posterior cerebral arteries. Each of these is joined to the corresponding internal carotid artery by a communicating branch (posterior
communicating arteries). This forms what is known as the Circle of
Willis, i.e. a circle of arteries consisting of the two posterior cerebral
arteries, the two anterior cerebral, the two internal carotid arteries
and the posterior and anterior communicating arteries (see Figure
1.19). Although the Circle of Willis provides a link between the major
arteries that supply the brain, under normal conditions there is little
exchange of blood between the main arteries through the slender
anterior and posterior communicating arteries, since the arterial
pressure in the internal carotid arteries is similar to that in the basilar
artery. The Circle of Willis, however, provides alternative routes for
blood when one of the major arteries leading into it is occluded. For
example, if one of the posterior cerebral arteries is occluded where it
branches from the basilar artery, the pressure distal to the occlusion
will drop allowing blood from the internal carotid on the same side to
flow into the posterior cerebral via the posterior communicating
artery. These anastomoses (an anastomosis is a connection between
two tubular organs), however, are frequently inadequate, especially in
the elderly where the communicating arteries may be narrowed by
vascular disease (atherosclerosis). Unfortunately the Circle of Willis
and its immediate branches are also common sites for aneurysms
(sacs in blood vessel walls).

The regions of the cerebral hemisphere supplied by the various

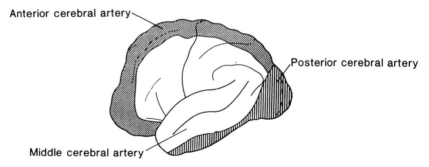

Figure 1.20 Lateral view of the left cerebral hemisphere showing the distribution of the major cerebral arteries.

cerebral arteries are shown in Figure 1.20. The middle cerebral artery, the largest branch of the internal carotid, travels laterally in the lateral fissure. Eventually it emerges from the lateral fissure onto the lateral surface of the cerebral hemisphere, as can be seen from Figure 1.20. Branches of the middle cerebral artery supply almost all of the lateral surface of the hemisphere, including the motor and sensory areas for the face, hand, arm, shoulder, trunk and pelvis. In the dominant hemisphere the region supplied by the middle cerebral artery also includes the major speech and language centre, making it the most important artery involved in pathologies associated with the occurrence of aphasia. While in the lateral fissure, the middle cerebral artery gives off several branches which supply structures such as the cortex of the insula and various sub-cortical structures. One branch, the lateral striate artery, supplies components of the basal ganglia such as parts of the caudate and lenticular nuclei and the internal capsule. (The blood supply to sub-cortical structures is described in further detail in Chapter 3.)

The anterior cerebral artery branches off the internal carotid artery near the olfactory tract. It travels along the corpus callosum in the longitudinal fissure and supplies all the medial surface of the cerebral cortex as far back as the parieto-occipital sulcus including the foot and leg areas of the motor strip. A branch of the anterior cerebral artery called the medial striate artery supplies part of the caudate nucleus, lenticular nucleus and internal capsule. The anterior cerebral artery also supplies the undersurface of the frontal lobe.

The posterior cerebral artery branches off the basilar artery at its terminal bifurcation and curves laterally around the mid-brain and then dorsally to the temporal and occipital lobes. It supplies the medial and inferior surface of the temporal lobe and the medial surface and pole of the occipital lobe. Branches of the posterior cerebral artery, which include the posterior choroidal arteries, the thalamo-perforating branches, and the thalamo-geniculate branches also supply parts of the mid-brain, the majority of the thalamus and part of the internal capsule.

Whereas the cerebrum is supplied primarily by branches of the internal carotid artery, the brainstem and cerebellum receive their arterial supply via branches of the basilar and vertebral arteries. The posterior spinal artery arises from the vertebral artery at the mid-medulla level and descends along the dorsal surface of the lower medulla and spinal cord. It helps supply the dorsal region of the medulla, including the nuclei cuneatus and gracilis and the dorsal portion of the spinal cord.

Two anterior spinal arteries arise from the vertebral arteries at the level of the olive. They unite almost immediately to form a single anterior spinal artery which descends in the anterior median fissure of the medulla and spinal cord. It supplies a medial wedge of the medulla including the hypoglossal nucleus, and the anterior portion of the spinal cord.

The posterior inferior cerebellar artery branches off the cerebral artery at the level of the mid-medulla and courses dorsally along the medulla and then curves upward along the inferior surface of the cerebellum. It supplies the dorso-lateral part of the medulla which contains the spino-thalamic and rubro-spinal tracts, the nucleus ambiguus, the dorsal motor nucleus of the vagus and the inferior cerebellar peduncle. It also supplies parts of the inferior vermis as well as the inferior surface of the cerebellar hemisphere and the deep nuclei of the cerebellum. Branches of this artery also supply portions of the choroid plexus of the fourth ventricle.

The anterior inferior cerebellar artery branches off the basilar artery at the level of the caudal pons and passes caudally and laterally to reach the inferior surface of the cerebellum. It helps to supply the tegmentum of the caudal pons, parts of the inferior vermis and the inferior surface of the cerebellar hemisphere and deep nuclei of the cerebellum.

Numerous small branches of the basilar artery form the major supply to the pons. These branches include paramedian branches, that supply the medial portion of the pons, excluding the tegmentum and circumferential arteries that curve backward to supply the lateral and dorsal portions of the pons.

The superior cerebellar artery branches off the basilar artery at the level of the mid-brain just below the point where it bifurcates into the two posterior cerebral arteries. It then passes back to the superior surface of the cerebellum. It contributes branches to the circumferential arteries that supply central and lateral parts of the crus cerebri and substantia nigra and lateral parts of the mid-brain tegmentum. It also supplies the superior vermis, superior surface of the cerebellar hemisphere and the deep nuclei of the cerebellum.

1.6.2 Venous blood supply

The brain is drained by two sets of veins, both of which empty into the dural venous sinuses which, in turn, empty into the internal jugular veins. These two sets of veins are known as the deep or great cerebral veins and the superficial cerebral veins. Before dealing with each of these, however, it is necessary to have an understanding of the dural venous sinuses.

The major dural venous sinuses include the superior sagittal sinus, which runs along the superior portion of the longitudinal fissure at the junction of the falx cerebri and the cranial dura and the inferior sagittal sinus, which runs deep within the longitudinal fissure along the deep margin of the falx cerebri. These two sinuses are joined together by the straight sinus, which runs along the mid-line crest of the tentorium cerebelli at its junction with the falx cerebri. The straight sinus empties into the transverse sinus. The two transverse sinuses pass laterally from the junction of the straight sinus and superior sagittal sinus and course downwards to leave the cranium through the jugular foramina and become the internal jugular veins. Another important venous sinus, the cavernous sinus, is a large, rather diffuse sinus located around the sella turcica. It drains eventually into the transverse sinuses and jugular veins via the superior and inferior petrosal sinuses. A sinus known as the occipital sinus runs in the falx cerebelli. It opens into the confluence of sinuses where the straight, transverse and superior sagittal sinuses meet.

The cerebral veins themselves have very thin walls, no muscular layer and no valves. As suggested above, they are found on the surface of the brain or in the substance of the central nervous system. The superficial cerebral veins lie in the sulci and are more external than the arteries. Those draining the cerebral hemispheres can be divided into superior, middle and inferior cerebral veins. A variable number of superior cerebral veins drain the superior surface of the cerebral cortex and empty into the superior sagittal sinus. The middle cerebral veins drain most of the lateral and inferior surface of the cerebral hemispheres. These vessels are found in the lateral fissure and empty into the cavernous sinus. The inferior cerebral veins drain the lateral occipital gyrus and part of the temporal lobe. These vessels drain into the transverse sinuses. All dural sinuses receive blood from their immediate vicinity.

The deep cerebral veins conduct blood from the centre of the cerebrum and converge upon a single large vein called the great cerebral vein or vein of Galen. The vein of Galen empties into the straight sinus. The veins that drain the brain stem and cerebellum also drain into the dural sinuses, some via the great cerebral vein.

Although most of the venous blood coming from the brain eventually leaves the cranial cavity via the internal jugular veins, other exits from the cranial cavity also exist. Some dural sinuses connect with the veins superficial to the skull by emissary veins. For example, the cavernous sinus is connected with emissary veins, including the ophthalmic vein, which extends through the orbit. A number of such communications via emissary veins exist between the dural sinuses and extra-cranial veins.

1.6.3 The blood–brain barrier

There is a free and rapid passage of substances between the brain tissue and the cerebrospinal fluid, but there is a barrier between the blood and brain tissue, the blood–brain barrier. This maintains a constant milieu for brain metabolism and is a protection against noxious substances present in the circulation (e.g. urea). The barrier is equally effective against antibiotics except when inflammation changes its characteristics.

The capillary network of the central nervous system is extensive, especially in the grey matter. The capillaries in the central nervous system have permeability characteristics that are fundamentally different, however, from those of capillaries elsewhere in the body. The endothelial cells in brain capillaries form a continuous rather than a fenestrated layer. The capillaries form the blood–brain barrier which is a major obstacle to the free movement of substances from the blood to the brain. In fact, the diffusion of most substances is definitely limited, except for lipid-soluble compounds and water. The importance of the barrier is that it may prevent potentially therapeutic drugs from reaching the brain. In such cases, these drugs may be administered directly into the cerebrospinal fluid via lumbar puncture.

1.7 SPEECH AND LANGUAGE CENTRES OF THE BRAIN

Within the dominant hemisphere, two major cortical areas have been identified as having specialized language functions. These two areas

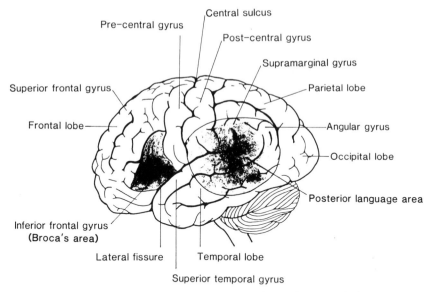

Figure 1.21 Lateral view of the brain showing the major speech–language centres.

are located in the peri-sylvian region (region surrounding the Fissure of Sylvius) and include the anterior or motor speech–language area and the posterior or sensory speech–language area (see Figure 1.21). The two major speech–language areas have been largely identified by the study of patients in whom these areas were damaged by either occlusion of blood vessels or by war injuries. Until recently the most reliable information concerning the areas of the brain important for language has come from the results of long-term studies of language-disordered patients whose lesions were identified at post-mortem examination. Since the mid-1970s, computer tomography has been extensively used to localize lesions associated with various types of language disorders. Using this technique, investigators were able for the first time to study the relationship between regions of brain damage and disturbances in language function in the living subject, particularly in those cases where the lesions involved deep cerebral structures below the level of the cortex. More recently introduced brain imaging techniques such as positron emission tomography (PET) scanning and magnetic resonance imaging (MRI) scanning have further expanded the ability of researchers to localize lesions associated with specific language disorders in living subjects.

The anterior speech–language area was first identified by Paul Broca in the mid-nineteenth century and is, therefore, commonly referred to as Broca's area. Broca's area occupies the pars opercularis and pars triangularis of the inferior frontal gyrus (also called the third frontal convolution) which represents Brodmann areas 44 and 45 and lies immediately in front of that part of the cortical motor strip devoted to the peripheral organs of speech.

The posterior speech–language area lies posterior to the Fissure of Rolando. The existence of this area was first indicated by Carl Wernicke in the 1870s. Reports in the literature on the location and extent of the posterior speech–language area vary widely. Wernicke originally indicated that the auditory association cortex (Brodmann area 22) of the dominant hemisphere (Wernicke's area) acts as a language centre. Subsequent authors have modified and extended this area to include a greater part of the temporal lobe and parts of the parietal lobe. Currently most descriptions of the posterior speech–language area include within its boundaries the lower half of the post-central gyrus, the supra-marginal and angular gyri, the inferior parietal gyrus and the upper part of the temporal lobe, including parts of the superior and second temporal gyri and Wernicke's area.

In the majority of people (approximately 96%) the language areas are located in the left cerebral hemisphere. The anterior and posterior speech–language areas communicate with one another via the arcuate fasciculus, a bundle of association fibres that travel as part of a long association tract called the superior longitudinal fasciculus. The arcuate fasciculus sweeps around the insular region and its fan-shaped ends connect parts of the temporal and frontal lobes.

It appears from observations of the speech and language deficits of patients with known lesion sites, that the posterior language area is devoted to those tasks having to do mainly with the recognition, comprehension and formulation of language. As this region of the cerebral hemisphere also deals with the reception of sensory stimuli through the auditory, visual and somatosensory (body sensations) systems, it is believed that language data, which are transmitted through these same modalities, are processed in this area of the hemi-sphere. In contrast to the posterior area, the anterior language area is involved with the programming and execution of overt acts, such as those that result in speaking, writing or gesturing.

Obviously, the anterior and posterior language areas do not function separately. Rather, for normal language function to occur, the anterior and posterior areas must be in communication. As described above, the two areas are connected primarily by the arcuate fasciculus, although other connections through sub-cortical structures such as the thalamus also exist. It is believed, therefore, that information concerning the spoken or written word is decoded and translated in the posterior language area and as a consequence of this, damage to the posterior area results in impairment of the ability to comprehend written or spoken word. Comprehension of speech takes place when auditory impulses are transmitted to the auditory cortex of both hemispheres and subsequently relayed to the posterior language area in the dominant hemisphere for translation. Comprehension of written words, on the other hand, takes place when visual impulses are transmitted to the visual cortex in the occipital lobes of each hemisphere and are subsequently relayed to the posterior language area. Following translation in the posterior area, information is then passed to the anterior area via the arcuate fasciculus for the complex programming of the speech organs in order to make a verbal response. Damage to the anterior language area, therefore results in language production problems involving planning and execution. The anterior area feeds information to the primary motor cortex of each hemisphere and from here instructions are sent via the motor pathways to the muscles of the speech mechanism to produce a verbal response.

In addition to the above two areas, a third cortical area has been identified which may be involved in speech and language functions. This third area is small and lies mainly in the medial surface of the frontal lobe (i.e. within the longitudinal fissure) immediately anterior to the foot region of the primary motor strip. It is known as the secondary speech area or the supplementary motor area. Lesions in this area often lead to temporary aphasias and difficulty in producing rapidly alternating movements such as required in the oral region during speech. The entire secondary speech area can be excised, however, without causing a permanent language disorder.

Although the centres described above, including the anterior and posterior language areas and the secondary speech area, comprise the primary speech–language centres of the brain, it is evident that other brain structures also play a role in language function. In particular these other areas include the parietal, temporal and occipital association areas and various sub-cortical structures such as the basal ganglia.

1.8 NEUROLOGICALLY-BASED COMMUNICATION DISORDERS – DEFINITIONS

Speech sounds are produced by regulating the exhaled air stream as it passes from the lungs to the atmosphere. This regulation is brought about by movements of the jaw, lips, tongue, soft palate, pharynx and vocal cords which vibrate the air column and alter the shape of the

Table 1.4 Basic processes involved in speech production

---

1. A concept of the speech output has to be formed and symbolically formulated for expression as speech – disruption at this level is associated with APHASIA
2. The symbolically formulated concept of speech output has to be externalized as speech through the concurrent motor functions of respiration, phonation, resonance, articulation and prosody – disruption at this level is associated with DYSARTHRIA
3. Prior to externalization as speech, a programme has to be developed which determines the sequence of muscle contractions required to produce individual sounds and words that comprise the intended speech output – disruption at this level leads to APRAXIA OF SPEECH

---

vocal tract. The movements are brought about by the contraction of skeletal muscles, which in turn are regulated by nerve impulses. The entire process of speech production is, of course, controlled by the central nervous system.

The efficient execution of speech production requires the smooth sequencing and co-ordination of three basic neurological processes (see Table 1.4). Impairment of each of these three processes results in a distinctive communication disorder. Impairment in the first process involving the organization of concepts and their symbolic formulation and expression is caused by pathological processes that damage the cerebral hemisphere that contains the speech–language centres, thereby leading to aphasia. Aphasia has been defined as the loss or impairment of language function caused by brain damage. It is an impairment, due to brain injury, of the capacity to interpret and formulate language symbols. Aphasia is a multi-modality disorder (i.e. it manifests in difficulties in speaking, reading and writing) and involves a reduction in the capacity to decode (interpret) and encode (formulate) meaningful linguistic elements (i.e. words [morphemes] and larger syntactical units such as sentences). The aphasic patient is impaired in the comprehension, formulation and expression of language although the relative amount of loss in each of these areas varies between one type of aphasia and another (see Chapter 2). All aphasic patients do, however, show some loss in all three of these areas.

Impairment in the second process involving the motor production of speech is associated with dysarthria, a group of speech disorders resulting from interference with any of the basic motor processes involved in speech production (a more complete definition of dysarthria is given in Chapter 8). Damage located at a number of different sites in the nervous system including the cerebrum, brainstem or cerebellum can be associated with dysarthria, in each case a different type of dysarthria resulting (see Chapters 8, 9 and 10).

Impairment of the third process involving the programming of motor actions involved in speech production is caused by damage to those circuits located in the cerebrum devoted to determining the sequence of muscle contractions required to produce speech. Such impairment leads to a communication disorder called apraxia of speech (verbal apraxia). Apraxia of speech is, therefore, a disorder of motor speech programming which manifests primarily as errors in the articulation of speech and secondarily by what are thought by many researchers to be compensatory alterations of prosody (e.g. pauses, slow rate, equalization of stress). It is a disorder in which, although the muscles of the speech mechanism are neither paralysed nor weak, the individual has difficulty speaking because of a cerebral lesion that prevents executing voluntarily and on command the complex sequence of muscle contractions involved in speaking.

Whereas aphasia is considered to be a language disorder, dysarthria and apraxia of speech are motor speech disorders involving disruption of the motor control of speech. Although each of these three disorders are distinctive, it should be remembered that they can occur in combination and consequently a neurologically disordered patient may exhibit characteristics of more than one of these disorders. Many aphasic patients, for instance, may exhibit some apraxic elements and also some type of dysarthria.

1.9 NEUROPATHOLOGICAL SUBSTRATE OF NEUROGENIC SPEECH AND LANGUAGE DISORDERS

Any type of neuropathology capable of producing structural alterations in an appropriate portion of the brain, whether that be the cerebral cortex, sub-cortical structures, brainstem or cerebellum, is capable of producing a communication deficit in the form of either a speech or language disorder or both. Widely diverse disease processes affecting particular brain structures may produce similar abnormalities in brain function. Consequently, it is the neuroanatomical location of the brain damage rather than the causative agent that largely determines the nature of the communicative deficit. The specific causative disease, however, can usually be identified by certain characteristics of the patient's history, the specific pattern of neurological dysfunction and by appropriate laboratory and/or clinical examinations.

The major diseases of the nervous system that produce speech and language disorders are cerebrovascular disease, neoplastic disorders, head trauma, degenerative disease, toxic conditions, demyelinating disorders and infectious diseases.

1.9.1 Cerebrovascular disorders

Disorders in which one or more of the blood vessels of the brain are primarily involved in the pathological process are the most common form of neurogenic disease. Consequently, in peacetime cerebrovascular disorders are the most common cause of brain damage relating to the occurrence of neurogenic speech and language disorders.

When the blood supply to the brain is seriously disturbed spontaneously (i.e. not due to trauma or surgical ligation of cerebral vessels) the condition is referred to as a cerebrovascular accident or stroke. The three major characteristics of cerebrovascular accidents include; (1) an abrupt onset of focal brain dysfunction; (2) the disability produced (including any speech or language deficit) is worst at onset or within a short period of onset; (3) if the patient survives, the disability tends to improve, in some cases partially, in others almost totally.

Cerebrovascular accidents can be divided into two major types, ischaemic strokes and haemorrhagic strokes. Ischaemic strokes occur when the supply of blood to part of the brain suddenly becomes inadequate for the brain cells to function. Haemorrhagic strokes occur when a blood vessel ruptures and blood either rushes through the brain tissue destroying it (intra-cerebral haemorrhage) or collects outside the brain in one of the spaces between the meninges causing compression of the brain within the skull.

Ischaemic strokes can arise in two ways – firstly through occlusion of the vessel by thrombus formation (cerebral thrombosis) and secondly through occlusion of the vessel by an embolus (cerebral embolism). Thrombosis is most commonly associated with atherosclerotic changes in the blood vessel wall. However, it can also be associated with inflammatory disorders which affect the blood vessels such as giant-cell arteritis, syphilitic endarteritis and systemic lupus erythematosus. Thrombotic stroke usually develops abruptly, often during sleep or shortly after rising. In some cases, however, it may be preceded by transient 'warning' signs in which case it has a step-wise onset over several hours or days. Thrombotic strokes are the most common type of cerebrovascular accident.

Embolic strokes are almost always abrupt in onset and the patient is only rarely forewarned by transitory symptoms. Embolism is now well recognized as a frequent and important source of stroke. The potential sources of emboli are remarkably widespread. For a long time it was believed that almost all emboli came from the heart, as a result of small pieces of mural thrombosis (from the walls of the heart) becoming dislodged from the cardiac wall by atrial fibrillation or other cardiac arrhythmia. Angiography has demonstrated that calcified plaques associated with atherosclerosis particularly in the carotid vessels, are also frequent sources of cerebral emboli. Cardiac surgery and bacterial endocarditis are less common but also real sources of emboli. Consequently, disorders such as rheumatic heart disease with atrial fibrillation, atrial fibrillation with coronary heart disease, recent myocardial infarction with mural thrombus formation, or bacterial endocarditis all predispose to brain damage as a result of cerebral embolism. Occasionally, emboli emanate from the lungs or even the great veins and on rare occasions emboli of tumour cells may become lodged in the vessels of the brain.

Of central importance in both types of ischaemic stroke is the fact that they deprive brain tissue of needed oxygen. Both thrombosis and embolism cause acute ischaemia in the tissues receiving their vascular supply from the occluded vessel which, in turn, produces an area of cell death (infarct). Embolic infarctions develop much more rapidly than thrombotic infarcts. Both neurones and the myelinated pathways are affected but the white matter is considerably less sensitive to ischaemia than grey matter (i.e. the cortex). The centre of an infarct will be totally destroyed, but towards the periphery there may be preservation of white matter pathways and there is often a surrounding zone of lesser ischaemia in which cells cease to function on a temporary basis, but cell death does not occur. In time, some of these injured neurones recuperate sufficiently to resume function and many white-matter pathways survive to carry impulses again. This delayed return of function to certain areas within an infarct provides one explanation (but not the only one, e.g. reduction in degree of associated oedema, etc. is another) of the spontaneous recovery so often seen in many types of aphasias. The final outcome of an infarct is a cyst-like area from which both neurones and white matter have disappeared, surrounded by a scarred, sclerotic zone of glia.

Ischaemic attacks vary in their severity. At one extreme a major vessel may be almost totally occluded by thrombosis or by a major embolism. At the other extreme the ischaemic attack may be only transient and therefore may not deprive the brain tissue of oxygen for long enough to cause permanent brain damage. Transient ischaemic attacks tend to involve repetitive stereotyped attacks of focal neurological function followed by complete recovery. Formerly considered to be caused by episodic narrowing or 'spasm' of blood vessels, it is now thought that transient ischaemic attacks are produced by repeated embolization of small particles from proximally-located atherosclerotic plaques in the large vessels of the neck (e.g. the internal carotid arteries).

Haemorrhagic stroke may have a sudden onset with evolution to maximum deficit occurring in a smooth fashion over several hours. Cerebral haemorrhage, when the result of vascular disease (as opposed to trauma, etc.), is most often associated with hypertension but it may occur with a variety of pathologies affecting the cerebral vessels such as aneurysm, angioma, arterio-venous malformation, blood dyscrasia or arteritis. Anti-coagulant therapy (e.g. warfarin therapy) is acknowledged as a frequent cause of cerebral haemorrhage which can lead to the production of speech and language disorders.

Most haemorrhages occur during activity and without warning. Onset therefore is abrupt and is associated with severe headache, vomiting and often loss of consciousness. The most common site for intra-cerebral haemorrhages is the region of the internal capsule in which case the patient suddenly complains of something wrong in the head, followed by headache, dysarthria and/or aphasia, paralysis down the opposite side of the body and variable alterations in consciousness. With brainstem haemorrhage there is usually rapid loss of consciousness and often death in a short time. Cerebellar haemorrhages are associated with vertigo, nausea and ataxia followed by coma and often death. Overall the prognosis for recovery for haemorrhagic strokes is poorer than for ischaemic strokes.

Intra-cerebral haemorrhages usually involve deeper structures of the brain than the cerebral cortex and produce brain damage both by local destruction and by compression of surrounding brain tissue. The force of blood coming from a ruptured blood vessel directly damages the brain tissue. This extravasated blood forms a clot called a haematoma which increases in size and displaces surrounding brain tissue. As the skull is a fixed box, the intra-cranial pressure increases as the clot develops causing compression of the brain tissue. Secondary rupture into the ventricular system or sub-arachnoid space may also occur. Emergency evacuation of the intra-cerebral clot is of value in aiding the relief of symptoms in some cases.

In addition to hypertension, rupture of the intra-cranial aneurysms is another major cause of haemorrhagic strokes. An aneurysm is a thin-walled enlargement of a blood vessel usually found in the Circle of Willis or its major branches. Aneurysms tend to occur at junctions or bifurcations and are believed to represent congenital deficiencies in the development of the vessel wall. They tend to increase in size and may produce cranial nerve palsies or focal seizures by compression of adjacent structures prior to rupture. Rupture usually occurs during activity and produces severe headache, collapse and unconsciousness. Generally, bleeding occurs into the sub-arachnoid space but may also occur into the brain tissue forming an intra-cerebral haemorrhage. In the latter case prolonged unconsciousness and focal signs such as hemiplegia, hemianaesthesia and aphasia may also occur.

1.9.2 Neoplasms

Intra-cranial tumours (neoplasms) are the third most common disorder of the nervous system after cerebrovascular diseases and infections. Although they are, in general, a less frequent cause of speech and language disorders than cerebrovascular accidents, intra-cranial tumours are nonetheless not uncommon as aphasia-producing lesions. Such tumours may be either benign or malignant. Tumours affecting the central nervous system are said to be primary tumours if they grow from cells within the cranial cavity itself, or secondary (metastases) if they travel to the brain from a primary tumour elsewhere in the body (e.g. breasts, lungs, etc.)

Brain tumours produce symptoms in three ways. First, because tumours are 'space-occupying lesions', as they develop they cause the intra-cranial pressure to rise, leading to compression and distortion of surrounding brain structures. Secondly, as tumours grow they may disrupt the blood supply to specific regions of the brain or may interrupt the circulation of cerebrospinal fluid, such as by compressing the ventricles or occluding the cerebral aqueduct, thereby leading to increased intra-cranial pressure. Thirdly, the tumour may directly damage the brain tissue in a localized area. The direct effect produces symptoms and signs (e.g. paralysis down one side of the body, epileptic fits, etc.) which become gradually worse and more extensive as the tumour grows, in complete contrast to the sudden onset of a cerebrovascular accident. Tumours growing in the dominant hemisphere may cause progressively increasing aphasia.

Intra-cranial tumours can be divided into two major types, namely intra-cerebral tumours and extra-cerebral tumours. Intra-cerebral tumours are those that directly involve the cerebral tissues while extra-cerebral tumours arise from tissues outside the brain itself (e.g. the meninges and skull bones). By far the majority of intra-cerebral neoplasms are gliomas which develop from the supporting tissue of the brain (i.e. the neuroglial cells), tumours of nerve cells being rare. The various types of glioma take their names from the particular neuroglial cells involved and include astrocytomas, oligodendro-cytomas and microgliomas. Some intra-cerebral tumours called ependymomas develop from the cells lining the ventricles (ependymal cells) while others called medulloblastomas develop from primitive cells in the roof of the fourth ventricle. Any variety of intra-cerebral tumour is capable of producing a speech and/or language disturbance dependent upon its location in the brain. On the other hand, language disorders are rarely caused by extra-cranial tumours which include among others those growing from the meninges (meningiomas), sheaths of peripheral nerves (neurofibromas, e.g. acoustic neuromas), the skull bones (ostoemas) and the pituitary gland (e.g. various adenomas). These tumours are mostly benign and do not directly cause destruction of cerebral tissues as in the case of intra-cerebral tumours but instead may produce abnormal neurological signs as a result of distortion or displacement of cerebral tissue.

Although intra-cerebral tumours cause language disorders more often than extra-cerebral tumours, in neither variety does aphasia usually become a major complaint until late in the course of the disease. The reason why aphasic symptoms usually only appear late in the disorder is that intra-cerebral neoplasms infiltrate the cerebral tissues widely before producing focal destruction. Further, extra-cerebral tumours tend to develop slowly, allowing considerable accommodation by the cerebral tissues with only minimal disruption of functions until late in the course of the disorder. If aphasic symptoms do appear early in the development of a tumour, it is usually because the tumour has either disrupted the cerebral blood supply or interfered with the circulation of cerebrospinal fluid. Although the particular neurological signs, including any speech or language disorder, associated with the presence of a tumour may give some indication as to the location of that tumour in the brain, due to the local effects of the tumour, it must be remembered that distortion and/or compression of cerebral tissue may actually occur at a distance from where the tumour is located. Consequently the par-ticular speech–language deficit exhibited may have no direct relationship with the location of the tumour itself.

Some intra-cranial tumours occur more frequently in persons belonging to a particular age group. Others produce characteristic syndromes because of their predilection for certain sites. In the first two decades of life, gliomas of the cerebellum, brainstem and optic nerves are common. Meningiomas, neurofibromas of cranial nerves, gliomas of the cerebral hemispheres and pituitary tumours are more common in the middle decades. Metastatic tumours are most common in the later decades of life.

Surgical removal of tumours may also be the cause of speech and/ or language deficits. Often such surgery requires destruction of both the grey and white matter that has been infiltrated by the tumour. In addition, there is evidence to suggest that radiotherapy often given as part of the treatment of intra-cranial tumours may cause damage to the nervous tissue which may manifest several years later as impaired language and cognitive abilities.

1.9.3 Head trauma

Traumatic head injury is a common cause of speech and/or language disorders, particularly in young adult males. Although head injury can result from a variety of different incidents, in peacetime the majority of head injuries are caused by motor-vehicle accidents. Over the years, traumatic head injury cases, particularly subjects with brain injury resulting from war wounds, have provided an important source of language-disordered patients for academic study. The nature of the speech and language deficits seen in association with traumatic head injury is discussed in detail in Chapter 4.

1.9.4 Degenerative disorders

Degenerative diseases of the nervous system include a broad range of disorders all of which are characterized clinically by progressive deterioration of neurological function and pathologically by cellular depletion with atrophy of nervous tissue. Those affecting the cerebrum and particularly the cerebral cortex are characterized by progressive dementia in the middle or later decades of life and include disorders such as Alzheimer's disease and Pick's disease. Both of these conditions are associated with an initial dulling of intellectual abilities with impairment of memory and confusion. Language impairment has also been reported to be a common occurrence in these disorders. Progressive deterioration occurs in months or years, leading to profound dementia, immobility and death from secondary infections. Focal signs such as hemiparesis, hemianaesthesia, cranial nerve palsies and increased intracranial pressure do not occur. The language disorders associated with the major clinically-encountered forms of dementia are discussed in Chapter 6.

In addition to those disorders characterized by atrophy of the cerebral cortex, some degenerative disorders are associated with degeneration primarily of the region of the basal ganglia. Examples of the latter conditions include Huntington's chorea and Parkinson's disease. Huntington's chorea is a dominantly inherited disorder characterized by mental deterioration and choreiform movements which may involve the muscles of the speech mechanism causing a speech disturbance called hyperkinetic dysarthria. The characteristics of hyperkinetic dysarthria are described in Chapter 9. Parkinson's disease is considered a degenerative disease with prominent motor system involvement and minimal organic mental symptoms. Tremor, muscular rigidity, bradykinesia (slowness and lack of movement), a mask-like face, stooped-flexed posture, a shuffling gait and hypokinetic dysarthria are characteristic features of the disorder (see Chapter 9).

1.9.5 Toxic disorders

Toxins are poisons which may be either produced within the body (e.g. when the kidneys fail) or may be introduced from outside. A wide range of different substances may interfere with the normal functioning of the nervous system including a large number of drugs (e.g. barbiturates, tranquillizers, some antibiotics, some antidepressants, etc.), heavy metals (e.g. lead, mercury, arsenic, etc.), organic phosphates (widely used in insecticides) and alcohol. Some toxic disorders of the nervous system are capable of producing a speech or language deficit as part of their overall neurological impairment. For example, tardive dyskenesia (see Chapter 9), a toxic disorder resulting from long-term treatment with anti-psychotic drugs, has as one of its symptoms a hyperkinetic dysarthria.

Probably the best known toxic disorder of the nervous system is Wernicke–Korsakoff syndrome (Chapter 6). This is a well-known complication of chronic alcoholism characterized by paralysis of eye movements, ataxia, variable alterations of consciousness, confusion, disorientation and memory loss. These patients also have a tendency to confabulate, often in an elaborate and colourful manner. Most of these patients also have a polyneuritis (inflammation of many nerves).

1.9.6 Demyelinating disorders

Demyelinating diseases comprise a group of chronic disorders in which spontaneous degeneration of the myelin sheaths of nerve fibres in the central nervous system is the primary pathological alteration. Multiple (disseminated) sclerosis (see Chapter 10) is the most important disorder in this category of disease. In a typical case of multiple sclerosis various symptoms of focal damage to the central nervous system appear and disappear over a prolonged period of time due to numerous scattered areas of demyelination in almost any area of the central nervous system, including the cerebrum, brainstem and cerebellum.

Multiple sclerosis only rarely causes a language deficit and when it does, the language problem usually takes the form of a word-finding difficulty and almost always occurs in combination with severe amnesia and/or dementia. Patients with multiple sclerosis do, however, more often exhibit a speech deficit in the form of a mixed dysarthria (see Chapter 10).

1.9.7 Infectious disorders

The nervous system and its coverings can be infected by the same microorganisms that affect other organs of the body. Infections of the nervous system are classified according to the major site of involvement and type of infecting organism. Infection of the meninges (usually the leptomeninges) is called meningitis while inflammation of the brain is referred to as encephalitis. In some cases both the meninges and brain may be infected, a condition called meningo-encephalitis. Inflammation of the spinal cord is known as myelitis.

There are three major types of meningitis. These include: pyogenic meningitis, caused by a pus-forming bacteria (e.g. meningococci, pneumococci and the influenza bacillus); tuberculous meningitis, caused by the tubercle bacillus; and viral meningitis, caused by a variety of different viruses (e.g. polio, mumps, etc). Meningeal infections by common bacteria produce obvious systemic and neurologic symptoms which include pyrexia (fever), headache, nausea, vomiting, photophobia (avoidance of bright light), neck stiffness or rigidity, a positive Kernig's sign and alterations in the level of consciousness. Signs of focal damage to the nervous system are rare. Viral meningitis produces a similar but less severe clinical picture.

As in the case of meningitis, encephalitis may be caused by either pyogenic bacteria or viruses. In addition, in some regions of the world encephalitis may also be due to various forms of parasite acquired from animals (e.g. hydatid disease). The general features of encephalitis include moderate headache, vomiting, confusion, delirium and increasing drowsiness eventually leading to coma. Kernig's sign is negative and, unless the meninges are also involved, there is little neck stiffness.

Most varieties of intra-cranial infection produce rather widespread neurological symptomatology and any associated language disorder is, therefore, liable to be lost amongst other neurobehavioural and cognitive dysfunctions. Occasionally, however, a significant aphasia can be traced to a central nervous system infection. Currently the most common infection reported to give rise to aphasia syndromes is herpes simplex encephalitis.

Aphasia can also result from the formation of intra-cerebral abscesses. A cerebral abscess is a pus-filled cavity in the brain which develops around a localized bacterial infection. Abscesses most commonly develop in the frontal and temporal lobes. Prior to anti-biotic drugs, temporal-lobe abscesses were a frequent source of aphasia secondary to chronic ear infections. In a manner similar to other types of 'space-occupying lesions' such as intra-cerebral tumours, as it grows an abscess can produce symptoms by compres-sing and distorting surrounding brain structures and by interrupting the vascular supply or the flow of cerebrospinal fluid.

CHAPTER TWO

Bostonian and Lurian aphasia syndromes

Aphasia has been defined as the loss or impairment of language caused by brain damage (see Chapter 1). Although aphasic disability is complex, many aphasic patients are clinically similar and fall into recurring identifiable groups. Over the years, a bewildering amount of nomenclature has been used to describe and classify the various aphasia syndromes. This vast amount of terminology serves to confuse the student of aphasiology in that, dependent upon their own individual concept and model of language, different authors have often used different terms to describe the same aphasic disturbance. For instance, at various times the aphasic disturbance associated with damage to the anterior language centre has been referred to as Broca's aphasia, motor aphasia, efferent motor aphasia and verbal aphasia.

In an attempt to unravel at least some of the terminological tangles associated with aphasia, prior to looking at the major contemporary aphasia classifications systems, it is useful to briefly review the history of aphasia in terms of the models of language that have been proposed by various authors. A brief look at the history of aphasia also allows a better appreciation of some of the controversy that still surrounds many aspects of aphasia, including arguments regarding the role of various speech–language areas and other brain structures in language.

2.1 MODELS OF LANGUAGE – A BRIEF HISTORY

Although the existence of a wide variety of language disorders has been recognized since as far back as Hippocrates in 400 BC, much of what is currently known about aphasia has come from the large amount of research into this area that has taken place since the middle of last century. By the early nineteenth century, two clearly separate schools of thought regarding the brain's function in language had developed. One school was comprised of those investigators who believed that specific (mental) functions were subserved by specific areas of the brain. Investigators supporting this viewpoint became known as 'localizationists'. By correlating specific language disorders exhibited by individuals during life with lesions of specific brain structures determined at post-mortem, the localizationists designated certain regions of the brain as centres responsible for specific language functions. Some localizationists even went as far as to denote specific areas of the brain as centres for mental functions such as love, pride and greed as well as speech. Opponents to the localizationist viewpoint, known as 'holists', believed that mental function was the product of the entire brain working as a unit and that mental ability was a reflection of total brain volume. This latter viewpoint became known as the holistic viewpoint.

Aphasia researchers over the past century or so can generally be divided into one or other of the above two schools. Many of the problems associated with the study of aphasia, such as the diverse terminology, has resulted from the controversy generated between the localizationists versus the holists. One of the early localizationists was Gall (1758–1828). He was the first to relate speech to a particular area of the brain. In a description of what he called two-speech disordered patients, Gall postulated the existence of an organ for words and language in the anterior portions of the brain (i.e. the frontal lobes). Bouillaud (1825) collected clinical as well as pathological evidence to support Gall's hypothesis. In a study of 850 cases, he found lesions in the frontal lobes of 116 patients who were speech defective. He insisted that it was necessary to distinguish two different phenomena in the act of speech: first, the power of creating words (internal speech) as a sign of our ideas; and secondly the power of articulating these same words (external speech). On this basis, Bouillaud suggested that there are two causes which can lead to loss of speech, each in its own way, one by destroying the organ of memory of words, the other an impairment in the nervous principle which directs the movements of speech. The validity of Bouillaud's division of aphasic disorders into an articulatory and an amnesic category is still generally accepted under the rubric of non-fluent and fluent types of aphasia.

Opposed to this early localization point of view was Flourens (1824) who stated that all parts of the brain were equipotential and that specific areas for specific purposes did not exist. According to him, if an area of the brain was damaged due to disease or injury, any other area of the brain could take over that function. Although we now know that this may happen to some degree, it clearly does not occur, especially in adult patients, to the extent predicted by Flourens. For instance, most global aphasics remain severely language impaired and do not exhibit the degree of recovery that could be expected if Flourens' hypothesis was correct.

Another opponent of Gall and Bouillaud was Andral (1840). In a study of 37 patients with lesions in the frontal lobes (found at autospy), Andral reported only 21 having speech disorders. In addition, he also reported having seen 14 cases of aphasia with lesions not involving the frontal lobes, but rather confined to post-rolandic areas. He concluded that loss of speech is not a necessary result of lesions in the anterior lobes, and that speech disorders can also result from lesions not involving the frontal lobes.

The major stimulus for the revolution in aphasia research was provided in 1861 by Paul Broca. On the basis of the autospy findings of two aphasic patients that Broca had cared for in the last months of their lives, Broca reported that the 'motor speech area' was located in the third frontal convolution. Broca interpreted his findings as supporting the Gall–Bouillaud thesis that the seat of language was in the frontal lobes. He emphasized, however, that he did not mean to imply that all forms of aphasia were related to frontal lobe disease, only the motoric type which he called aphemia (loss of speech) and which was essentially the same as the articulatory aphasia proposed by Bouillaud. After further study in 1865 Broca was the first to draw attention to the fact that language is a function of the left cerebral hemisphere, having noted that in all of his aphasic patients, the lesion was located in the left hemisphere. This discovery led to a major revolution in medical and physiological thinking. From a medical standpoint, aphasia was transformed from a minor curiosity to an important symptom of focal brain disease. (It is now generally accepted that the link between the left hemisphere and language was first reported by Dax (1836); however, Dax's findings were not published until after the publication of Broca's reports.)

The next major advance in aphasiology came from a German neuropsychiatrist, Carl Wernicke, who demonstrated that the occurrence of the other major type of aphasic disorder (i.e. the amnesic type) postulated by Bouillaud, was related to disease of the left temporal lobe. In a publication in 1874, Wernicke described the major features of what he called sensory aphasia. These features were fluent but disordered speech, with analogous disturbances in writing and reading (both oral and silent) and impaired understanding of oral speech. Wernicke determined that the lesion most commonly associated with this disorder was situated in the posterior part of the

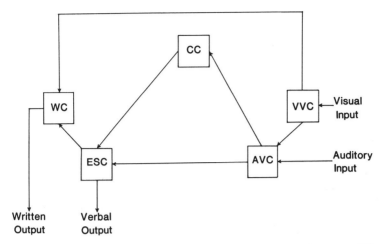

Figure 2.1 The Wernicke–Lichtheim model. AVC, audio-verbal centre; CC, concept centre; ESC, expressive speech centre; VVC, visuo-verbal centre; WC, writing centre.

superior temporal gyrus (first temporal gyrus) of the left hemisphere (i.e. the auditory association cortex of the dominant hemisphere).

Wernicke further stated that the anterior portions of the brain were devoted to motor functions and the posterior areas to sensory functions. He accepted the existence of Broca's area and, on this basis, proposed the existence of two distinct types of aphasia – motor aphasia and sensory aphasia. Not only did Wernicke emphasize the importance of cortical centres in language performance, he also stressed the role played by the association fibre tracts which connect the cortical language centres. He postulated the existence of a bundle of fibres (now known to be the arcuate fasciculus) connecting Broca's area to the first temporal gyrus and suggested that lesions here would produce conduction aphasia (an entity now well recognized). In this way Wernicke was the first to propose that aphasia could be caused by cortical–cortical disconnections. From his simple neural model, Wernicke therefore not only accounted for aphasic syndromes known at his time, but also correctly predicted the existence of syndromes that had not been described at the time.

Following the publication of Wernicke's work, a large number of schemes and aphasia classification systems, based mainly on localizationist theories, were proposed by various authors. Among the most notable of these contributions were the works of Bastian (1898), Charcot (1877), Lichtheim (1885) and Nielsen (1936). As an example of these works, Lichtheim (1885) elaborated Wernicke's model and postulated the existence of five interconnected cortical centres (see Figure 2.1): a centre for the memory images of the movement patterns of oral speech (expressive speech centre) located in Broca's area in the frontal lobe; a centre for the memory images of word sounds (audio-verbal centre) located in Wernicke's area in the temporal lobe; a centre for the memory images of the movement patterns of writing (writing centre); a centre for the memory images of written words (visuo-verbal centre); and a centre in which concepts or ideas to be expressed are formulated (concept centre) located in the inferior part of the parietal lobe. According to Lichtheim's proposal, these centres were linked, not only with each other but also with other cortical and sub-cortical structures. For example, Lichtheim proposed that the audio-verbal and visuo-verbal centres were intimately associated with the primary cortical receptive areas for hearing and vision.

The model of the cortical speech centres developed by Lichtheim on the basis of Wernicke's proposals has become known as the Wernicke–Lichtheim model (or schema). Lichtheim agreed with Wernicke that a disruption of the connection between the audio-verbal centre and the expressive speech centre would lead to a conduction aphasia. Further, Lichtheim postulated that a break in the connection between the concept centre and the expressive speech centre would lead to a language disorder characterized by impoverished spontaneous speech, since formulated ideas would not be able to be expressed verbally. At the same time, however, since the basic mechanisms of speech would remain intact in such a case, strictly linguistic performance such as repetitive speech and reading aloud would be preserved. The understanding of speech would be spared, since both the audio-verbal centre and its connection with the expressive speech centre remain intact. The real existence of this syndrome, designated transcortical motor aphasia, together with conduction aphasia, has since been confirmed by more recent workers.

Using the Wernicke–Lichtheim model as an example, it is evident that the models of language proposed by the early localizationists were successful, at least in part, in predicting the existence of a number of now clinically-recognized aphasia syndromes. Aphasia, however, is a complex disorder and, as will become evident later, the localizationist models of language have been only partly successful in explaining some of the curious combinations of symptoms often encountered in aphasic patients.

Strongly opposed to the localizationist theories of the latter part of last century was the English neurologist, Hughlings Jackson. Jackson started publishing his thoughts in 1864 and considered aphasia from a dynamic, psychological viewpoint rather than as a static, neuroanatomical correlation. He was the founder of what is often called the 'cognitive school' in the field of aphasia and distinguished between two levels of speech: emotional ,(or automatic) and intellectual. He believed it was the intellectual level of utterance, involving the statement of what he termed 'propostions' that was impaired in the aphasic patients, considering that aphasics may show considerable preservation of automatic language in the form of interjections, oaths, clichés and recurring utterances. Jackson did not deny that Broca's area was frequently damaged in patients who suffered from aphasic disturbances, especially when motor speech involvements were manifest. He refused, however, to localize language function in Broca's area alone, and stressed the notion that language was a psychological rather than a physiological function. Jackson emphasized that for language as well as for other intellectual functions, the brain operates as a functional unit.

The non-localizationist viewpoint was supported by the work of Pierre Marie (1906). Marie presented a paper entitled 'The third frontal convolution plays no special role in the function of language' in which he rejected the notion of two distinct language centres as claimed by Wernicke, suggesting that the superior temporal gyrus was the only true language centre. Marie suggested that the language disorder known as Broca's (motor) aphasia is actually a Wenicke's (sensory) aphasia combined with anarthria. Further he proposed that there is only one true form of aphasia, Wernicke's (sensory) aphasia which is characterized by poor comprehension, paraphasia, jargon and reading and writing difficulties. For a lesion in Broca's area to cause aphasia, Marie suggested it would be necessary for the lesion to extend posteriorly to involve Wernicke's area or its connections to the thalamus. Marie agreed with Jackson that, while localized lesions could not be held responsible for language and speech disturbances, lesions in certain cortical areas can more readily disturb speech than can lesions in other areas. Both he and Jackson emphasized the point that a knowledge of pathological conditions which disturb and impair language function does not provide information *per se* as to how the function is normally controlled in the healthy individual. Jackson, in particular, emphasized the importance of observing the live patient rather than studying autospy findings. Following the publication of Marie's work, the influence of the holistic approach on aphasia increased over the early part of the twentieth century causing Wernicke's disconnectionist model of language to fall into some degree of disrepute. Major advocates of the holistic approach included Critchley (1970), Head (1926), Luria (1970), Pick (1931), Schuell, Jenkins and Jimenez-Pabon (1964) and Wepman (1951).

Although, as discussed above, it is generally possible to divide models of language proposed over the past century into those belonging to the anatomically-based localizationist school and those belonging to the psychologically-based holistic school, most clinicians at some time have observed language behaviours supportive of each of these two viewpoints. In addition, most contemporary aphasia researchers make use of both approaches in explaining and classifying the aphasia exhibited by their subjects. For example, the Soviet psychologists headed by Luria tend to lie between the strict localizationists and the holists promoting equipotentiality for all areas of the brain. Luria, sometimes described as a dynamic localizationist, considered that the material basis of the higher nervous processes (including language) is the brain as a whole. Further, however, Luria stated that he believed the brain to be a highly differentiated system whose parts are responsible for the different aspects of the unified whole. The work of Luria and his associates, based on a moderate localization position, features careful descriptions of aphasia syndromes often correlated with an anatomical localization of pathology. (Luria's aphasia classification system is discussed later in this chapter.) In this way a better integration of the anatomical and psychological approaches to aphasia has resulted.

In 1965 Norman Geschwind re-introduced the disconnectionist approach to the study of language and aphasia first proposed by Wernicke. The publication of Geschwind's work (1965a,b) subsequently provided a new impetus to the anatomical study of aphasia syndromes and has led to the revival of the Wernicke–Lichtheim model of language. Added to this, the introduction of new brain-imaging techniques since the mid-1970s, including computed tomography scanning (p. 91), magnetic resonance imaging (p. 93), positron emission tomography (p. 95) and regional cerebral blood-flow studies, which allow the localization of aphasia-producing lesions in the living subject, have better enabled clinico-pathological studies of aphasia to be carried out. The results of studies which have utilized these techniques have, in general, tended to vindicate the Wernicke–Lichteim model of language. The new brain-imaging techniques have, however, also verified the need expressed by many researchers over the years, to expand the classical concept of the 'speech–language area' of the dominant hemisphere to include sub-cortical structures such as the components of the striato-capsular region and the thalamus. Consequently, in addition to the specific cortical centres and cortical–cortical connections (e.g. the arcuate fasciculus) specified in older language models, recently proposed models have been extended to acknowledge the importance of sub-cortical structures as centres relevant for language processes (Crosson, 1985). (The role of sub-cortical structures in language is discussed in Chapter 3.)

It should be stressed that the intent of the above brief history of aphasia was to provide some insight into the major approaches to aphasia over the past century or so and the controversies surrounding them. Consequently, only a very small proportion of the contributors to aphasiology have been mentioned. The works of many of the more recent researchers in aphasiology are referred to either later in this chapter or in subsequent chapters.

2.2 CLASSIFICATION OF APHASIA

As a consequence of the complexity and variability of aphasia, over the years an extraordinarily large number of systems have been developed to classify the various aphasia syndromes. In general, individual aphasia researchers have tended to develop their own terminology to differentiate the various types of aphasia based on their own idea of the nature of aphasia. Although the various authors have tended to agree to a greater extent than may be readily apparent on the major criteria that distinguish one aphasia type from another, they have used a diverse array of terminology to define the aphasia syndromes that they recognized. Consequently, it is difficult for the student of aphasiology to determine how the various aphasia syndromes as defined in the different classification systems are related. Further, because of the variability of the language disturbance seen from one aphasic patient to the next, clinicians often have difficulty when making a diagnosis as to where the boundaries between one aphasia type and another exist.

To avoid the confusion associated with a broad classification system, many authors make use of simple dichotomies to classify aphasia. The two most widely used dichotomies are the expressive–receptive division proposed by Weisenberg and McBride (1935) and the motor–sensory division originally introduced by Wernicke (1874). Two additional dichotomies also frequently used in the literature include the fluent–non-fluent (Benson, 1967) and anterior–posterior (Goodglass and Kaplan, 1972) dichotomies.

Expressive or motor aphasia is presumably associated with lesions involving the anterior language centre of the dominant hemisphere (i.e. Broca's area) and is therefore primarily related to an inability to translate speech concepts into meaningful sounds. In other words, the primary disturbance here is an inability to execute speech production. As a result, speech is non-fluent, with pauses between words or phrases. Receptive or sensory aphasia is presumably associated with lesions in the posterior language area of the dominant hemisphere. It therefore is related to problems in the comprehension, formulation etc. of speech. The primary disturbance in the receptive aphasic patient is lack of comprehension of language. In that the expressive aspects of speech are dependent upon normal functioning of the receptive speech processes, however, the expression of language may also be disturbed in persons with sensory aphasia, with the production of jargon, substitution of words (paraphasia), and other expressive disturbances being evident in their spoken output. Speech, however, is generally fluent, although full of mistakes. Thus the major differences between motor and sensory aphasia are that comprehension of language is only mildly affected in motor but severely affected in sensory aphasia, and that speech is usually non-fluent in motor aphasia but fluent in sensory aphasia.

As a modification of the expressive–receptive dichotomy, some authors, including Wernicke (1874) and Geschwind (1969, 1971), have divided aphasics into those with fluent speech (fluent aphasia) (able to effortlessly produce well-articulated, long phrases with a normal grammar, melody and rhythm) and those with non-fluent speech output (non-fluent aphasia) (able to produce only sparse speech that is uttered slowly with great effort and poor articulation). Wernicke (1874) suggested an anatomical correlation for the two different aphasic outputs. The significance of the dichtomy in aphasic verbal output has been re-emphasized by a number of recent studies which have confirmed Wernicke's observations and shown that non-fluent aphasic patients have lesions involving structures anterior to the Fissure of Rolando while fluent aphasic patients on the other hand have lesions posterior to the Fissure of Rolando (Benson, 1967; Poeck, Kerschensteiner and Hartje, 1972; Wagenaar, Snow and Prins, 1975).

Although each of the dichotomies has a validity and a usefulness, none adequately characterizes the distinguishing features of most varieties of aphasia. For instance, pure forms of aphasia (i.e. pure motor or pure sensory) are not commonly encountered. This is especially the case with sensory aphasia which is accompanied by some loss of motor speech. Thus most aphasic patients will have some symptoms of both these types. Further, posterior lesions can produce disorders of expression and anterior lesions can produce disorders of comprehension. In addition, aphasic disturbances can be caused by brain lesions which do not directly involve the cortical language centres of the dominant hemisphere. Studies in recent years, for instance, have shown that sub-cortical and even right hemisphere structures contribute to language function.

Although the fluent–non-fluent dichotomy does offer significant anatomical localizing information, it should be pointed out that exceptions to the anterior–posterior correlation do exist. First, not all aphasics can be classified as fluent or non-fluent (Benson, 1967; Karis and Horenstein, 1976). Secondly, almost all children who become acutely aphasic are non-fluent (often they are mute) (see Chapter 11) and even with recovery, almost never does a child develop a fully fluent, paraphasic, jargon type of output. Thus the acquired aphasia of childhood does not fit into the fluent–non-fluent division. Finally, evaluation of fluency in the early stages of aphasia may be misleading. Many patients with freshly acquired aphasia show non-fluent characteristics at the onset. While some aphasics with lesions posterior to the Fissure of Rolando rapidly become fluent, others may be non-fluent for weeks before altering to a fluent output (Benson, 1979).

It is evident, therefore, that aphasic disturbances need to be divided into a greater number of types rather than simply being divided into one of the categories specified by the above dichotomies. Two major classification systems above all others have gained some degree of clinical acceptance in recent years. These include the Boston Aphasia Classification System (developed by the Speech Pathology Section, Aphasia Research Unit, Veterans Administration Hospital, Boston) and the Lurian Aphasia Classification System proposed by Luria (1970).

2.2.1 The Boston Classification System

According to a modified version of the Boston Classification System (Benson, 1979), there are eight clinically recognizable aphasia syndromes. These include: (1) Broca's aphasia; (2) Wernicke's aphasia; (3) conduction aphasia; (4) global aphasia; (5) transcortical motor aphasia; (6) transcortical sensory aphasia; (7) isolation (mixed transcortical) aphasia: (8) and anomic aphasia.

The suggested lesion sites, according to the Wernicke–Lichtheim model (see Figure 2.1) for each of these aphasia types are as follows.

1. Broca's aphasia is associated with a lesion involving the expressive speech centre.
2. Wernicke's aphasia is caused by a lesion of the audio-verbal centre.
3. Conduction aphasia results from a lesion involving the pathways connecting the audio-verbal and expressive speech centres.
4. Global aphasia is produced by an extensive lesion involving both the audio-verbal centre and the expressive speech centre.

5. Transcortical motor aphasia is associated with disruption of the pathways connecting the concept centre to the expressive speech centre.
6. Transcortical sensory aphasia results from lesions of the pathways connecting the audio-verbal centre to the concept centre.
7. Isolation aphasia is caused by lesions which disconnect the concept centre from both the audio-verbal centre and the expressive speech centre.
8. Anomic aphasia is produced by a lesion involving the pathways which connect the concept centre to the expressive speech centre or, in those cases where comprehension is also disturbed, by a lesion of the concept centre.

Each of the Bostonian aphasia syndromes is characterized by a particular group or cluster of language signs and symptoms. Consequently allocation of language-disordered patients to each of these syndromes is carried out on the basis of the particular cluster of language symptoms exhibited by the patient. It is important to realize, however, that even though each syndrome has its own set of language characteristics, the particular set of language signs and symptoms associated with each syndrome is not fixed. Certainly not every member of the set needs to be present in the patient before a diagnosis of that syndrome can be made. Likewise, occasionally language findings not normally characteristic of a particular aphasia syndrome may be present among a group of signs and symptoms more typical of that aphasia syndrome. Consequently, speech–language pathologists need to allow some degree of flexibility in the membership of the language signs and symptoms they take to be indicative of specific aphasia syndromes.

The various Bostonian aphasias can be grouped a number of ways in order to assist localization. Lesions in the central speech areas that bank the Fissure of Sylvius give rise to the central aphasias, which have in common loss of repetition. Central aphasias include Broca's aphasia, Wernicke's aphasia, conduction and global aphasia. The peri-central aphasias which include transcortical motor, transcortical sensory, anomic and isolation aphasias are caused by lesions surrounding the central speech areas and have in common good repetition. Thus a central aphasia can be distinguished from peri-central aphasia by testing the repetition abilities of the patient. In fact, if repetition is not tested, the examiner will have difficulty in distinguishing between the following central and peri-central syndromes: Broca's from transcortical motor; Wernicke's from transcortical sensory; anomic from conduction; and global from isolation.

Differentiation between the peri-sylvian and the border-zone (peri-central) types of aphasia as determined by the ability to repeat is not only a valid localizing feature, but it also offers a clue as to the nature of the underlying pathology. When vascular disease is the aetiology, most aphasias with normal repetition are based on occlusive disease of the left internal carotid artery. Following acute occlusion of the carotid vessel, the limited arterial circulation available through the Circle of Willis may be sufficient to perfuse only the immediate peri-sylvian cortex, allowing this area to remain viable but not providing enough oxygenated blood to maintain border-zone tissues. In contrast, a cerebrovascular accident producing an aphasia with repetition disturbance is most likely based on thrombotic or embolic vascular problems with involvement of one or more of the branches of the middle cerebral artery (Romanul and Abramowitz, 1961). The various Bostonian aphasia syndromes are discussed in terms of their symptoms and lesion sites later in this chapter).

2.2.2 The Lurian Classification System

Based on his studies of soldiers who had suffered head wounds in war, Luria distinguished seven main types of aphasia: (1) sensory (acoustic) aphasia; (2) acoustico-mnestic aphasia; (3) semantic aphasia; (4) efferent (kinetic) motor aphasia; (5) pre-motor aphasia; (6) afferent (apraxic) motor aphasia; and (7) frontal dynamic aphasia. (The symptoms and lesion sites of each of the Lurian aphasia syndromes are described later in this chapter.)

2.3 BOSTONIAN APHASIA SYNDROMES

2.3.1 Broca's aphasia

Broca's aphasia is characterized by non-fluent speech output and poor repetition abilities. Auditory comprehension, however, is relatively spared. Speech output is slow, effortful, agrammatic and often tele-grammatic (i.e. containing a predominance of content words such as nouns and action verbs and a paucity of adjectives, adverbs and prepositions giving the patient's speech a telegraphic style). The vocabulary of patients with Broca's aphasia is generally restricted in range. These patients may use words repetitively (perseveration) and long pauses may occur between words or phrases. A motor speech disorder in the form of an apraxia of speech (see Chapter 7) and/or dysarthria often accompanies Broca's aphasia.

As the more lateral portions of the primary motor strip are usually damaged by the causative lesion, Broca's aphasia is most often associated with either a right hemiplegia (paralysis down one side of the body) or right hemiparesis (muscle weakness down one side of the body) affecting the lower half of the face and the right arm more severely than the right leg. Hyperactive reflexes and pathological reflexes (e.g. a positive Babinski sign) are frequently present on the right-hand side. Ideomotor dyspraxia (see Chapter 7), an inability to carry out on command a task that can be performed spontaneously, frequently involves the 'non-pathological' left hand of Broca's aphasics. Although sensory deficits such as loss of pain and temperature sense are occasionally present, they are inconsistent findings in Broca's aphasia. Likewise, persistent visual field deficits occur less frequently in association with Broca's aphasia than most other aphasia syndromes.

Writing in Broca's aphasia is usually similarly impaired as speech. Typically the written material contains multiple misspellings and the omission of letters. Individual letters tend to be oversized and poorly formed, a situation made worse by the fact that because most Broca's aphasics have a concomitant hemiplegia, they are required to use their left hand for writing. Typically these patients are better able to copy written material than they can write either to command or to dictation. Most Broca's aphasics read aloud poorly, while reading comprehension is usually similar to auditory comprehension.

The localization of lesions associated with Broca's aphasia has been an area of controversy since Broca's first description of the area to bear his name in 1861. For a long time, we were taught to associate the occurrence of Broca's aphasia to damage in Broca's area. Clinico-pathological studies, using computed tomography, however, have shown us that the localization of Broca's aphasia is less clearly defined, Broca's area being damaged in some cases but unharmed in others.

Mohr *et al.* (1978) carried out an extensive study of Broca's aphasia in which the lesions in 20 cases were documented by autopsy, computed tomography scan or arteriography. They found that a lesion confined to Broca's area produced not aphasia, but transient speech apraxia. Their findings suggest that infarction affecting Broca's area and its immediate environs, even deep into the brain, causes a mutism that is replaced by a rapidly improving dyspraxia and effortful articulation, but that no significant disturbance in language function persists (Broca's area infarction syndrome). Rather than being confined to Broca's area, it would appear that Broca's aphasia requires a large lesion to the Sylvian region, encompassing much of the opercula, insula and subjacent white matter in the territory of the upper division of the middle cerebral artery, including Broca's area. As emphasized by Marie (1906) and Mohr (1976), the extensive lesion in Paul Broca's original case included much more than the posterior portion of the left frontal gyrus (Broca's area). Indeed, clinco-pathological studies have shown that Broca's aphasia usually follows extensive cortical–subcortical damage of the left fronto-parietal operculum (including both the inferior frontal gyrus and pre-central gyrus) with the lesion, in some cases, also involving the anterior parts of the parietal and, sometimes, the temporal lobe (Naeser and Hayward, 1978; Mazzocchi and Vignolo, 1979; Murdoch et al., 1986a). Mazzocchi and Vignolo (1979) also noted that all of their Broca's aphasics had lesions involving the lenticular nucleus and insula.

The involvement of Broca's area in Broca's aphasia remains somewhat controversial. Certainly from Mohr's work, lesions restricted to Broca's area itself do not appear to produce Broca's aphasia but rather Broca's area infarction syndrome (centred on apraxia of speech). (The relationship between apraxia of speech and Broca's aphasia is discussed further in Chapter 7.) Whether or not a lesion needs to involve Broca's area to produce a Broca's aphasia, however, is less clear.

2.3.2 Wernicke's aphasia

The major feature of the language disorder exhibited by patients with Wernicke's aphasia is an impairment in language comprehension. Consequently, these patients have a problem in understanding what is said to them and what they read. Some researchers have reported a bipolar tendency in the comprehension deficit of Wernicke's aphasia. Some Wernicke's aphasics have their greatest problem in understanding spoken words (word deafness) while others have more difficulty comprehending written words (word blindness). Elements of deficit in both areas are always present to some degree, however, in Wernicke's aphasics.

In addition to impaired comprehension abilities, Wernicke's aphasics also exhibit poor naming and poor repetition abilities. Unlike Broca's aphasics, however, the speech output of Wernicke's aphasics is fluent, the speech being well articulated with phrases of normal length and melody. Despite being fluent, however, the content of the spoken language is abnormal and is contaminated by the substitution of words (verbal paraphasias) or parts of words (literal paraphasias). As these patients are not capable of monitoring their own verbal expression due to their comprehension difficulty, they often unknowingly invent new words (neologisms) as they speak and, although they speak in full sentences, the sentences mean nothing to others. Therefore, despite their fluent speech, Wernicke's aphasics fail to communicate their ideas to others and are commonly frustrated by their inability to make themselves understood. In severe cases, verbal confusion and the presence of paraphasic errors may result in a very disjointed language disorder in which the patient produces meaningless jargon.

Unlike patients with Broca's aphasia, those with Wernicke's aphasia rarely have a concomitant hemiplegia. Consequently, the handwriting of Wernicke's aphasics is usually satisfactory in terms of motor ability, the written output consisting of well-formed legible letters. As in the case of their spoken language, however, the content of the written language of Wernicke's aphasia is also abnormal. Although in their writing the individual letters are often arranged in the appearance of words, the letters are often combined in a meaningless manner. In some instances the written output resembles the paraphasic spoken output.

Reading aloud and reading comprehension are both disturbed in Wernicke's aphasics although, as indicated above, reading comprehension may be better than auditory comprehension. Likewise the naming abilities of these patients are also impaired, Wernicke's aphasics often failing totally or producing grossly paraphasic responses when asked to name objects, body parts, etc.

The results of a neurological examination are typically negative in the majority of Wernicke's aphasics, there being no indication of the presence of any associated neurological deficit. As indicated above, there is usually no concomitant hemiplegia or hemiparesis and disturbances in general sensation are only rarely present. A visual field deficit in the form of a quadrantanopsia (blindness in one-fourth of the visual field) is present in some Wernicke's aphasics.

Classically, Wernicke's aphasia is believed to result from a temporo-parietal lesion which invariably involves Wernicke's area itself. A number of studies based on computed tomography have provided evidence to support this suggestion. For example, using computed tomography scanning, Naeser and Hayward (1978) found the lesions associated with four cases of Wernicke's aphasia to primarily involve the post-rolandic and temporo-parietal areas and included Wernicke's area in all cases. There was no pre-rolandic overlap or extension into Broca's area in any of their cases. Kertesz, Harlock and Coates (1979) also reported the case of a chronic Wernicke's aphasic (> 12 months) who had a post-rolandic lesion without extension into the frontal operculum.

Some authors, however, have reported that Wernicke's aphasia can be associated with lesions involving both pre- and post-rolandic structures (Kertesz, Harlock and Coates 1979; Mazzocchi and Vignolo, 1979). Lesions restricted to areas anterior to the Rolandic fissure, however, do not appear from most reports to be capable of supporting a Wernicke's aphasia.

Basso *et al.* (1985) reported eight cases of Wernicke's aphasia (characterized by both comprehension and repetition disorders co-existing with fluent jargon) associated with extensive peri-sylvian lesions which might have been expected to produce global aphasia. These authors did not, however, document the involvement of sub-cortical structures in these lesions and the extent of sub-cortical damage is therefore unknown. The documentation of sub-cortical involvement in these cases would appear to be important in the light of the findings of Naeser *et al.* (1982) that Wernicke's aphasia can result from capsular/putaminal lesions with posterior white-matter lesion extension across the auditory radiations in the temporal isthmus. In stark contrast to classical expectations, Basso *et al.* (1985) also reported four cases of Wernicke's aphasia associated with anterior-only lesions. Such cases, however, seem to be rare exceptions and, although not commented on by Basso *et al.* (1985), their respective computed tomography scans showed involvement of the sub-cortical structures deep to Broca's area in the lesion in all these cases. Mohr *et al.* (1978) also reported that one of their patients (case 3 in their study) with a lesion in the left inferior frontal region had Wernicke's aphasia characterized by a comprehension disorder, impaired repetition and fluent jargon. As in the four cases reported by Basso *et al.* (1985), however, the lesion also involved the sub-cortical structures deep to Broca's area and the depth of the sub-cortical white-matter lesion may, therefore, be more critical to the occurrence of the language impairments observed in these cases than the cortical damage in Broca's area.

Post-rolandic lesions are recognized as being associated with fluent aphasias. Although fluent aphasias are associated with comprehension deficits (hence the terms 'receptive' and 'sensory' aphasias), the different sub-types of fluent aphasias are defined largely by the relative occurrence of two major errors in speech production: lexical semantic and phonological (phonemic) errors. There is some evidence that lexical errors occur primarily with lesions of the infero-lateral temporal (middle and inferior temporal gyri) and parietal 'integration' cortices, and that lexical processing is more bilaterally organized (Cappa, Cavallotti and Vignolo, 1981). Phonological errors, on the other hand, occur with lesions of the left Wernicke area proper (posterior part of the superior temporal gyrus and supramarginal gyrus) (Cappa, Cavallotti and Vignolo, 1981). These findings raise the possibility of there being several different forms of Wernicke's aphasia.

Pure or sub-cortical word-deafness (auditory aphasia) is a very rare and fractional language disorder closely related to Wernicke's aphasia. Although these patients distinguish words from other sounds, they cannot understand them. Consequently their own speech sounds like a foreign language. Patients with this disorder cannot repeat words or write to dictation although spontaneous speech, writing and reading are unimpaired. The condition is thought to be caused be a lesion of the white matter deep to the posterior part of the left (dominant) superior temporal gyrus.

2.3.3 Conduction aphasia

Conduction aphasia is characterized by a disproportionate impairment in repetition relative to spontaneous speech and oral and written comprehension. Although the spontaneous speech output of conduction aphasics is relatively fluent, these patients have a problem in choosing and sequencing their phonemes so that their speech is contaminated by numerous paraphasic errors, typically of the literal type (i.e. involving use of incorrect phonemes within words). In addition there may be a slight impairment of articulation and frequent pauses, and hesitations associated with word-finding difficulties are usually present causing the speech output to be dysprodic.

The impairment in repetition is the most outstanding feature of this disorder and is most marked for multi-syllabic words and sentences. It is during performance of repetition tasks that the literal paraphasic errors are most prominent in conduction aphasics. Most of these patients also demonstrate difficulty in confrontation naming. Auditory comprehension abilities are good.

Written spontaneous language is impaired similarly to spoken spontaneous language, although the motor aspects of handwriting are usually normal. Conduction aphasics are usually able to produce well-formed letters, but spelling is poor with omission, reversal and substitution of letters being present in their written output. Also words in a sentence are frequently interchanged, misplaced or omitted. The ability to read aloud is also impaired, oral reading being paraphasic. These patients, however, often retain relatively good reading comprehension abilities.

The neurological examination of conduction aphasics shows considerable variation from case to case. Often no associated neurological abnormally can be demonstrated. Significant hemiparesis is rare. When present, however, muscle weakness almost always involves the arm to a greater extent than the leg. Sensory findings are also variable, sensory abnormalities being totally absent in some patients and present in the form of a limited right-sided hemianaesthesia (loss of sensation down one side of the body) in others. Visual-field defects involving either a hemianopsia (loss of one-half of the visual field) or quadrantanopsia are present in some conduction aphasics. Ideomotor dyspraxia is frequently, but not constantly, present. When asked to perform bucco-facial or limb movements on command conduction aphasics may fail, even while protesting that they know what they want to do. Often such commands result in an improper movement which nevertheless demonstrates that the patient has comprehended the command (e.g. waving the hand near the face when asked to salute). (Ideomotor dyspraxia is discussed more fully in Chapter 7.) The variation in neurological findings in conduction aphasia probably reflects involvement of different sites neighbouring the locus of pathology underlying this condition.

Wernicke (1874) proposed that conduction aphasia results from separation of the posterior language comprehension area and the anterior motor speech area of the left hemisphere. This proposal has since been supported by Geschwind (1965a,b) and the arcuate fasciculus has been the most frequently suggested site of such a disconnection. In general, the findings of clinico-pathological studies that have identified the location of the aphasia-producing lesion by computed tomography have tended to support the disconnection theory (Naeser and Hayward, 1978; Damasio and Damasio, 1980). For example, in the cases of conduction aphasia reported by Naeser and Hayward (1978) the lesions were either sub-cortical with lesions deep to, but not including, Wernicke's area or cortical with lesions extending from the surface of the supramarginal gyrus area to the body of the left lateral ventricle. In addition, the lesions were primarily post-rolandic but did not involve either Broca's or Wernicke's areas and were consistent with involvement of the posterior portions of the arcuate fasciculus.

Although involvement of the arcuate fasciculus has been proved in a number of cases of conduction aphasia, some authorities insist that involvement of the supramarginal cortex, not the underlying white matter, is the essential finding (Levine and Calvanio, 1982). Wernicke's and Geschwind's interpretation of the disturbance as a disconnection syndrome, therefore, has not been universally accepted. The arcuate fasciculus theory has been further challenged by reports of conduction aphasia occurring subsequent to lesions in the dominant Wernicke's area, suggesting that a cortical lesion, not a disconnection, is the crucial factor (Kleist, 1962 ; Mendez and Benson, 1985).

Mendez and Benson (1985) reported three atypical cases of conduction aphasia in which the lesions did not lie in the arcuate fasciculus. Two of their patients had left temporo-parietal lesions, the other was a right-handed case with a right temporo-parietal lesion (i.e. a 'crossed aphasic'). They concluded, however, that even these atypical conduction aphasias are best explained by the disconnection concept, which they agree remains the most tenable model for this language disturbance.

Kertesz (1979) proposed that there are actually two types of conduction aphasia, one he called 'efferent conduction aphasia' and the other 'afferent conduction aphasia'. Kertesz based this proposal on radioisotope localization studies (Kertesz, Lesk and McCabe, 1977). In the efferent type of conduction aphasia, the patients are less fluent and have more anterior lesions while, in the afferent type, the subjects are more fluent with more posterior lesions.

2.3.4 Global aphasia

In global or total aphasia, all major language functions are seriously impaired, including both the expressive and receptive components of language. In its most severe form the patient does not communicate and verbal output is limited to expletives or a stereotypic repetitive utterance. Occasionally, however, these utterances are said quite fluently, with inflection and associated emotional expression conveying some meaning. Comprehension abilities are severely impaired although they are frequently reported to be better than verbal output, possibly because global aphasics become skilled at interpreting non-verbal communication through gesture, facial expressions, etc. It is possible that this non-verbal comprehension may be mistaken by clinicians for comprehension of the spoken word. Repetition, naming, reading and writing are all severely, usually totally, disturbed.

In the majority of cases, global aphasics exhibit a range of con-comitant neurological signs indicative of severe brain damage. These neurological signs may include hemiplegia, sensory loss, visual-field defects and often an attention disturbance.

Most studies reported in the literature indicate that global aphasia results from an extensive left-hemisphere lesion involving both Broca's and Wernicke's areas (Hayward, Naeser and Zatz, 1977; Naeser and Hayward, 1978; Kertesz, Harlock and Coates, 1979; Murdoch et al., 1986a). Although most studies have found the lesions associated with global aphasia to be large, involving the entire peri-sylvian region and sub-cortical structures in the frontal, parietal and temporal lobes, some authors have found that exceptions do exist (Mazzocchi and

Vignolo, 1979; Basso *et al.*, 1985; Naeser *et al.*, 1982; Vignolo, Boccardi and Caverni, 1986). These last authors suggest that global aphasia does not necessarily result from large lesions involving both Broca's and Wernicke's areas. In particular, Wernicke's area may be spared even in chronic global aphasia with persisting comprehension deficit.

Global aphasia occurring without lesions of Wernicke's area was first described by Mazzocchi and Vignolo (1979). They suggested that the critical anatomical difference between global and Broca's aphasia might be related to the greater size of the lesions rather than to actual damage to Wernicke's area. However, this does not seem to be invariably true, since the lesions found in global aphasics are not always larger than those found in Broca's aphasics. An example of this can be found in the work of Basso *et al.* (1985) who reported 10 cases of global aphasia with spared Wernicke's area. Vignolo, Boccardi and Caverni (1986) reported a further eight cases of global aphasia with anterior lesions sparing Wernicke's area and three cases with posterior lesions sparing Broca's area. Further, these authors described four cases of global aphasia with deep lesions centred on the insula and lenticular nucleus. It is apparent, therefore, that there is more than one type of lesion underlying global aphasia.

Naeser *et al.* (1982) reported the occurrence of global aphasia in association with small sub-cortical lesions. They described three cases of lasting global aphasia subsequent to capsular/putaminal lesions with both anterior–superior and posterior lesion extension and suggested that the severely limited speech output observed in these patients was probably the result of isolation of Broca's area due to disruption of the afferent and efferent pathways. A global aphasia was also described in four cases with striato-capsular lesions involving both the anterior and posterior limbs of the internal capsule by Murdoch *et al.* (1986b). It is evident therefore that, in addition to the large combined cortical and sub-cortical lesions reported in most studies, damage to sub-cortical structures such as the white-matter pathways (including the internal capsule, extreme capsule, genu of the corpus callosum and temporal isthmus) can also be the cause of a global aphasia. Consequently, damage to both cortical and sub-cortical structures should be documented in any descriptions of lesions associated with global aphasia. Unfortunately, a number of authors, such as Basso *et al.* (1985), have concentrated their lesion descriptions primarily on involvement of the cerebral cortex and have largely ignored damage to sub-cortical structures.

2.3.5 Transcortical motor aphasia

The term 'transcortical aphasia' is used to describe a group of aphasic syndromes characterized by retention of repetition out of all proportion compared to other language functions. Three types of transcortical aphasia are recognized: transcortical motor aphasia; transcortical sensory aphasia; and mixed transcortical aphasia.

Transcortical motor aphasia is characterized by a marked reduction in the quantity and complexity of spontaneous speech in the presence of a retained ability to repeat. Preserved repetition is the most striking feature of this condition. The repetitions of transcortical motor aphasics, however, are not mandatory and therefore these patients cannot be regarded as echolalic (echolalia is the automatic repetition by patients of what is said to them). Although transcortical motor aphasics may echo a word or phrase, they will correct grammatically incorrect statements they are asked to repeat and at the same time will reject nonsense syllables.

The limited spontaneous speech output of transcortical motor aphasics is non-fluent, the speech often being described as stumbling, repetitive and 'stuttering-like'. Conversational verbalization is produced only with considerable effort and is in most cases agrammatic and highly simplified. Series speech is usually performed well once the patient is started. Often these patients may require prompting for the first few numbers in a series but can then continue unhindered.

Comprehension of spoken and written language is relatively preserved. Reading aloud is almost invariably defective, a poorly articulated output being produced. The ability to write is affected in the majority of transcortical motor aphasics, the written language output featuring large, clumsily produced letters, poor spelling and agrammatic output. Transcortical motor aphasics usually perform poorly in confrontation naming tasks.

In general the associated neurological signs exhibited by transcortical motor aphasics are similar to those found in Broca's aphasics. The majority of these patients have a right hemiplegia. Similarly ideomotor dyspraxia is a common finding in the non-paralysed left hand of these patients. Neither sensory loss nor visual field defects, however, are characteristic of transcortical motor aphasia.

Most textbooks tell us that transcortical aphasias are associated with lesions in the arterial border zone of the left hemisphere. Lesions in the anterior parts of the border zone are usually said to be associated with transcortical motor aphasia, while those in the posterior parts are said to produce transcortical sensory aphasias (Benson,

1979). Isolation syndrome (mixed-transcortical aphasia) results from lesions involving both the anterior and posterior parts of the border zone (Benson, 1979).

Few studies have investigated lesions associated with transcortical motor aphasia (Naeser and Hayward, 1978; Massocchi and Vignolo, 1979; Ross, 1980b). Naeser and Hayward (1978) reported the lesions associated with four cases of transcortical motor aphasia. In general the lesions tended to be small and agree with the border-zone theory and reports based on radionuclide scans (Rubens, 1976) in that they were scattered primarily anteriorly and superiorly to Broca's area in the frontal lobe. Two patients, however, had lesions that included the superior portion of Broca's area. None of the four patients had lesions directly involving Wenicke's area. A single case of transcortical motor aphasia was also reported by Mazzocchi and Vignolo (1979). This case had a typical clinical picture with a computed tomography scan showing a lesion in the frontal lobe anterior and superior to Broca's area. Two further cases of transcortical motor aphasia associated with infarction of the anterior cerebral artery were reported by Ross (1980b). In both cases, the lesions were confined to the medial aspects of the left frontal lobe.

Overall, it would appear that in transcortical motor aphasia, Broca's area may remain intact or be only slightly damaged. Transcortical motor aphasia has also been reported to follow unilateral sub-cortical damage, such as a lesion near the anterior horn of the left lateral ventricle (Damasio, 1981) or in the anterior capsular–putaminal region, interrupting the thalamic–frontal connections (Sterzi and Vallar, 1978). (Aphasias associated with sub-cortical lesion sites are discussed more fully in Chapter 3.)

2.3.6 Transcortical sensory aphasia

Transcortical sensory aphasia is characterized by impaired comprehension abilities occurring in conjunction with preserved repetition and a fluent speech output. Comprehension of spoken language is severely disturbed in transcortical sensory aphasia, often to the point of total non-comprehension. This group of patients often incorporate words and phrases uttered by the clinician into their ongoing speech output while at the same time failing to comprehend the meaning of these words and phrases. In fact, the most outstanding feature of transcortical sensory aphasia is the presence of echolalia. Unlike the situation in transcortical motor aphasia, the repetition of statements by transcortical sensory aphasics often appears to be mandatory, the patients apparently being unable to omit from their utterances the statements made by the examining clinician. In contrast to patients with transcortical motor aphasia, those with transcortical sensory aphasia repeat syntactically incorrect statements, nonsense words and even foreign phrases without apparent awareness of what is said and without appropriate correction.

Spontaneous speech, although fluent, is often contaminated by paraphasic errors, including both neologistic and semantic (verbal) substitutions and by pauses associated with word-finding difficulties. Series speech, if initiated by the examiner, is good. Confrontation naming is seriously defective.

The ability to read aloud is better preserved in transcortical sensory aphasics than reading for comprehension. The latter is almost invariably defective in a way similar to that seen in Wernicke's aphasia.

The concomitant neurological problems evidenced at neurological examination of transcortical sensory aphasics vary from case to case. The majority of patients with this condition show no elementary neurological deficit. Others, however, show a mild and usually transient hemiparesis. Sensory deficits, although not common, are found in some transcortical sensory aphasics.

Transcortical sensory aphasia is a syndrome characterized by poor comprehension but excellent repetition. The most extensive localization study of transcortical sensory aphasia, to date, is that carried out by Kertesz, Sheppard and Mackenzie (1982). Their findings were in agreement with the border-zone theory, the lesions being primarily located in the inferior parietal–temporal–occipital area. They found that the lesions separated into two groups, one group being located in a more medial, inferior and posterior position in the territory of the posterior cerebral artery. In the other group, the lesion was located in a relatively more lateral, superior and anterior position in the watershed (border zone) area between the distributions of the middle and posterior cerebral arteries. It appears, therefore, that transcortical sensory aphasia is most often seen in association with infarction in the territory of posterior cerebral artery or subsequent to watershed area lesions that involve the territory between the posterior cerebral and middle cerebral arteries, in the posterior temporal–parietal region.

2.3.7 Isolation aphasia (mixed-transcortical aphasia)

Isolation aphasia is a rare aphasia syndrome characterized by preserved repetitional abilities occurring in association with a marked reduction in spontaneous speech and impaired comprehension of language. The most outstanding feature of this disorder is the loss of the voluntary aspects of language, including spontaneous speech and the ability to initiate and actively participate in conversation. The verbal output of patients with isolation aphasia is almost entirely limited to what has been said to them. Consequently, these patients tend to speak only when spoken to. Isolation aphasics, however, usually exhibit the completion phenomenon (i.e. if told the beginning of a common phrase, the patient may not only repeat what has been said but also continue the phrase to completion). Although the ability of isolation aphasics to repeat is dramatically preserved compared to other language functions, it is limited compared to the repetition abilities of normal individuals. For instance, the number of words in a phrase that can be repeated may be limited to only three or four.

The articulation of phonemes during repetition is good and isolation aphasics demonstrate preserved series speech (e.g. counting) abilities once they are started. Reading aloud as well as reading comprehension and writing are all severely disturbed in this disorder. Isolation aphasics also have severe difficulties in naming, often producing no response at all but occasionally producing neologisms or semantic paraphasias in naming tasks. In many ways, therefore, isolation aphasics resemble global aphasics, with the major exception that the former group are able to repeat what has been said.

The results of neurological examination of these patients is variable. Some exhibit bilateral upper motor neurone paralysis producing a severe quadriplegia or quadriparesis. Others exhibit unilateral motor disturbances such as right hemiplegia. A significant sensory loss is also frequently present. A visual-field deficit in the form of a hemianopsia has been reported in some cases.

Only a limited number of studies have documented the results of a comprehensive clinico-pathological analysis of this condition (Geschwind, Quadfasel and Segarra, 1968; Whitaker, 1976; Chenery and Murdoch, 1986). Only two studies using computed tomographic localization have been reported (Ross, 1980b; Chenery and Murdoch, 1986). The studies by Geschwind, Quadfasel and Segarra (1968) and Whitaker (1976) relied on autopsy evidence which suggested involvement of both the anterior and posterior aspects of the vascular border zone between the distributions of the major cerebral arteries. However, in the case reported by Chenery and Murdoch (1986), computed tomography scans revealed no focal lesions to be present, only mild cerebral atrophy indicated by prominent cortical sulci and dilated ventricles. Ross (1980b) documented a case of mixed transcortical aphasia with infarction involving the left anterior cerebral artery. In the case in question, the lesion extended beyond the left motor and sensory cortices to involve the anterior precuneus lobule of the left parietal lobe. This finding led Ross to suggest that the supplementary sensory area of the left medial parietal lobe participates in receptive language functions.

The full spectrum of the anatomical basis of isolation syndrome remains unknown. It has been argued that comprehension and spontaneous speech are lost in this syndrome because the central speech area no longer has access to other cortical areas necessary for these functions, whereas those functions that can be carried out by the central speech area such as repetition and completion of well-learned phrases are preserved.

2.3.8 Anomic aphasia

Anomia, the name given to a word-finding difficulty in confrontational naming tasks and in spontaneous speech, is a symptom common to all types of aphasia. When, however, anomia is the most prominent feature of an aphasic disorder, the condition is referred to as an anomic aphasia.

Anomic aphasia is a commonly encountered form of aphasia in which the patient has little expressive or receptive difficulty. Spontaneous speech is produced easily and fluently, although at times there is an emptiness resulting from a lack of substantive words. In addition, the speech output at times is very circumlocutory (circuitous) as a result of replacement of specific words (names) by generalizations (e.g. thing, it, them, etc.) which often fail to communicate the message satisfactorily. As a result the speech of anomic aphasics is often described as being vague. The degree of word-finding difficulty varies widely from case to case, some patients showing only a mild naming disturbance on confrontation naming tasks while in others confrontation naming is severely disturbed.

Anomic aphasics have good repetitional abilities and near normal auditory and written comprehension. Their ability to read aloud is also relatively good in most cases. Writing is near normal in some cases and impaired in others. Anomic aphasia is often the end result of recovery from other aphasic syndromes such as Wernicke's or conduction aphasia. Anomia remains the complaint of many well-recovered aphasics.

The associated neurological findings vary widely in anomic aphasia. Many cases of this disorder exhibit no associated neurological signs at all. On the other hand, hemiparesis, hemiplegia, hemisensory loss and visual-field defects may occur in some of these patients.

Anomic aphasic is regarded as a non-localizing aphasia syndrome. In general, the lesion site cannot be readily localized to a particular cortical area. Gloning, Gloning and Hoff (1963) in a study of the location of pathology in patients with various aphasia syndromes, found that 60% of their patients with anomic aphasia had a dominant hemisphere parietal–temporal junction lesion. The other 40% of their anomic patients, however, had lesions scattered over a wide area. Computed tomographic studies have come up with similar findings. For example, Hayward, Naeser and Zatz (1977) reported that, although there was some concentration of lesions in the region of the angular gyrus, in general the lesions in anomic aphasia were scattered about the left hemisphere with a lack of specific location involvement.

2.3.9 Sub-cortical aphasia syndromes

Since the introduction of computed tomography in the 1970s with its ability to demonstrate deep structured brain pathology in the living subject, it has become increasingly recognized that aphasia syndromes can be caused by damage to sub-cortical structures as well as cortical lesions. In particular, it has been suggested that aphasia can be produced by lesions involving the region of the basal ganglia or thalamus. For this reason, in recent years an increasing number of researchers has advocated the addition of sub-cortical aphasia syndromes to the list of clinically recognized aphasic disturbances. The various 'sub-cortical aphasias' are discussed in Chapter 3.

2.4 LURIAN APHASIA SYNDROMES

Based on an extensive study of patients who exhibited speech–language disorders subsequent to traumatic brain injuries, Luria (1970) described seven major types of aphasia including: (1) efferent (kinetic) motor aphasia; (2) frontal dynamic aphasia; (3) pre-motor aphasia; (4) afferent (apraxia) aphasia; (5) sensory (acoustic) aphasia; (6) acoustico-mnestic aphasia; and (7) semantic aphasia. By careful mapping of the sites of injury in wounded soldiers and examining their associated speech–language disorder, Luria was able to establish with some confidence the correlation between the above aphasia syndromes and the territory in which the brain was damaged. Unfortunately, few studies using modern localization methods such as computed tomography have been carried out in order to either confirm or refute Luria's proposals. Remembering that Luria formed his classification on shrapnel wounds, such studies are needed to confirm the relevance of his classification system to cerebrovascular accident cases.

Luria proposed that there are three principal functional units of the brain whose participation is necessary for any type of mental activity (including speech–language function). These three units include: (1) a unit for regulating tone and waking and mental states; (2) a unit for receiving, analysing and storing information; and (3) a unit for programming, regulation and verification of activity. Luria suggested that damage to either of the last two units could cause aphasia. Damage to the second functional unit, which comprises the primary receptive areas for vision, hearing and general senses and the association areas of the parietal, temporal and occipital lobes, may lead to the various receptive aphasias including sensory (acoustic) aphasia, acoustico-mnestic aphasia, afferent (apraxic) motor aphasia and semantic aphasia. On the other hand, lesions of the third functional unit, located in the cerebral hemispheres anterior to the pre-central sulcus, cause the various expressive aphasias including pre-motor, efferent (kinetic) motor aphasia and frontal dynamic aphasia.

2.4.1 Efferent (kinetic) motor aphasia

Patients with efferent motor aphasia have difficulty in formulating their thoughts in language. Although these patients are able to pronounce individual sounds easily, they fail when required to produce those same sounds as part of a whole word. In particular, efferent motor aphasics have difficulty in shifting from one articulatory position to another and therefore their speech disorder becomes most apparent when pronouncing multi-syllabic words or combinations of words.

According to Luria, there are two essential components to efferent motor aphasia, the first involving the loss of serial organization of speech and the second a disturbance in inner speech. Inner speech mediates the transition of thought to external speech. As part of the formulation of inner speech, the predicative structure or dynamic schema of the speech output to follow is determined. Disruption of inner speech in these patients, therefore, makes the formulation of sentences impossible. Sentences, therefore, are absent in the everyday speech of efferent motor aphasics. Likewise, the loss of inner speech causes disruption of the dynamic schema of words leading to a loss of the predicative significance or words and causing the articulation of words to also become impossible.

Efferent motor aphasics exhibit a writing disturbance similar to the disturbance in speech. In addition, most of these patients have a concomitant right hemiparesis. The lesion associated with this disorder involves the inferior pre-motor region of the left hemisphere.

2.4.2 Frontal (dynamic) aphasia

Although patients with frontal (dynamic) aphasia are able to utter words (e.g. repetition tasks) and distinguish speech sounds, they are deprived of spontaneous speech and seldom use it for purposes of communication. It is thought that these patients have lost the ability to formulate thoughts into sentences (propositionizing) and consequently show no spontaneity of speech and are unable to use speech for generalizing or expressing thoughts or desires.

Writing is disturbed in a manner similar to speech. Although these patients are able to follow simple instructions, comprehension is also disturbed. Reading aloud is disturbed. A right hemiplegia or hemiparesis is a common finding in this type of aphasia. According to Luria, the lesion is usually located in the inferior part of the left frontal lobe just anterior to Broca's area.

2.4.3 Pre-motor aphasia

Pre-motor aphasia is caused by lesions in the upper and middle portions of the pre-motor area. It is characterized by a loss of smoothness of speech, agrammatism, disturbed comprehension and intonation and perseveration.

Patients with pre-motor aphasia pronounce individual sounds and words haltingly. Long pauses occur in the transition from one word to another and each word is produced with considerable effort. Pre-motor aphasia, however, does not simply represent a motor speech disorder. Rather a disturbance of inner speech is present. In particular the patient appears unable to store the schemata of inner speech which in turn leads to the halting speech and a reduction in the grammatical complexity of sentences produced. The vocal and written speech of these patients is characterized by short sentence fragments. Pre-motor aphasics speak in monotone and are unable to suppress articulatory patterns once created, leading to perseveration. Comprehension is also disturbed and statements may need to be presented several times to these patients before they are understood. Even then, complex statements may only be understood in part.

Luria suggests that the basic disorder in patients with pre-motor aphasia is not an impairment in their ability to create new speech articulations but rather a disturbance in the automaticity of continuous speech as a result of a disturbance in the schema of speech. A right hemiparesis is a concomitant finding in the majority of these cases.

2.4.4 Afferent (apraxic) motor aphasia

The most outstanding feature of afferent (apraxic) motor aphasia is the patient's inability to determine immediately the positions of the components of the speech mechanism (e.g. lips and tongue) necessary to articulate the required sounds of speech. Consequently a given individual sound may be articulated differently depending upon the syllables in which it occurs. This difficulty is present in both spontaneous speech and repetitive speech.

According to Luria, the disturbance in expressive speech involves difficulty finding the articulatory movements necessary for the pronunciation of individual sounds and sound sequences which go to make up words and phrases. In other words the disorder involves an apraxic disturbance of the speech organs. Unlike the three Lurian syndromes associated with pre-motor lesions described above, the dynamic aspects of speech are not disturbed in afferent motor aphasia. Although in severe cases the patient may not know where to put the tongue, lips etc. to produce necessary sounds, in more mild cases the patient may only confuse the positions for similar articulemes, leading to substitution of phonemes within words (i.e. literal paraphasias).

Deficits in writing and reading are also present in this disorder. An ideomotor dyspraxia of the oral, lingual and pharyngeal musculature is a common finding. A right hemiparesis affecting the arm more than the leg is present in some cases. According to Luria, the lesion associated with afferent motor apraxia is located in the lower parts of the post-central area of the left hemisphere (i.e. the left inferior parietal region).

2.4.5 Sensory (acoustic) aphasia

Sensory (acoustic) aphasia is associated with lesions of the superior parts of the temporal lobe. The major symptoms include a disturbance of phonemic hearing, loss of meaning of words, difficulty naming objects, the presence of literal and verbal paraphasias and a writing disturbance.

The major factor underlying this condition is a disturbance in phonemic hearing which in turn leads to a breakdown in those linguistic tasks requiring sound discrimination. These tasks include the understanding of speech, the naming of objects and the recalling of words. Difficulty in understanding spoken speech (i.e. the loss of the meaning of words) arises from the loss of the ability to discriminate between closely sounding phonemes. Likewise, because they no longer have a differentiated phonemic system of language, the sensory aphasic finds it difficult to name objects and to recall necessary words, thereby leading to the substitution of incorrect words or phonemes (i.e. paraphasia errors). As a result of a deficient phonemic system of language these patients are unable to monitor their own speech and are therefore unaware of the defects in their speech. Consequently, they are unable to correct for these deficits so that their speech becomes converted into empty jargon in which the nominative components (the substantives) are lacking, and their output is reduced to consisting of interjections or habitual expressions. The melodic and intonational aspects of their speech, however, remain intact.

In addition to the loss of phonemic hearing, the semantic aspect of language is also profoundly disturbed in these patients.

2.4.6 Acoustico-mnestic aphasia

Lesions in the posterior–inferior part of the temporal lobe cause a disturbance in audio-verbal memory which in turn gives rise to a condition called acoustico-mnestic aphasia. Patients with acoustico-mnestic aphasia have difficulty in retaining word series. Although typically these patients can retain single words and repeat them after several minutes, they cannot retain a short series of words presented aloud, being able to repeat only the first or last word in most cases.

Although in the acute stage post-onset these patients may have difficulty understanding the speech of others, this symptom usually clears rapidly. Unlike patients with sensory (acoustic) aphasia, phonemic hearing is preserved in acoustico-mnestic aphasia and literal paraphasias are absent. Writing is also usually intact.

2.4.7 Semantic aphasia

Semantic aphasia is caused by lesions of the temporal–parietal–occiptal region of the left hemisphere. The language disorder seen in this condition is based on neither a hearing nor memory deficit. Rather, it represents a disturbance in the understanding of logico-grammatical relationships. Semantic aphasia is characterized by an inability to synthesize isolated simultaneous events into a meaningful unity.

Patients with semantic aphasia experience an inner loss of the semantic structure of words. They have no difficulty in hearing or understanding individual words and speak fluently. Their phonemic hearing is intact and they show no signs of having forgotten the primary meaning of words. Often the only language disturbance noted by an outside observer is an impairment in naming objects. Careful testing, however, reveals that even though patients with semantic aphasia appear to understand the general meaning of speech, they are unable to see the grammatical relationship between words because they can no longer see the entire complex of associations of words. Consequently, these patients are unable to process or decode information according to the logico-grammatical rules of language. Therefore, although there is a preserved ability to understand isolated words, semantic aphasics are unable to grasp the meaning of an entire sentence.

Semantic aphasics often exhibit a concomitant disturbance in spatial orientation. Difficulties with computation (acalculia) are also a frequent finding. Writing is disturbed in a similar manner to speech.

Hier *et al.* (1980) documented the lesions associated with three cases of semantic aphasia using computed tomography. In all cases, the lesions involved the left temporal–parietal–occiptal junction in agreement with Luria's suggestion (one case actually had bilateral damage involving both the left and right temporal–parietal–occiptal junction).

2.5 METHODS OF LESION LOCALIZATION IN APHASIA

Since the middle of the last century, many investigators have attempted to prove or disprove the theory that the various aphasia syndromes and lesions of specific anatomic sites of the brain are correlated. The general methods used to investigate this theory have been to obtain indices of the site and size of brain lesions, to measure the features and characteristics of the aphasic patient's clinical presentation, and to attempt a correlation of these two parameters.

While these basic procedures have been followed, many of the studies have used different kinds of specific instrumentation to measure both the pathological and the clinical parameters of the various aphasia syndromes. Until recently, the most commonly used methods of ascertaining lesion localization have been post-mortem examination, post-traumatic skull defects, cerebral angiography, cortical stimulation, electroencephalography, regional cerebral blood flow and radioisotope brain scans.

Unfortunately, all of these localization techniques have important limiting factors which restrict their validity for making correlations between site of pathology and aphasia type. Post-mortem examination, although an accurate localizing technique, by its very nature, prevents direct correlation of clinical findings with the location of pathology in the living subject. Post-traumatic skull defects do not necessarily reflect specific site or extent of brain lesion and, indeed, may be quite misleading, due to *contre coup* effect. Similarly, cerebral angiography indicates only the vessel occluded, not the specific part of the brain infarcated. Electroencephalograms also lack localizing precision. Cortical stimulation has provided important information about the localization of some language functions, for example, naming, repetition, reading and writing (Ojemann and Whitaker, 1978). It does not, however, give information about the co-ordination and integration of these functions. Regional cerebral blood flow provides accurate two-dimensional information, but gives no indication of depth of lesion, nor, as in the case of angiography, is it conducted without some risk to the patient. In recent years a number of studies have correlated language and other cognitive behaviours with regional cerebral blood flow. The findings of these studies suggest that, in addition to the classic language areas, many other parts of the brain are also activated during speech production and comprehension.

Radiosotope brain scanning is safe and relatively effective, and is probably the best method of localization of those mentioned above. However, it also has some technical limitations. An accurate three-dimensional picture, that is, information about depth of lesion, cannot be obtained because the area of isotope uptake may not necessarily conform to the boundaries of the lesion and all areas of damaged brain tissue are not necessarily shown. Radioisotope scans lack clear definition and do not differentiate between ischaemic and haemorrhagic lesions. In addition, positive readings for identifying brain lesions are only evident in radioisotope scans taken within a limited time post-onset.

Since the mid-1970s, the most commonly used method for localizing aphasia-producing lesions has been computed tomography.

2.5.1 Computed tomography

Computed tomography or CT has a number of advantages compared to the previously used methods of lesion localization. Most importantly it was the first technique to allow visualization of deep-structured pathology (i.e. sub-cortical lesions) in the living subject. Further, computed tomography is useful for viewing a wide variety of different brain pathologies including tumours, haematomas, infarcts and the effects of trauma.

Briefly, computed tomography is the product of applying computer technology and techniques of image reconstruction to modern radiological equipment. An X-ray source produces a narrow beam which transverses the head of the patient. This source is coupled to a radiation detector system, and they both move across the head in a linear fashion. Each point in the brain is investigated from 180° angles. That is to say, the X-ray beam has to complete a semi-circle so that the object is transversed 180 times. The computer then analyses the information obtained and generates an image on the cathode-ray tube. The brightness of the image is proportional to the density of the brain tissue at that point.

The examination starts at the base of the skull (the plane may be horizontal (axial) or angled, often parallel to the orbito-meatal line) and extends up the cranium in a series of about ten slices. Scan slices are usually 8–10 mm thick. Within the brain, computed-tomography scans can define structures such as the ventricles, the sulci, the cisterns, the putamen and thalamus, the choroid plexus, the eye, optic nerve, eye muscles, pineal gland, pituitary fossa, nasopharynx, cranial bone structures, venous structures, foramen magnum, etc.

As an aphasia localization technique, computed tomography has been used in research investigating clinico-pathological correlations of aphasia since the mid 70s. Although this technique has provided considerable evidence to support an anatomically-based model and classification of aphasia, it is evident that we cannot assume that the area of brain damage identified on computed tomography scans is solely responsible for concomitant disturbances in language function. There are many cases of aphasia reported in the literature with unexpected computed tomography lesions which point to limitations of anatomical analysis to fully explain the neuropathology of language. In addition, the reported findings of studies based on more recently introduced brain-scanning methods have indicated that structural brain damage, as indicated by computed tomography, is associated with modifications to brain metabolism in areas distant to the lesion. It is possible that these distant effects may also influence language behaviour.

The findings of studies based on computed tomography have forced a revision of the classical concept of language function. Although in previous years it was thought that the occurrence of aphasia was associated with damage to the 'speech–language area' (including Broca's and Wernicke's area) of the dominant hemisphere, computed tomography has demonstrated the need to extend that area to include sub-cortical as well as cortical structures.

In recent years, two further brain scanning techniques have been developed, each with its own set of advantages and disadvantages compared to computed tomography. These two methods are magnetic resonance imaging and positron emission tomography.

2.5.2 Magnetic resonance imaging

Magnetic resonance imaging or MRI used to be called nuclear magnetic resonance imaging. Overall, the magnetic resonance scanner bears a fair resemblance to the computed tomography scanner. In this technique, the objects of study are the nuclei of atoms that have an odd number of protons (e.g. hydrogen-1, phosophorus-31, carbon-13, sodium-23). The nucleus that yields the strongest magnetic resonance signal is hydrogen. It is also the most abundant atom in biological tissue and has consequently received the greatest attention in magnetic resonance techniques. These atomic nuclei are of interest because they have a property called 'spin' (or magnetic moment) that makes them behave like small bar magnets spinning around their axes. Normally these nuclei assume a random alignment. When we place the head in the strong magnetic field of the magnetic resonance unit, the spinning protons become aligned with the magnetic field (i.e. reach a state of equilibrium). A short burst of radio waves at specific frequencies is then introduced which causes the aligned protons to be 'tipped' in a controlled way. After the burst of radio waves, the protons return to their equilibrium point but, as they do, emit electromagnetic signals whose frequency and duration can be measured by a receiver and, subsequently, used to generate spatial co-ordinates and a visual image using computed methods similar to those employed for computed tomography. The images produced largely represent the distribution of hydrogen (or other elements, e.g. sodium) through the brain or other body region being examined. Different tissues have characteristic differences in hydrogen atom (water) concentrations. Abnormalities can be detected by observing differences between the concentrations of hydrogen atoms (water) found and those expected.

Magnetic resonance imaging provides an image which correlates with anatomy and can distinguish grey from white matter to a greater extent than the computed tomography scanner. Magnetic resonance imaging can also detect the presence of tumour, oedema, atrophy, arterio-venous malformation, haematoma and infarcts, and can identify the effects of demyelinating disease. It also has the potential to measure brain metabolism and cerebral blood vessel flow.

Magnetic resonance imaging has a number of advantages and disadvantages compared to computed tomography. These are summarized in Table 2.1. Because of the newness of this technique, there is little literature to cite which demonstrates the application of magnetic resonance imaging to our understanding of speech and language pathologies. However, it is likely to have an influence on speech–

Table 2.1 Summary of the relative advantages and disadvantages of magnetic resonance imaging

Advantages	Disadvantages
• Magnetic resonance imaging is non-invasive • Does not require use of ionizing radiation • Has very good resolution in transverse, coronal and sagittal planes • Unlike computed tomography, bone does not induce artefacts in the image. The posterior cranial fossa can therefore be viewed in detail • Distinguishes grey and white matter better than computed tomography scans • Sensitive to early changes associated with stroke • Sensitive to changes associated with demyelinating disease • Has the potential for early detection of diseases in which chemical and physiological alterations precede structural (anatomical) changes	• High capital outlay and running costs • Lengthy data acquisition time • Narrow gantry causes claustrophobia in some patients • Potential hazards related to ferromagnetic surgical clips or prostheses and effects on patients with cardiac pacemakers • Because of the high magnetic field, ferrous instruments, etc. in proximity to the magnetic resonance imaging unit may become airborne missiles • Because it is a new technique, at present magnetic resonance imaging scans may be difficult to interpret. It is possible that it may measure changes in brain tissue that are difficult to associate with specific language functions

language research in the following ways. First, the ability of magnetic resonance imaging to distinguish grey from white matter and deep sub-cortical structures, as well as locate infarcts and associated oedema may significantly increase and modify our current understanding of the anatomical correlates of aphasia, apraxia of speech and other related disturbances. This may be particularly useful in helping to sort out the controversy regarding 'sub-cortical aphasias'. A study by DeWitt *et al.* (1985) on two aphasics showed that apparently sub-cortical lesions, as evidenced on computed tomography scans, may be demonstrated by magnetic resonance imaging to involve the cortex as well, thereby suggesting that computed tomography may be inadequate to define the actual extent of lesions. However, because magnetic resonance imaging is a relatively new technique, interpretation at present is often difficult and identification of the nature of pathology is not always possible. If the magnetic resonance imaging scan represents differences in water content, then it would be expected that the magnetic resonance imaging lesions would be larger than infarcts on computed tomography because of changes in water content in the penumbra and other adjacent brain regions. The influence of such changes on language would be hard to associate with specific language deficits and it is possible, therefore, that magnetic resonance imaging may be too sensitive for lesion localization in aphasia. Whether or not this is the case will only be answered by future research and development of the magnetic resonance imaging technique.

The sensitivity of magnetic resonance imaging to changes associated with demyelinating disease is rapidly being established. In this way, it may increase our knowledge about the anatomical correlates of dysarthria in multiple sclerosis. Also, the capacity of magnetic resonance imaging to image, without significant artefact, the posterior fossa opens the door to confirming what we have been assuming about the anatomical bases for many of the dysarthrias. Finally, the non-invasive nature of magnetic resonance imaging and its potential focus on biochemistry raises the possibility of its use in the never-ending search for neurological correlates of childhood language and learning disabilities and so-called developmental verbal dyspraxia.

2.5.3 Positron emission tomography

Positron emission tomography or PET is a non-invasive brain imaging technique which utilizes positron-emitting radionuclides to measure brain metabolism. It is a technique which estimates functional damage rather than structural damage as in computed tomography. Positron emission tomography involves computerized techniques the same as computed tomography to generate visual images, and uses the same brain slices. The method involves measurement of the metabolism of radioactively labelled molecules (positron emitters) such as ^{18}FDG (fluorodeoxyglucose), $^{15}O_2$, etc. In most reported studies so far, ^{18}FDG has been the most commonly utilized molecule.

The radioactive material (e.g. ^{18}FDG) is injected intravenously into the patient and accumulates in the brain over a 45-minute period. Accumulation of ^{18}FDG is proportional to the rate of glucose metabolism. A major disadvantage with the technique is the short half-life of the radioactive compounds (e.g. for ^{18}FDG this is about 2 hours). A cyclotron is needed nearby, therefore, to generate these compounds for immediate use.

On the visual images generated, areas of high metabolic activity show up as dark areas and areas of low metabolic activity as light areas. A number of studies have been reported which have utilized positron emission tomography scans to localize lesions associated with aphasia (for a review see Metter, 1987). Most of these have been conducted by Metter and his co-workers in California. In general, their data suggest that reliance on computed tomography in delineating the extent of brain lesions in aphasia or other neuropsychological defects can be misleading. The reason for this is that positron emission tomography scans usually reveal that cerebral metabolic activity is diminished in an area larger than that of the infarction demonstrated by computed tomography. This suggests that function (metabolism) in non-structurally damaged tissue is abnormal and that consequently the observed language impairment may not be solely attributable to the structural brain damage evident in the computed tomography scan.

In addition, it has been found that in some cases, positron emission tomography scans reveal metabolic lesions to be present that might account for aphasia in patients where computed tomography scans revealed no structural deficit to be present. The full implications of positron emission tomography scan findings on speech–language are as yet unknown.

Sub-cortical aphasia syndromes

Aphasia has traditionally been described as a language disorder resulting from damage to the cerebral cortex. The classical models of language function, based on post-mortem findings, proposed by Wernicke (1874) and Lichtheim (1885) predicted that sub-cortical lesions could only produce language deficits if they disrupted the association pathways that connect the various cortical language centres. Such disruption has been suggested to cause conduction aphasia (Geschwind, 1965a,b ; Damasio *et al.*, 1979). In recent times, however, the belief that the cortex operates as a closed neuronal circuit responsible for language functioning has been challenged by a number of authors (Ojemann, 1976; Kornhuber, 1977; Brunner *et al.*, 1982; Glosser, Kaplan and LoVerme, 1982).

Although a relationship between sub-cortical lesions and aphasia was first postulated at the beginning of this century (Marie, 1906), it is only since the introduction of the computed tomography (CT) scanner that more conclusive scientific evidence has been offered to support this hypothesis. The sub-cortical structures that have received the greatest attention as possible lesion sites linked with aphasia are the thalamus and the striato-capsular region. The evidence available suggests that lesions involving the thalamus produce language deficits through different pathophysiological mechanisms than other sub-cortical lesions (Graff-Radford *et al.*, 1984).

3.1 NEUROANATOMY OF THE STRIATO-CAPSULAR REGION AND THALAMUS

A large amount of neuroanatomical terminology is contained in the literature relating to sub-cortical aphasia syndromes. Consequently, it is important that the reader be provided with a basic understanding of the neuroanatomy of the striato-capsular region and thalamus prior to any discussion of the relationship between specific sub-cortical lesions and language pathology.

3.1.1 Neuroanatomy of the striato-capsular region

The striato-capsular region occupies the deep, central portion of each cerebral hemisphere and is comprised of the basal ganglia and internal capsule. The basal ganglia are a collection of sub-cortical nuclei which process motor information in parallel with the cerebellum. Anatomically, the basal ganglia consist of the caudate nucleus, the putamen, the globus pallidus and the amygdaloid nucleus. Some neurologists also include another nucleus, the claustrum, as part of the basal ganglia. Although a number of brainstem nuclei, including the sub-thalamic nuclei, the substantia nigra and the red nucleus are functionally related to the basal ganglia, they are not anatomically part of it. Collectively, the globus pallidus and the putamen are referred to as the lenticular nucleus (lentiform nucleus).

The relative positions of the basal ganglia to other structures within the cerebral hemispheres are shown in Figures 3.1 and 3.2. The caudate nucleus is the most medial part of the basal ganglia. It is an elongated mass of grey matter, which is bent over on itself and

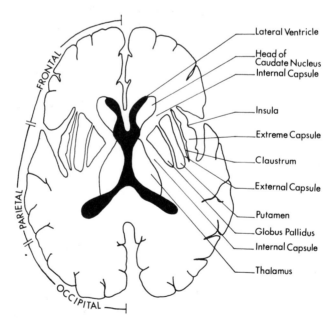

Figure 3.1 Horizontal section of the cerebral hemispheres showing the anatomy of the striato-capsular region.

throughout its length follows the lateral ventricle. The nucleus is divided into a head, body and tail. The head of the caudate nucleus bulges into the anterior horn of the lateral ventricle and lies rostral to the thalamus. The body extends along the dorso-lateral surface of the thalamus. The remainder of the caudate nucleus is drawn out into a highly-arched tail which, conforming to the shape of the lateral ventricle turns into the temporal lobe and terminates in relation to the amygdaloid nucleus. Throughout much of its extent, the caudate nucleus is separated from the lenticular nucleus by the internal capsule.

The lenticular nucleus is located in the midst of the cerebral white matter. Its shape is somewhat similar to that of a biconvex lens, hence the name lenticular or lentiform. The largest portion of the lenticular nucleus is the putamen, which is a rather thick, convex mass, located just lateral to the globus pallidus and internal capsule. Its lateral surface is separated from the cortex by the claustrum, the external capsule and the extreme capsule. The globus pallidus is the smaller and most medial part of the lenticular nucleus. It is traversed by numerous bundles of white fibres which make it appear lighter in colour than the putamen. The globus pallidus is sub-divided into medial and lateral parts by a small band of white fibres called the medial medullary lamina. The medial pallidal segment in turn is divided by the accessory medullary lamina into outer and inner

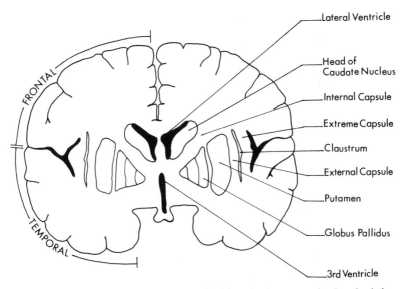

Figure 3.2 Coronal section of the cerebral hemispheres at the level of the optic chiasma showing the anatomy of the striato-capsular region.

portions. The lenticular nucleus combined with the caudate nucleus make up what is known as the corpus striatum, so named because of the striated (striped) nature of this region. The function of the corpus striatum is concerned with somatic motor functions. The term 'extra-pyramidal motor system' is used by neurologists to group together the corpus striatum and certain brainstem nuclei considered to subserve these somatic motor functions. The extrapyramidal system is discussed more fully in Chapter 9.

The claustrum is a thin layer of grey matter which lies between the insular cortex and the lenticular nucleus. It is separated from the more medial putamen by the external capsule and from the insular cortex by the extreme capsule. Both the external and extreme capsules carry association fibres. The amygdaloid nucleus (body) is a small, spherical grey mass located in the temporal lobe in the roof of the inferior horn of the lateral ventricle.

Connections exist between the various individual nuclei of the basal ganglia and between the nuclei and other brain structures which include the cerebral cortex, thalamus, red nucleus and reticular for-mation. These connections are extremely complex and as yet have not been fully determined in humans. Briefly, the afferent inflow to the corpus striatum is mainly from the massive cortico-striatal pathways which project fibres to the caudate nucleus and putamen from nearly all parts of the cerebral cortex, but especially from the motor areas. From the caudate nucleus and putamen, the input is relayed to the globus pallidus. Other less significant inputs come from the thalamus via the thalamo-striatal pathways and brainstem nuclei including the substantia nigra (nigro-striatal tracts). Output from the basal ganglia occurs primarily through the globus pallidus. It sends massive bundles of inhibitory fibres mainly to the thalamus (pallido-thalamic tracts), particularly to the ventral anterior nucleus of the thalamus, but also to the brainstem. Since the ventral anterior nucleus of the thalamus projects its output to the motor and pre-motor cortex, an important circuit is established between the motor cortex, basal ganglia, thalamus and motor cortex again. The importance of this circuit to language function is discussed below. Lesions of the basal ganglia are associated with several extrapyramidal syndromes including Parkinson's disease, chorea, athetosis, dystonia and hemiballismus. These syndromes are discussed in Chapter 9 in relation to hypo- and hyperkinetic dysarthria.

The internal capsule is composed of afferent fibres to and efferent fibres from the cerebral cortex, diverging from, and converging toward the brainstem. In horizontal section (see Figure 3.1), the internal capsule in each hemisphere is 'V'-shaped with the apex pointing

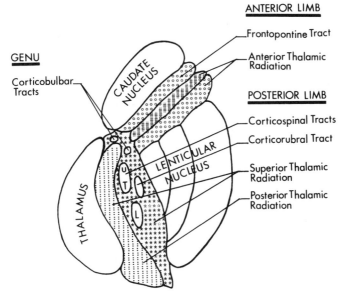

Figure 3.3 Schematic diagram of the right internal capsule as seen in a horizontal section similar to that shown in Figure 3.1. U, upper limbs; T, trunk; L, lower limbs.

medially. Two distinct parts of the internal capsule are evident in horizontal sections, an anterior limb and a posterior limb which meet at the genu (the apex of the 'V'). The larger limb, the posterior limb is bordered medially by the diencephalon and laterally by the lenticular nucleus. The anterior limb lies between the caudate nucleus and the putamen.

Efferent fibres in the internal capsule arise from cells in various regions of the cerebral cortex and project to specific nuclei in the brainstem and spinal cord. In so doing they form a number of tracts which include the cortico-thalamic, cortico-rubral, cortico-pontine, cortico-reticular, cortico-bulbar and cortico-spinal tracts. On the other hand, the afferent fibres in the internal capsule arise mainly from the thalamus and project to almost all regions of the cerebral cortex. These afferent fibres are referred to as the thalamo-cortical radiations and together with the cortico-thalamic fibres form the thalamic radiations. The location of the various efferent and afferent tracts in the internal capsule is shown in Figure 3.3.

The anterior thalamic radiation (or peduncle) and the fronto-pontine tract are located in the anterior limb. The genu contains cortico-bulbar and cortico-reticular fibres. Within the posterior limb are found the cortico-spinal fibres, the superior and posterior thalamic radiations and relatively smaller numbers of cortico-tectal, cortico-rubral and cortico-reticular fibres.

Because of the high concentration of nerve fibres, more widespread disability is produced by lesions in the internal capsule than in any other region of the nervous system. Most injuries to the internal capsule are of vascular origin arising, in most cases, from either thrombosis or haemorrhage of branches of the middle cerebral artery. Unilateral lesions of the posterior limb may result in contralateral hemianaesthesia of the head, trunk and limbs due to injury of the thalamo-cortical fibres *en route* to the sensory cortex. A contralateral hemiplegia is also present due to involvement of the cortico-spinal tracts.

3.1.2 Neuroanatomy of the thalamus

The thalamus is a large mass of grey matter located above the mid-brain. It forms part of the diencephalon. Although almost completely separated by the third ventricle into right and left thalami, the thalamic mass in each hemisphere is connected in most cases to that in the opposite hemisphere by a band of grey matter called the inter-mediate mass or interthalamic adhesion. In horizontal section (see Figure 3.1) the thalamus can be seen to lie medial to the posterior limb of the internal capsule.

The thalamus has multiple connections with both higher and lower structures in the nervous system such as the cerebral cortex and spinal cord and acts as an important relaying and integrating centre for both sensory and motor impulses. The majority of sensory impulses arriving at the cerebral cortex (with the exception of ol-factory impulses) have travelled through one or more nuclei in the thalamus. The thalamus receives sensory stimuli from the peripheral receptors via the sensory pathways, integrates and organizes this information and then relays the stimulus to the appropriate sensory area of the cerebral cortex. Visual stimuli, for instance, are relayed by the thalamus to the cortex of the occipital lobe for interpretation.

The thalamus also contributes to emotional responses to sensory experience. The posterior end of the thalamus is an expanded prom-inence called the pulvinar which overhangs the superior colliculus of the mid-brain. It is the largest nucleus in the thalamus. Attached to the pulvinar are two smaller prominences; the lateral geniculate body, which receives the optic tract and projects to the visual cortex, and the medial geniculate body, which receives the ascending auditory fibres and projects to the auditory cortex of the temporal lobe. Internally, the grey matter of the thalamus is divided by a 'Y'-shaped

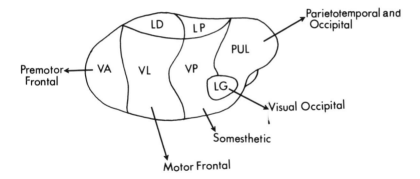

(a)

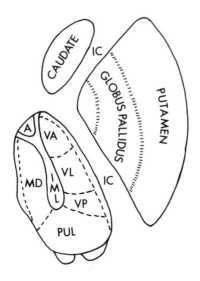

(b)

Figure 3.4 (a) Left lateral view of the thalamus showing the positions of the larger nuclei and their cortical projections; (b) schematic diagram showing the location of the major thalamic nuclei as seen in a horizontal section of the right cerebral hemisphere. VA, ventral anterior nucleus; VL, ventral lateral nucleus: VP, ventral posterior nucleus; LG, lateral geniculate body; PUL, pulvinar; LP, lateral posterior nucleus; LD, lateral dorsal nucleus; A, anterior nucleus; MD, Dorsomedial nucleus; IML, internal medullary lamina; IC, internal capsule.

vertical sheet of white matter, the internal medullary lamina, into three parts. The anterior part of the thalamus lies between the two limbs of the 'Y' and the medial and lateral parts lie on the sides of the stem of the 'Y'. The internal medullary lamina consists of nerve fibres which connect the various parts of the thalamus.

The thalamus contains more than 30 nuclei some of which have been indicated as important in language. The location of some of the major thalamic nuclei mentioned in the speech–language literature together with their cortical projections are shown in Figure 3.4. The pulvinar interconnects with many other thalamic nuclei and is connected reciprocally with the supramarginal gyri, the angular gyri, the superior parietal lobule and the occipital and posterior temporal regions. The ventral lateral nucleus and the ventral anterior nucleus are specific motor nuclei in the sense that they receive information from the cerebellum, corpus striatum and substantia nigra and project to motor areas in the frontal lobe via the anterior limb of the internal capsule. The ventral posterior nucleus is a specific sensory nucleus and functions as a thalamic relay for general sensations. It receives input from ascending sensory systems and projects fibres to the primary somesthetic cortex located on the post-central gyrus via the posterior limb of the internal capsule. Where appropriate the thalamic nuclei are discussed below in relation to models proposed that have attempted to explain the role of sub-cortical structures in language.

Lesions in the thalamus are, in most cases, the result of thrombosis or haemorrhage of one of the branches of the posterior cerebral artery. Neurological signs of damage to the thalamus include contra-lateral hemianaesthesia (especially if the ventral posterior nucleus is damaged), contralateral hemiplegia and emotional disturbances. Some or all of these concomitant neurological signs may be evident in patients with thalamic aphasia. Although impairment of all forms of sensation may occur, some pain and temperature sensation from the contralateral side may be retained. Some patients may experience 'thalamic syndrome', a condition in which there is an over-reaction to sensation such that mild sensory stimuli (e.g. light touch) may trigger exaggerated sensory responses which may lead to intractable pain.

A contralateral upper motor neurone paralysis (spastic paralysis) usually accompanies thalamic lesions due to involvement of the adjacent internal capsule. Emotional instability may also occur with spontaneous laughing and crying being evident in some patients. Neurosurgical destruction of the ventral anterior thalamic nuclei inter-rupt connections between the basal ganglia and motor areas of the cerebral cortex and serve to decrease rigidity and tremor in patients with Parkinson's disease.

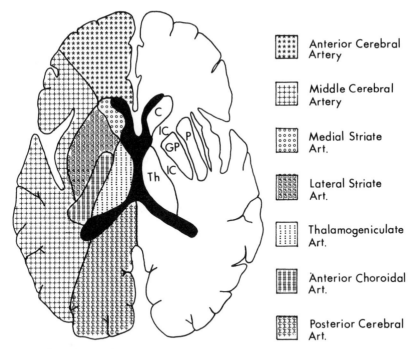

Figure 3.5 Horizontal section of the cerebral hemispheres showing the sub-cortical distribution of the anterior, middle and posterior cerebral arteries and branches. Th, thalamus; GP, globus pallidus; C, caudate nucleus; P, putamen; IC, internal capsule.

3.2 BLOOD SUPPLY TO THE STRIATO-CAPSULAR REGION AND THALAMUS

In most reported cases of sub-cortical aphasia, the language disturbance has occurred secondary to a cerebrovascular accident, usually a haemorrhage. As the particular vessel involvement determines which of the sub-cortical structures will be involved in the area of lesion and consequently influences the expected language disturbance, it is important that the reader has some knowledge of the vascular supply to the striato-capsular region and thalamus prior to looking in detail at the variety of language disorders reported to occur in association with lesions in these two areas.

The distribution of the anterior, middle and posterior cerebral arteries and their deep branches to the sub-cortical region is shown in Figure 3.5. The medial and lateral striate (also called lenticulo-striate) and the anterior choroidal arteries provide the blood supply to the internal capsule. The anterior limb is supplied primarily by the lateral striate branches of the middle cerebral artery, the medial striate (also called the recurrent artery of Heubner) from the anterior cerebral supplying the rostral–medial tip. Blood supply to the genu comes via some direct branches from the internal carotid or from the lateral striate arteries. The posterior limb of the internal capsule is supplied partly by the anterior choroidal artery (internal carotid) and partly by the lateral striate arteries.

Arterial supply to the corpus striatum is mainly provided by the lateral striate arteries. Part of the head of the caudate nucleus, however, is provided with nourishment by the medial striate artery. Most of the putamen and body of the caudate is supplied by the lateral striate arteries while the tail of the caudate and posterior portion of the putamen receive blood from the anterior choroidal artery. The majority of the globus pallidus is also supplied by this latter artery.

Whereas the striato-capsular region is supplied by branches of the middle and anterior cerebral arteries, branches of the posterior cerebral artery provide the major blood supply to the thalamus. The medial and anterior regions of the thalamus are supplied by the thalamo-perforating branches while the pulvinar (posterior portion of the thalamus) and lateral aspects of the thalamus are nourished by the thalamo-geniculate branches. The dorsal part of the thalamus is supplied by the posterior choroidal branches of the posterior cerebral artery. Some branches of the anterior choroidal artery are thought to pass to the lateral geniculate body.

3.3 APHASIA ASSOCIATED WITH THALAMIC LESIONS

In recent years, a number of reports documenting thalamic involvement in individuals with significant language disturbance have appeared in the literature (for reviews see Jonas, 1982; Lhermitte, 1984). Despite the increasing number of reports, the role of the thalamus in normal language function, and the existence of the so-called 'thalamic aphasia' remains controversial.

Evidence to support a role for the thalamus in speech–language functions has come from studies of language changes due to surgical destruction and stimulation of thalamic targets as well as from observations of language disturbances following spontaneous thalamic lesions, including neoplastic, infectious and vascular lesions. In most reported cases of thalamic aphasia, the language disorder has been associated with a thalamic haemorrhage (Bugiani, Conforto and Sacco,

1969; Ciemens, 1970; Fazio, Sacco and Bugiani, 1973; Mohr, Watters and Duncan, 1975; Reynolds *et al.*, 1978; Cappa and Vignolo, 1979; Reynolds *et al.*, 1979; Kirshner and Kistler, 1982; Chesson, 1983; Murdoch, 1987). A smaller number of cases have also been reported in association with thalamic tumours (Smyth and Stern, 1938; Arseni, 1958; Cheek and Taveras, 1966) and thalamic abscesses (Panchal, Parikh and Karapurkar, 1974). Because these disorders (i.e. thalamic haemorrhage, tumour and abscess) are in most cases associated with a considerable mass effect, a number of authors have suggested that the accompanying language deficit may not be directly attributable to localized damage of the thalamus *per se* (Mohr *et al.*, 1975; Van Buren, 1975; Horenstein, Chung and Brenner, 1978). Rather the observed language abnormalities may be the products of the effects of pressure and oedema on other areas of the brain, particularly the left cerebral hemisphere where the known speech and language centres are located. In other words, the aphasia associated with localized thalamic lesions reported in many studies may not be the direct products of thalamic destruction, but rather the secondary product of pressure effects on other parts of the brain. Further, aphasia does not always follow left thalamic haemorrhage (Cappa *et al.*, 1986).

It is interesting to note that those structures most likely to suffer compression effects as a result of thalamic haemorrhage are those structures that lie adjacent or in close proximity to the thalamus. These structures include the isthmus of the temporal lobe (through which are considered to pass fibres connecting the posterior language zone to Broca's area and to the remaining brain, including the arcuate fasciculus). Damage to the temporal isthmus produces a syndrome inseparable from Wernicke's aphasia (Nielsen, 1962) while interruption of the arcuate fasciculus produces conduction aphasia (Geschwind, 1965a,b). If the aphasia associated with thalamic haemorrhage, etc. is, in fact, due to compression of these adjacent structures rather than damage to the thalamus itself, it could be expected that the aphasic syndrome would resemble either a conduction or a Wernicke's aphasia, both of which have among their characteristics impaired repetition.

To some extent, the influence of extraneous factors such as pressure effects can be reduced by studying patients with ischaemic infarcts in the thalamus rather than haemorrhagic lesions. Although such cases of infarction are rare, several authors have documented aphasia subsequent to ischaemic infarcts in the thalamus (Cohn, Gelfer and Sweet, 1980; McFarling, Rothi and Heilman, 1982; Gorelick *et al.*, 1984; Graff-Radford *et al.*, 1984). These findings lend support to the suggestion that aphasia subsequent to thalamic lesions is the direct result of thalamic damage rather than the product of associated pressure effects.

A language function for the thalamus is also indicated by the results of studies of patients following surgical intervention in the thalamus. The effects on speech and language functions have been studied in cases of surgically induced thalamic lesions (Waltz *et al.*, 1966; Bell, 1968; Ciemens, 1970; Riklan and Levita, 1970; Darley, Brown and Swenson, 1975b) and in cases of thalamic stimulation (Fedio and Van Buren, 1975; Ojemann, 1977; Cappa and Vignolo, 1979). These studies have shown that, although aphasia does not inevitably follow left-sided thalamic ablations or stimulations, it does occur often. Although highly suggestive, as with the haemorrhage and tumour cases, surgical manipulation of the left thalamus cannot be said to provide conclusive evidence that the thalamus participates in language. The procedures are always performed on brains with pre-existing lesions, and both stimulation and ablation procedures may affect neighbouring structures. If the thalamus does have a language function, stimulation studies suggest it to be selectively related to the ventro-lateral nucleus and the antero-superior pulvinar (Ojemann, 1975).

Based on conclusions drawn from thalamic stimulation studies and post-operative thalamotomy data, several authors have concluded that thalamic involvement with speech and language is entirely a left thalamic function. Other more traditional researchers advocate a dominant hemisphere thalamic speech–language function. The lesion has primarily involved the left thalamus rather that the right in most reported cases of acquired thalamic aphasia (Riklan and Cooper, 1975; Reynolds *et al.*, 1978; Cappa and Vignolo, 1979; Reynolds *et al.*, 1979; Cohen, Gelfer and Sweet, 1980; McFarling *et al.*, 1982). Neither ablation nor stimulation of the right thalamus has been reported to produce effects on language (Lhermitte, 1984). Language disturbances have also, in recent years, been reported after right thalamic haemorrhage in right-handed (Murdoch, 1987) and left-handed (Kirshner and Kistler, 1982; Chesson, 1983) patients, thereby lending support to the suggestion that thalamic involvement in language is a dominant hemisphere function.

The clinical picture as described in the literature relating to the language deficits arising from thalamic lesions is fairly homogeneous. The language disorders associated with thalamic lesions, however, do not readily fall into any of the currently accepted cortical aphasia syndromes such as Broca's and Wernicke's aphasia (Alexander and LoVerme, 1980). Overall, the language disturbance appears to resemble a transcortical aphasia (Cappa and Vignolo, 1979; Alexander and LoVerme, 1980; McFarling *et al.*, 1982; Chesson, 1983; Gorelick *et al.*, 1984).

The features of thalamic aphasia most commonly reported include preserved repetition, variable but often relatively good auditory comprehension, a reduction in spontaneous speech, a predominance of semantic paraphasic errors, and anomia. In particular, it is the preservation of repetition that indicates similarities between thalamic aphasia and transcortical aphasia. McCarthy and Warrington (1984) have shown that there are at least partially separated semantic and phonological functions in language production. Jonas (1982) concluded that the thalamus does not play a major role in several aspects of language, including syntactic structuring. The relatively intact repetition abilities of thalamic aphasics in the presence of semantic impairment (semantic paraphasias) suggests that the system for semantic monitoring involves the thalamus, while that involved in the transmission of phonological linguistic information for repetition is separate and does not involve thalamic structures. Crosson (1985) suggests that the arcuate fasciculus is involved in phonological processing. Preservation of repetition further suggests that the aphasia associated with thalamic lesions is not due to compression of adjacent or surrounding brain structures such as the temporal isthmus, as damage to these structures is known to cause aphasia syndromes in which repetition abilities are not preserved, e.g. Wernicke's and conduction aphasia. Because of the similarities to transcortical aphasia (particularly transcortical sensory), Alexander and LoVerme (1980) raise the possibility that the language deficits noted in cases of thalamic aphasia may be due to compression of the internal carotid artery intracranially, or its middle and anterior cerebral branches, or both. Such restriction of the internal carotid is known to produce diffuse ischaemic damage in the arterial border zone of the left hemisphere, a condition known to be associated with transcortical aphasias. This theory, however, does not explain why the auditory comprehension abilities of thalamic aphasics is often reported to be relatively good whereas in transcortical sensory aphasia, comprehension of spoken language is severely defective. Although similar, therefore, to transcortical aphasia, thalamic aphasia does show some differences especially in comprehension abilities. Thalamic aphasics exhibit greater retention of the phonemic level of verbal behaviour than the semantic level, as evidenced by the high number of semantic paraphasic errors present in their speech. The preserved repetition abilities of thalamic aphasics suggests that integrity of the thalamus is not crucial for intentional speech and the phonemic level of verbal behaviour.

The prognosis for thalamic aphasia varies, some authors reporting a good prognosis for some of their patients and a poor prognosis for others. Riklan and Levita (1970) found that speech disorders associated with left unilateral thalamotomy for Parkinson's disease tend to recover well. Benson (1979, p. 96), suggests that the language findings following thalamic haemorrhage are transient, 'recovery often beginning within days or weeks and except in cases complicated by widespread damage, the course is usually one of consistent improvement over a few weeks or months'. Although several cases of aphasia following left thalamic haemorrhage have been reported to recover completely, or have been left with only mild deficits, some reports suggest that persistent language deficits, particularly naming problems, may occur (Fazio *et al.*, 1973; Mohr *et al.*, 1975; Reynolds *et al.*, 1978; Chesson, 1983; Murdoch, 1987).

3.4 APHASIAS ASSOCIATED WITH STRIATO-CAPSULAR LESIONS

A number of authors have documented the occurrence of language disturbances subsequent to lesions in the basal ganglia (Yarnell, Monroe and Sobel, 1976; Sterzi and Vallar, 1978; Alexander and LoVerme, 1980; Brunner *et al.*, 1982; Damasio *et al.*, 1982; Cappa *et al.*, 1983; Leader, 1983; Wallesch *et al.*, 1983; Murdoch *et al.*, 1986b). Damasio *et al.*, (1982) suggested that critical sites for sub-cortical lesions associated with aphasia are the putamen, the head of the caudate nucleus and anterior limb of the internal capsule. Aphasia has also been reported in association with lesions in the left caudate nucleus (Barat *et al.*, 1981).

In general, there has been considerable variation reported, both within and between studies, in the pattern of language disturbances exhibited by patients with lesions of the basal ganglia. Consequently, the clinical picture still defies a definition and the pattern of language behaviours exhibited cannot be readily classified into one of the classical cortical aphasia syndromes. Some characteristics, however, are frequently reported: language disturbances are generally mild, presumably transient and repetition is preserved. This description, as outlined above, also fits the clinical picture of thalamic aphasia and, in fact, Alexander and LoVerme (1980) stress that putaminal and thalamic aphasias are not distinguishable from one another. Basso, Sala and Farabola (1987), however, could not identify any picture typical of sub-cortical aphasias and concluded that the term 'sub-cortical aphasia' covers patients identified by site of lesion and not by specific clinical syndromes.

A number of authors have suggested that a difference exists between the type of aphasia associated with anterior subcortical lesions compared to posterior sub-cortical lesions (Damasio *et al.*, 1982; Naeser *et al.*, 1982; Cappa *et al.*, 1983; Murdoch *et al.*, 1986b). Based upon nine cases, Naeser *et al.* (1982) described three aphasia syndromes resulting from lesions involving the putamen and internal capsule. Subjects with left capsular–putaminal lesions extending into the anterior–superior white matter exhibited good comprehension, grammatical but slow dysarthric speech and a persisting right hemiplegia. Capsular–putaminal lesions with posterior white matter extensions were associated with poor comprehension, fluent Wernicke-type speech and a persisting right hemiplegia. Global aphasia was present in those cases where both anterior–superior and posterior white-matter extensions were observed. It should be noted, however, that in addition to sub-cortical structures such as the putamen, internal capsule and corona radiata, the lesions described by Naeser *et al.* (1982) were not purely sub-cortical in that they also involved cortical areas such as the insula and components of the cortex buried in the sulci. Although Cappa *et al.* (1983) suggested that damage to the cortex of the insula may be the cause of speech and language disturbances, Alexander, Naeser and Palumbo (1987) produced evidence that the insular cortical lesions are not the cause of the aphasia observed in patients with predominantly striato-capsular lesions that also involve the cortex in the region of the insula.

Cappa *et al.* (1983) also distinguished between aphasias associated with anterior sub-cortical lesions (putamen and anterior limb of internal capsule) and posterior sub-cortical lesions (putamen and posterior limb of internal capsule). They described an atypical non-fluent aphasia to be associated with anterior sub-cortical lesions and a mild fluent aphasia with posterior sub-cortical lesions. The lesions described by Cappa *et al.* (1983) were smaller than those reported by Naeser *et al.* (1982) and presumably spared the cortex. Overall, the language deficits exhibited by their subjects were less severe and atypical if compared with the classical cortical aphasia syndromes.

Similar findings have been reported by Damasio *et al.* (1982) and Murdoch *et al.* (1986b). A right hemiparesis, dysarthria or dyprosody and an atypical aphasia not readily classified according to the Boston System for Diagnostic Classification (see p. 69) was described by Damasio *et al.* (1982) in association with left, non-haemorrhagic lesions involving the anterior limb of the internal capsule, head of the caudate nucleus and putamen. Patients with similar lesions were described by Murdoch *et al.* (1986b) as exhibiting an atypical non-fluent aphasia. Although these patients exhibited features in common with Broca's aphasia, they also exhibited non-typical features such as semantic paraphasias usually associated with fluent aphasia

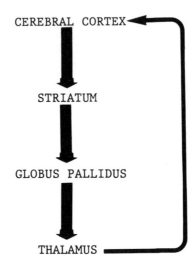

Figure 3.6 Connections between the cerebral cortex, basal ganglia and thalamus.

syndromes. Murdoch *et al.* (1986b) reported that posterior sub-cortical lesions involving the posterior limb of the internal capsule and posterior portion of the putamen to be associated with mild, fluent Wernicke's-type aphasia while lesions involving both anterior and posterior sub-cortical structures were associated in most cases with global aphasia.

In addition to clinico-anatomical correlation studies, anatomical evidence also suggests a role for sub-cortical structures in language. The basal ganglia are connected with almost all cortical areas by an anatomical loop: the cortex–striatum–pallidum–thalamus–cortex (Kemp and Powell, 1971; Krauthamer, 1979) (see Figure 3.6).

Wallesch (1985) suggested that this loop may have a role in language function. As described above, the caudate nucleus and putamen both receive input from most regions of the cerebral cortex. However, although the connections from the cortex to the caudate nucleus and putamen are prolific, there are no fibres of any consequence which travel in the opposite direction, i.e. from the caudate nucleus or putamen back to the cortex.

The main outputs from the basal ganglia are channelled through the globus pallidus which receives fibres from both the caudate nucleus and the putamen. The globus pallidus projects to the ventral lateral and ventral anterior nuclei of the thalamus. The ventral lateral thalamus, in turn, projects to the motor cortex and also receives input from the cerebellum as well as from the motor cortex. The ventral anterior thalamus sends output to the pre-motor and pre-frontal cortex. The functional correlate of this organization is most likely the participation of the basal ganglia in cortical functions, including language. The particular connections between sub-cortical structures just outlined, however, tells us something about the way in which they may be organized in language functions. Clearly, the cortical language centres may influence the caudate nucleus or putamen. However, due to an absence of direct connections, the caudate nucleus and putamen can have no direct influence on cortical language centres. Rather, any influence of the basal ganglia on cortical language centres must be mediated by thalamic centres. The implications for this on the participation of sub-cortical structures in language will be discussed more fully shortly.

Disruption of the loop connecting the cortex, basal ganglia and thalamus may occur with lesions involving the internal capsule, or following damage to the grey-matter neuronal operators located in the various nuclei such as the putamen and caudate nucleus. Lesions in these areas could, therefore, be expected to cause language deficits (Damasio et al., 1982).

Reports in the literature vary as to the prognosis of aphasic disturbances resulting from basal ganglia lesions. Some studies have shown that the aphasia is transient and resolves quickly (Brunner et al., 1982; Peach and Tonkovich, 1983; Olsen, Bruhn and Oberg, 1986) while others have indicated that long-lasting deficits result from left basal ganglia lesions (Naeser et al., 1982; Wallesch et al., 1983; Murdoch et al., 1986b).

3.5 ROLE OF SUB-CORTICAL STRUCTURES IN LANGUAGE

Several theories have been proposed to explain the role of sub-cortical structures in language. Penfield and Roberts (1959) introduced the suggestion that the thalamus, specifically the pulvinar, is primarily responsible for mediation between the anterior and posterior language centres (i.e. Broca's and Wernicke's areas). Their argument rested chiefly on the presence of strong fibre connections between the pulvinar and the temporo-parietal cortex and the parallel evolutionary expansion of the pulvinar and speech cortex over the mammalian species.

It has been suggested that the thalamus acts as an 'alerting system' for the cortical language areas (Ojemann, 1975; Mateer and Ojemann, 1983) and that the aphasia associated with thalamic lesions is the outcome of deficient arousal of otherwise intact cortical language mechanisms. The clinical features of such an aphasia would be expected to involve selective loss of the more complex, cortical-based, language functions. Such aphasias are classified transcortical aphasias, which are characterized by loss of more complex language functions with relative sparing of repetition, often to the point of echolalia. The above-mentioned 'transcortical-like' features of thalamic aphasia noted by various authors (Cappa and Vignolo, 1979; McFarling et al., 1982) tend to suggest, therefore, that Ojemann's conclusions may be correct. Further, anomia, found in all cases of thalamic aphasia, can be explained by the arousal hypothesis. Based on their findings that loss of verbal memory in both cortical and sub-cortical aphasias is correlated with decreased metabolic rates in the thalamus, Metter et al. (1983) concluded that abnormal metabolic functioning of the thalamus interferes with specific arousal mechanisms for memory.

A number of authors have suggested a semantic role for the thalamus in language (Cappa and Vignolo, 1979; Alexander and LoVerme, 1980). Crosson (1985) suggested that the thalamus provides a mechanism through which the temporo-parietal area of the cerebral cortex, involved in semantic and phonological decoding, monitors the encoding of language in the anterior language area prior to execution of encoded material in speech. In particular, Crosson proposed that reciprocal connections between the anterior cortical and temporo-parietal areas which pass through the thalamic structures (the anterior superior lateral pulvinar, the ventral anterior nucleus and internal medullary lamina) provide the means by which the temporo-parietal cortex checks the language encoded in the anterior cortex for semantic accuracy. When semantic errors are discovered during the monitoring process, information is carried back to the anterior language area via the thalamic pathway, and a process of semantic refinement is initiated. The predominance of semantic paraphasias observed in cases of thalamic aphasia (Cappa and Vignolo, 1979; Alexander and LoVerme, 1980) supports the idea of an interruption in semantic monitoring.

Aphasia subsequent to thalamic lesions may be a consequence of direct thalamic involvement in language processing (Riklan and Levita, 1965; Brown, 1975). Evidence for thalamic participation in the processing of complex visual stimuli was provided by Phillips and Singer (1974). Rolls et al. (1982) have recorded potentials from neurones in the thalamus during visual recognition and memory tasks. Alternatively, the aphasia seen with thalamic lesions might be the result of distance effects on the cerebral cortex. Metter et al. (1983) using positron emission tomography (PET) found that metabolic depression of the cortex occurs with structural lesions of the thalamus and basal ganglia. Patients with sub-cortical lesions may show blood-flow reduction in the ipsilateral cortex (Olsen *et al.*, 1983; Perani *et al.*, 1985). Olsen, Bruhn and Oberg (1984) showed that aphasic patients with sub-cortical lesions have a low cortical blood flow, while non-aphasics with sub-cortical lesions have no detectable blood flow abnormalities. Robin and Schienberg (1983) believed that the aphasia associated with sub-cortical lesions may occur as a result of a combination of factors including abnormal language processing at a sub-cortical level, decreased arousal mechanisms and distance effects. Other theories have emphasized that the thalamus might regulate access to stores of language information (Reynolds *et al.*, 1979).

Of the structures involved in language, the least data exist on the basal ganglia. Nonetheless, the data that are available do allow the formulation of tentative hypotheses regarding their role in language. Not least among these data is the anatomical position of the basal ganglia relative to the cortex and the thalamus. As previously highlighted, the putamen and caudate nucleus receive input primarily from the cortex, but these structures do not send any output directly to the cortex. Most output from the basal ganglia is mediated through the globus pallidus which sends numerous fibres to the ventral anterior and ventral lateral nuclei of the thalamus.

Although the exact role played by the basal ganglia in speech and language mechanisms is unknown, there is evidence to suggest that, as a group, these structures function as a programme generator for precise, sequential motor behaviour, including speech and language functions (Kornhuber, 1977, 1980; Marsden, 1982). Researchers believe that complex motor movements are the result of basal ganglia motor programmes working in conjunction with cortical activation. The basal ganglia receive increased blood flow during complex motor acts prior to the cortex (Roland *et al.*, 1982). Thus, speech production may be the result of initiation of motor programmes at the level of the basal ganglia. The thalamus may also be responsible for accurate transmission of information from the basal ganglia to the cortex (Kornhuber, 1974). It is possible, therefore, that for language to be processed efficiently and correctly, the specific neurones in the cortex, thalamus, basal ganglia and their connections, need all to be intact or error patterns will result.

As described above, the basal ganglia are in a position to receive input from various parts of the cortex and, on the basis of these inputs, the basal ganglia can in turn influence outputs from the thalamus to the cortex. Crosson (1985) proposed that the basal ganglia are involved in two mechanisms which influence language production by integrating inputs from the cortex and subsequently influencing thalamic mechanisms. First, the basal ganglia influence activity in the anterior cortical language areas by regulating the flow of excitatory impulses from the ventral anterior thalamus. The ventral anterior nucleus of the thalamus is thought to activate the anterior cortical language mechanisms for formulation of meaningful language by regulating the flow of excitatory impulses from the reticular formation. Inhibitory influences exercised on the anterior nucleus from the globus pallidus determine how much excitation is allowed to pass to the anterior cortical mechanism. If cortical excitation is maintained at too high a level, extraneous material will enter into the encoding (language formulation) and motor-programming processes. If the level of excitation is too low, language formulation will be inefficient or not occur spontaneously at all. The second mechanism is a motor release mechanism which allows language segments to be released at the proper time, after semantic monitoring has taken place.

The mechanism directly affecting the activation of the anterior language cortex by the ventral anterior nucleus of the thalamus is thought by Crosson to be located in the globus pallidus. Fluent language output was described after surgical lesions of the globus pallidus (Svennilson *et al.*, 1960). Further, the arrest of ongoing language during stimulation of the globus pallidus was reported by Hermann *et al.* (1966). The most obvious explanation for these two pallidal phenomena is that the globus pallidus maintains an inhibitory influence over the ventral anterior thalamus in such a way as to regulate the amount of excitation conveyed to the anterior language cortex. Thus, lesion of the globus pallidus would create disinhibition of the ventral anterior thalamus, leading to over-activation of the anterior cortical language zones. This over-activation would, in turn, result in the programming of extraneous material such as paraphasic errors in language (Svennilson *et al.*, 1960). Stimulation of the globus pallidus, however, would excite inhibitory mechanisms resulting in the inhibition of the ventral anterior nucleus and the interruption of on-going language, as noted by Hermann *et al.* (1966). An interpretation of Crosson's model, together with a summary of the predicted language segments being released as speech. Once the semantic infor- is presented in Figure 3.7.

As indicated above, pathways through the thalamus are responsible for the pre-verbal monitoring by the temporo-parietal cortex of the semantic content of language encoded by the anterior language area. During the semantic verification of language, Crosson proposed that inhibitory influences from the temporo-parietal cortex prevents language segments being released as speech. Once the semantic information has been verified to be correct, the temporo-parietal cortex

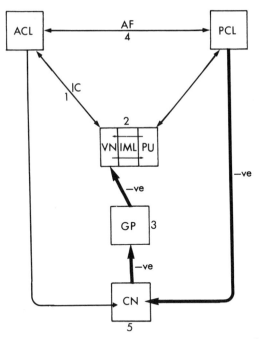

Figure 3.7 Sub-cortical participation in language: An interpretation of the model proposed by Crosson (1985) showing predicted language deficits with specific lesions. ACL, anterior cortical language area; CN, caudate nucleus; GP, globus pallidus; IML, internal medullary lamina; PCL, posterior cortical language area; PU, pulvinar; AF, arcuate fasciculus; VN, ventral anterior nucleus; IC, internal capsule; −ve, inhibitory pathways.

1. Lesion in the anterior limb of the internal capsule. May disrupt the reciprocal connections between the ventral anterior nucleus and the anterior language cortex leading to a decrease in the level of excitation of the anterior language cortex responsible for language formulation and motor programming – dysfluent speech (Damasio et al., 1982) and interruption of ongoing language (Hermann et al., 1966).

2. Lesion in the thalamus. Disrupts pre-verbal semantic monitoring leading to production of a predominance of semantic paraphasic errors (Mohr, Watters and Duncan, 1975; Cappa and Vignolo, 1979; Reynolds et al., 1979; Alexander and LoVerme, 1980). Repetition abilities remain relatively intact due to reliance on intact phonological pathways in the arcuate fasciculus (McCarthy and Warrington, 1984).

3. Lesion in the globus pallidus. Reduces inhibitory influence on ventral anterior nucleus leading to an increased level of excitation in the anterior language area – inclusion of extraneous materials in speech, e.g. semantic paraphasic errors (Svennilson et al., 1960).

4. Lesion in the arcuate fasciculus. Disrupts the transmission of phonological linguistic information used to perform the act of repetition – impaired repetitional abilities.

5. Lesion in caudate nucleus. Disinhibition of the globus pallidus leads to an increased inhibitory influence on the ventral anterior nucleus and a reduced level of excitation in the anterior language centre making it difficult to release language for motor programming – dysfluent speech.

reduces its normally inhibitory influence on the caudate nucleus. As a consequence of the reduced amount of cortical inhibitition, the caudate nucleus is free to send a greater number of inhibitory signals to the globus pallidus which responds by reducing the flow of inhibitory signals to the ventral anterior nucleus of the thalamus. The ventral anterior nucleus subsequently sends a greater number of excitatory signals to the frontal language mechanism causing initiation of motor programming of the semantically verified language and eventually leading to expression of the language in speech.

As can be seen from Figure 3.7, according to this model, semantic problems in the form of verbal paraphasic errors could result from lesions in either the thalamus or basal ganglia (particularly the globus pallidus). In agreement with this prediction, a number of authors have reported the occurrence of semantic paraphasias in association with both thalamic (Cappa and Vignolo, 1979; Alexander and LoVerme, 1980; Gorelick *et al.*, 1984) and basal ganglia (Damasio *et al.*, 1982; Wallesch, 1985) lesions. Also consistent with the model, a number of studies (Damasio *et al.*, 1982; Naeser *et al.*, 1982) have found that destruction of the anterior limb of the internal capsule produces dysfluent but often grammatical language. The model, however, cannot explain the presence of semantic paraphasias in association with a lesion located in the anterior limb of the internal capsule reported in a more recent study by Murdoch *et al.* (1986b). These authors did state, however, that although the resolution of their computed tomography scan was good, it was impossible to rule out with absolute certainty involvement of the adjacent basal ganglia. Wallesch (1985) proposed that two different types of neurones connect the basal ganglia to the frontal cortex. Lesions in one type result in non-fluent aphasia while lesions in the other, a functionally inhibitory neurone, produce paraphasic speech.

In summary, there is a growing acceptance of the existence of sub-cortical aphasic syndromes. In general, these syndromes appear to have features atypical of cortical aphasia syndromes and often appear to have a better prognosis for recovery. If sub-cortical aphasias do represent different disorders from cortical aphasic syndromes, it follows that they may require the development and implementation of different therapy strategies. Although in many forms (e.g. thalamic aphasia) they appear to have a better prognosis, long-term follow-up studies and more extensive language-testing techniques suggest that chronic-stage deficits may result.

To date, the relationship between lesions in specific areas of the striato-capsular region and the occurrence of particular language deficits remains uncertain. One reason for this is that critical differences in lesion extension of only a few millimetres may be responsible for major differences in language deficits (Naeser *et al.*, 1982).

Most of the evidence linking sub-cortical lesions to the occurrence of aphasia has come from studies based on computed tomography localization of lesion sites. Unfortunately, the extent of impairment of function cannot be exactly defined on the basis of computed tomography. Magnetic resonance imaging has been reported by DeWitt *et al.* (1985) to provide better differentiation of grey and white matter and a sensitivity in sub-cortical lesion detection as good as, and sometimes superior, to computed tomography. Using magnetic resonance imaging, these authors found that most sub-cortical lesions as identified by computed tomography, do in fact involve the cerebral cortex as well. In addition, positron emission tomography has shown that patients with sub-cortical lesions also exhibit a mild cortical metabolic depression (Metter *et al.*, 1983), a finding that has added to the controversy regarding the existence of sub-cortical aphasias.

Clearly, our understanding of the role of sub-cortical structures in language remains at best speculative. Most of the evidence currently available suggests that aphasias associated with sub-cortical lesions are different from the classical cortical aphasia syndromes. Further studies based on high-resolution computed tomography, magnetic resonance imaging and positron emission tomography scan findings are necessary to more precisely determine the relationship between lesions in particular sub-cortical structures and the occurrence of specific language deficits. Until such time that these studies are completed, formulation of individual treatment programmes based on careful detailed language assessment of sub-cortical aphasics is recommended.

CHAPTER FOUR

Speech–language disorders associated with traumatic head injury

Traumatic head injury can cause a variety of communication problems. Depending on the location of the damage in the nervous system, head injury may be associated with either speech disorders, language disorders or both (see Figure 4.1).

4.1 OPEN VERSUS CLOSED INJURY

Head injuries can be divided into two major types: open head injuries and closed head injuries. In an open head injury the brain or meninges are exposed. A closed head injury differs from an open head wound in that the meninges remain intact even though the skull may be fractured. Historically, most of what we know about the effects of head injury on speech and language function has come from studies of patients who have sustained penetrating missile wounds in wartime (Russell and Espir, 1961; Luria, 1970). In most cases these descriptions sought to connect a particular speech or language disorder to a specific lesion site in the central nervous system. Reports in the literature have indicated that the prevalence of communication disorders in large groups of patients with open head trauma ranges from 14 to 23% (Hillbom, 1959; Teuber, 1975). In contrast to the open head wounds characteristic of war injuries, however, the majority of traumatic head injuries in civilian life are of the closed type, stemming mainly from motor-vehicle accidents and affecting predominantly young adult males (Annegers, *et al.*, 1980). Due to the nature of the pathology, a number of authors have suggested that there may also be

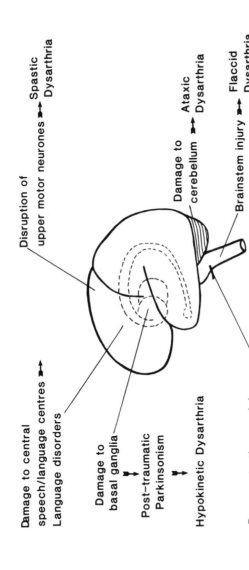

Damage to central
speech/language centres ➤➤
Language disorders

Disruption of
upper motor neurones ➤➤ Spastic
Dysarthria

Damage to
basal ganglia ➤➤
Post-traumatic ➤➤
Parkinsonism

Hypokinetic Dysarthria

Damage to the cranial nerves
in their peripheral course ➤➤ Flaccid Dysarthria

Damage to
cerebellum ➤➤ Ataxic
Dysarthria

Brainstem injury ➤➤ Flaccid
Dysarthria

Figure 4.1 Speech and language disorders associated with head injury.

differences in the associated communication deficit between open versus closed head injuries (Hagen, Malkmus and Burditt, 1979). Luria (1970) reviewed 800 head injury cases and compared the speech and language deficits exhibited by closed head injury cases with those observed in patients with open head injuries. He concluded that there was no significant difference in the language abilities of each group immediately post-trauma. Groher (1983), however, conducted a closer analysis of Luria's results and found they indicated that open-head injured patients exhibited language deficits for a longer period than did closed-head injured patients. Further the re-analysis showed that, in the initial period post-trauma, the patients with closed head injury as a group suffered fewer communication deficits than patients with penetrating head wounds. From this re-analysis it can be implied that closed-head injured patients with communication deficits exhibit greater and faster improvement than those with open head injuries. Darley (1982) also reported a better prognosis for the language disorder following closed head injury than for the communication deficit associated with open head wounds.

4.2 COMPLICATIONS OF HEAD INJURY

Following a head injury a patient may suffer from a number of complications which include: concussion, contusions, lacerations and skull fractures; vascular lesions (e.g. haemorrhage and thrombosis); infections (e.g. meningitis); increased intra-cranial pressure; rhinorrhea and otorrhea; cranial nerve lesions; focal brain lesions; post-traumatic epilepsy; and post-traumatic vertigo.

4.2.1 Skull fractures

The various different types of skull fracture associated with traumatic head injury are listed in Table 4.1. There is no direct relationship between the severity of any damage to the skull and the extent to which the brain is damaged. Although severe fractures of the skull are usually associated with severe cerebral injury, the brain may be extensively damaged without the skull being fractured. Alternatively, a fracture of the skull may occur without severe damage to the brain. Consequently, the fracture itself is of little importance in relation to occurrence of persistent neurological deficits following traumatic head injury. Rather the presence of neurological impairments is dependent on the location and extent of damage to the underlying structures, particularly the brain itself. Overall, depressed skull fractures are the

Table 4.1 The different types of skull fracture

Fracture type	Description
Simple	Cracks or fissures in the skull bone, skin intact
Compound	Cracks in the skull bone, scalp also breached
Comminuted	The damaged skull bone is broken into several pieces
Depressed	A piece of broken skull bone is driven inwards and may cause laceration or compression of the underlying brain tissue

most likely to produce severe and permanent neurological signs due to the possibility of the dislodged piece of bone causing lacerations of the brain tissue.

4.2.2 Concussion, contusions and lacerations

The most critical factor determining the neurological outcome of traumatic head injury is the degree of resultant brain damage. As indicated above, some head injuries involve penetrating wounds (e.g. bullet wounds, depressed skull fractures) which damage the brain tissue directly. The majority of head injuries in peacetime, however, do not involve penetration of the brain substance by foreign objects or materials. Even so, in these peacetime injuries the brain can be severely 'shaken up' leading to a temporary loss of function of the brain cells and a temporary loss of consciousness. This condition is referred to as concussion and need not be associated with permanent brain damage to any part of the brain. In fact the term 'brain concussion' by definition means a transient alteration of neuronal function without structural abnormality. Generally the term 'concussion' is used to apply to patients who show full neurological recovery in 1–2 hours.

A brain contusion (bruise) is a more-or-less diffuse disturbance of the brain following head injury which is characterized by multi-focal capillary haemorrhages, vascular engorgement and oedema. In the area of impact, small blood vessels of the cerebral cortex and/or underlying white matter may rupture resulting in extra-vascular extravasation of blood causing a bruise to form. In addition, the impact may cause the brain to strike the skull at a point opposite to the point of trauma, thereby resulting in additional vascular disruption and bruising at this site (the so-called *contre-coup* contusion). As the cerebral hemispheres are set into motion by the impact, the mid-brain may twist and buckle causing small blood vessels in this region to rupture producing a brainstem contusion.

The symptoms produced by a brain contusion depend very much on the size and location of the contusion. These symptoms commonly include language disturbances, speech disorders, hemiparesis and visual disturbances. Brainstem contusions can result in alterations of consciousness, disturbances of postural tone and cranial nerve palsies. In particular, where the functioning of cranial nerves is disrupted, speech disturbances may result (see Chapter 8).

When a brain contusion is sufficiently severe to cause a visible breach in the continuity of the brain, it is referred to as a laceration. In particular, lacerations are classically associated with penetrating head injuries and produce a more severe and prolonged unconsciousness and paralysis than contusions. As in the case of contusions, however, the specific signs of cerebral laceration depend upon the location of the tear in the brain tissue.

4.2.3 Increased intra-cranial pressure

An increase in intra-cranial pressure is seen in association with a number of different neurological conditions. It is, however, a particularly common finding following traumatic head injury. In adults the cranium acts as a fixed box that is filled with brain tissue, cerebrospinal fluid and blood. Intra-cranial pressure is dependent on the total volume of these three components. Depending upon the nature of the lesion causing an elevation in intra-cranial pressure and the time over which it develops, the intra-cranial contents can adjust in such a way as to minimize intra-cranial pressure changes. For instance, a slow-growing intra-cranial tumour can attain a considerable size without increasing the intra-cranial pressure because the cerebrospinal fluid can be shifted out of the cranial cavity, cerebral veins can be compressed and brain tissue can be either displaced slowly or atrophy. In contrast, rapidly accumulating intra-cranial masses (e.g. extra-dural haematomas) produce increased intra-cranial pressure even while small because they occur before the various compensatory devices can be effective. Consequently, these latter intra-cranial masses cause brain structures to be displaced by the pressure associated with the sudden increased volume of intra-cranial contents.

In the more severe traumatic head injury cases, increased intra-cranial pressure may cause part of the brain to be shifted to another cranial compartment. This phenomenon is called herniation. The major types of cerebral herniation seen in association with traumatic head injuries include transtentorial herniation, tonsillar herniation, and axial herniation.

(a) Transtentorial herniation

The tentorium cerebelli is a fold in the dura mater that separates the cerebral hemispheres from much of the brainstem and cerebellum. The upper part of the brainstem passes through a hole in the tentorium called the tentorial hiatus. An increase in intra-cranial pressure above the level of the tentorium, as might be caused by an extra-dural or sub-dural haematoma (see below), may cause the medial portions of the temporal lobes to be herniated through the hiatus alongside the brainstem. Herniation of this type causes compression of the brainstem and interferes with the functioning of the reticular formation, thereby leading to a deterioration in the level of consciousness. At the same time the third cranial nerve is also compressed causing pupillary dilation, first on the side of the herniation and later on the other side as well. Eventually, if untreated, compression of the brainstem will lead to death. The level of consciousness and the state of the pupils of the eyes are, therefore, critical factors that require monitoring following head injury.

(b) Tonsillar herniation

Tonsillar herniation occurs when the cerebellar tonsils are displaced inferiorly through the foramen magnum. The only structure compressed by tonsillar herniation is the medulla and the first sign of its presence is often respiratory insufficiency or apnoea.

(c) Axial herniation

A downward shift of the entire brainstem as a result of increased intra-cranial pressure is referred to as axial herniation. Although no structures are actually compressed by axial herniation, distortion of the brainstem may cause altered levels of consciousness and changes in respiration.

4.2.4 Vascular lesions

(a) Extra-dural haematoma

Extra-dural haematomas usually result from laceration of the middle meningeal artery by fractured bone and involve bleeding between the skull bones and the dura mater (see Figure 4.2). Haematomas of this type usually collect and enlarge rapidly and signs of increased intra-cranial pressure become evident within a short period post-injury.

Typically, although the patient may have been knocked uncon-
scious at the time of head injury, consciousness is quickly recovered
and then within 1–2 hours the patient becomes increasingly drowsy
and develops paralysis down one side of the body as a result of
compression of the ipsilateral cerebral hemisphere by the expanding
haematoma. Eventually the patient demonstrates pupillary dilation and
loses consciousness from compression of the third cranial nerve and
brainstem, respectively, as a consequence of herniation of the
temporal lobe through the tentorial hiatus.

Treatment of an extra-dural haematoma requires an emergency
operation which involves the drilling of a burr-hole over the bruise site
and evacuating the clot. If left untreated the patient will die as a result
of compression of vital centres (e.g. respiratory centres) in the brainstem. Although the speech–language centres in the dominant hemisphere may be compressed, because these haematomas develop
quickly and consciousness is lost rapidly, aphasia is a comparatively
rare occurrence in association with extra-dural haematomas (Benson,
1979).

(b) Sub-dural haematoma

Sub-dural haematomas result from bleeding between the dura mater
and arachnoid following disruption of the small blood vessels that
cross the sub-dural space (see Figure 4.2). This space is normally
filled with small amounts of a lymph-like fluid and has little capacity to
absorb blood.

Sub-dural haematomas develop much more slowly than extra-dural
haematomas and consequently, although the neurological signs and
symptoms resulting from the increasing intra-cranial pressure are the
same, they appear at a much later time, in some cases days, in others
weeks after the traumatic head injury. As a result of the slower development of this type of haematoma the patient has the time to complain of
increasing headache prior to losing consciousness. Further, the
occurrence of a language disturbance in the form of a mild word-
finding difficulty has also been reported in association with sub-dural
haematomas (Benson, 1979). In addition, as it develops, a
sub-dural haematoma may irritate the cerebral cortex, thereby trigger-
ing an epileptic fit. If the haematoma develops to the stage of causing
tentorial herniation, as in the case of extra-dural haematoma, surgical
evacuation of the clot is again required to prevent compression of the
brainstem.

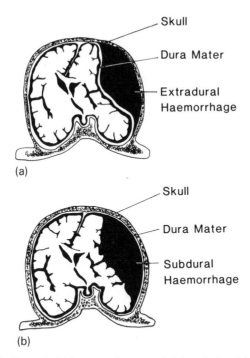

(a)

(b)

Figure 4.2 (a) Extra-dural haemorrhage and (b) sub-dural haemorrhage.

(c) Sub-arachnoid haemorrhage

A sub-arachnoid haemorrhage occurs following rupture of the blood vessels that cross the sub-arachnoid space between the arachnoid and the pia mater. Such haemorrhages are common occurrences after head trauma and can be detected by the presence of blood in the cerebrospinal fluid (Note: neither sub-dural nor extra-dural haemorrhages are associated with the presence of blood in the cerebrospinal fluid). Although patients with sub-arachnoid haemorrhages may experience severe headaches and stiffness of the neck for many days, they normally recover spontaneously.

(d) Intra-cerebral haemorrhage

In some cases, traumatic head injury can cause haemorrhages within the brain itself. Most commonly these lesions take the form of multiple small haemorrhages located in the contused region. Occasionally, large sub-cortical haemorrhages may also result from head injury. Depending on their location, these post-traumatic intra-cerebral haemorrhages can produce a variety of speech and/or language deficits.

4.2.5 Cranial nerve lesions

Traumatic head injuries can cause cranial nerve dysfunction either by damaging the cranial nerve nuclei in the brainstem or disrupting the nerves themselves in either their intra-cranial or extra-cranial course. Contusions of the brainstem can damage the cranial nerve nuclei leading to flaccid paralysis of the muscles innervated by the affected nerves. In particular, should these include muscles supplied by cranial nerves V, VII, X or XII, speech disorders may result. The effects of cranial nerve lesions on speech production are discussed in detail in Chapter 8.

Fracture of the base of the skull is the most common cause of damage to the cranial nerves in their intra-cranial course. The facial nerve (VII) is most commonly affected by this condition and as it supplies the muscles of facial expression such fractures have important implications for speech production. Branches of the facial (VII) and trigeminal (V) nerves may be damaged extra-cranially by trauma to the face. In general traumatic cranial nerve palsies are permanent, exceptions being those resulting from contusions of extra-cranial branches of the nerves.

4.2.6 Rhinorrhoea and otorrhoea

Subsequent to traumatic head injury cerebrospinal fluid may leak from either the nose (rhinorrhoea) or ear (otorrhoea). Rhinorrhoea occurs following fracture of the frontal bone with associated tearing of the dura mater and arachnoid. Otorrhoea, on the other hand, is caused by injuries to the base of the brain. As injuries in this region often damage the brainstem as well, otorrhoea is of more serious prognostic importance than rhinorrhoea. Infections and meningitis are potential hazards of both conditions.

4.2.7 Post-traumatic epilepsy and post-traumatic vertigo

These are two complications of head injury which, although not incapacitating, may have a profound effect on the lifestyle of the head-injured patient. Post-traumatic epilepsy occurs most commonly after a penetrating head wound through the skull, dura and brain. The epilepsy is triggered by the formation of scar tissue as a result of brain laceration. The scar may act as an irritating focus to trigger epileptic fits. In some cases convulsions may occur very shortly after impact (within 24 hours), especially in children. Where it occurs in adult head injury cases, however, the epilepsy usually develops within the first two years post-injury.

Some degree of vertigo accompanied by vomiting and unsteadiness is common after head injury. Post-traumatic vertigo may last for days or weeks or may persist in some cases for many months.

4.3 MECHANISMS OF HEAD INJURY

Brain damage following traumatic head injury may be either focal, multi-focal or diffuse in nature and may involve any part of the brain. Consequently, brain damage following head injury can be associated with a variety of communicative deficits depending primarily upon the location and extent of the lesion. Closed head injuries tend to produce more diffuse pathology (Brookshire, 1973; Wertz, 1978) while open head injuries are generally associated with more focal pathology (Hagen, Malkmus and Burditt, 1979).

The majority of traumatic head injuries in civilian life are due to closed head injuries. The force of the blow to the head is distributed to all parts of the brain. As a result all parts of the brain suffer to some extent (Brain and Walton, 1969). Brain contusions, lacerations and haemorrhages can result at the time of head injury from either direct trauma at the site of impact on the skull, acceleration of the brain against the bony shelves of the skull or from *contra-coup* trauma that occurs when the brain strikes the skull on the side opposite the point of insult. In all, three different destructive forces are applied to the brain at the moment of impact: compression, tension, and shearing (Brain and Walton, 1969). Compression forces the brain tissue together, tension pulls it apart and shearing, produced by rotational acceleration, develops primarily at those points where the brain impinges upon bony or ligamentous ridges (e.g. sphenoidal ridge) within the cranial vault.

The primary mechanism producing brain injury following closed head injury appears to be diffuse neuronal damage occurring at the time of impact (Strich, 1956; Adams *et al.*, 1977). In fact a number of neuropathological changes, including permanent microscopic alterations in both the white and grey matter, have been reported subsequent to closed head injury. These include: severe localized demyelination (Greenfield, 1938); widespread white-matter degeneration (Tomlinson, 1964; Adams *et al.*, 1977); and nerve cell damage (Horowitz and Rizzoli, 1966). Widespread injury to the cerebral white matter apparently results from shearing and stretching of the nerve fibres (Adams *et al.*, 1977; Levin *et al.*, 1981). The corpus callosum in particular has been reported to be vulnerable to diffuse shear strains (Adams *et al.*, 1977) and consequently hemispheric disconnection may occur if rotational acceleration of the skull is sufficient.

4.4 SPEECH AND LANGUAGE DISTURBANCES FOLLOWING HEAD INJURY

4.4.1 Nature of the language disturbance following head injury

There is general agreement that individuals who suffer a head injury may exhibit a communication deficit in the form of either a speech and/or language disorder. There is some controversy, however, regarding the nature of the communication impairments and the terms that should be used to describe them. In particular, the major controversy relating to the terminology applied to the speech–language deficits following closed head injury concerns the presence or absence of aphasia in closed-head injured patients.

A variety of aphasic syndromes have been identified post head injury. Glaser and Shafer (1932) reviewed 16 subjects with poor communication skills following closed head injury and classified them as follows: 1 case was a complete and 10 were partial motor aphasics, 4 were mixed aphasics and 1 was a sensory aphasic. These authors, however, did not account for their methods of evaluation and classi-fication, nor did they specify the point of time post-traumatically at which patients were assessed. More recently, Heilman, Safran and Geschwind (1971) studied 13 patients with acute aphasia secondary to closed head injury, and found 9 cases of anomic aphasia characterized by fluent speech with essentially normal comprehension and repe-tition, concurrent with verbal paraphasia and abnormal naming of objects. The remaining 4 subjects exhibited fluent paraphasic speech in conjunction with poor comprehension of both written and spoken language and deficient repetition skills, characteristic of Wernicke's aphasia; no subject was classified as either Broca's of global aphasics.

Other studies (Thomsen, 1975; Stone, Lopes and Moody, 1978) yielded results similar to those of Heilman, Safran and Geschwind (1971). Thomsen (1975) examined language disorders in the acute stage of severe closed head injury, and found amnestic (anomic) aphasia and verbal paraphasia to be the most frequent symptoms, with perseveration, dysgraphia and literal paraphasia occurring less frequently. On reassessment between 23 and 50 months later, none had recovered completely, 13 exhibited evidence of a language deficit when tested without showing clinical manifestations of a language disorder (sub-clinical aphasia), half the patients had persistent deficits of expression (mostly evident on naming tasks), 2 had resolved from mixed to sensory (Wernicke's) aphasia, and 1 had resolved from sensory to anomic aphasia. Thomsen (1975) concluded that aphasia following closed head injury is a receptive rather than an expressive disorder contributing to a neurophyschological syndrome often dominated by residual defects of general memory.

Levin, Grossman and Kelly (1976) examined language-disturbed patients with mild-to-severe closed head injuries. They found a trend towards expressive or anomic aphasia in mild–moderate closed-head injured patients and mixed deficits in the majority of severe closed-head injured cases, the latter group being more prone to linguistic disturbance than the former. It was also noted that closed-head injured patients who also had a period of prolonged coma often exhibited a general non-specific linguistic disturbance when consciousness returned.

Language and memory disorders following closed head injury were investigated by Groher (1977), using subjects averaging 17.1 days in coma and excluding cases involving skull fracture. Each patient was assessed as soon as possible after regaining consciousness, using the Porch Index of Communicative Ability (Porch, 1967) and the Wechsler Memory Scale Form I (Wechsler, 1945). Reassessments followed every 30 days for 4 months. In the acute stages, gestural skills were poorer than graphic skills, which were poorer than verbal skills. All subjects were reported to show marked anomia with literal and nominal paraphasic errors. Reading comprehension was poor and 9 of the 14 subjects had spastic dysarthria (from which 6 recovered). Oral apraxia was not observed in any subject. At the final assessment, all subjects could make their needs known and converse readily. However, 9 of the 14 subjects used inappropriate language and displayed confused thought content. Writing at this stage was just becoming functional, characterized by spelling errors, incomplete sentences and poor syntax. Reading comprehension was generally poor. By 4 months' post-trauma, graphic skills had become superior to both gestural and verbal abilities (in descending order of competence).

Sarno (1980) used four sub-tests of the Neurosensory Centre Comprehensive Examination for Aphasia (Spreen and Benton, 1969) to investigate the verbal impairments of 56 people following closed head injury. The closed-head injured subjects were classified into three groups on the basis of the scores obtained on visual naming, word fluency, sentence repetition and the Token Test (for auditory comprehension): those with aphasia, those with sub-clinical aphasia, and those with sub-clinical aphasia and dysarthria. It was noted that the aphasic group consisted of older individuals than either of the other two groups. A number of different aphasic syndromes were identified, including fluent aphasia (39%), non-fluent aphasia (38%), anomia (11%) and global aphasia (11%). These findings are in contrast to those of other studies (Heilman, Safran and Geschwind, 1971; Thomsen, 1975; Stone, Lopes and Moody, 1978) in which anomic aphasia was identified as the predominant type of aphasia following closed head injury. This discrepancy may, at least in part, be due to the greater severity of the closed head injury in Sarno's subjects.

Although a number of researchers have described the symptoms of aphasia in subjects following closed head injury other workers have questioned the validity of classifying all language disorders occurring secondary to closed head injury as aphasia (Russell, 1932; Russell and Smith, 1961; Halpern et al., 1973; Holland, 1982). Russell and Smith (1961) reported that initial language deficits following traumatic head injury are consistent with post-traumatic amnesia, implying that these language impairments result from faulty memory skills. Their proposal was modified by Akbarova (1972) who believed that aphasia due to closed head injury is accompanied by impairment in the storage of new information. Indeed, it has been found that defective short-term recognition memory for essentially non-verbal material is closely associated with linguistic deficits (Levin et al., 1976). Language following head trauma has also been described as 'confused', a term encompassing faulty short-term memory, mistaken reasoning, inappropriate behaviour, poor understanding of the environment and disorientation (Halpern et al., 1973).

Darley (1982) defined the language of confusion as impairment of language accompanying neurological conditions, often traumatically induced, characterized by reduced recognition and understanding of, and responsiveness to, the environment, faulty memory, unclear thinking and disorientation in time and space. Structured language events are usually normal and responses utilize correct syntax; open-ended language situations elicit irrelevance, confabulation. According to Darley (1982, p. 25), 'it is this pattern of a high degree of irrelevance of content, coupled with paradoxically adequate syntax and fluency that differentiates the language performance of confused patients from that of aphasic patients'.

Groher (1977) proposed that the initial communication problems exhibited by patients following closed head injury are comprised of both confused language and aphasic disabilities but that the aphasic component resolves leaving the patient with the language of confusion. Levin et al. (1981) adopted an eclectic approach, acknowledging that the diagnosis of aphasia following closed head injury may be confounded by the patient's confusion during the early post-comatose stage when amnesia is common. Thus, confabulation and profound memory impairment or reduplicative para-amnesia (mistaken identification of a person, place or event for one previously experienced) may be present in addition to a deficit in the formulation and/or interpretation of linguistic symbols (i.e. aphasia).

The language deficit following closed head injury has been classified as 'sub-clinical aphasia' by Sarno (1980, 1984). By definition, sub-clinical aphasics evidence linguistic processing deficits on testing in the absence of clinical manifestations of language impairment. Holland (1982), however, believes that the term 'sub-clinical aphasia' does nothing to describe the language problems of the head-injured patient and argued that, although aphasic patients exist in the head-injured population, in the majority of head injury cases, the associated language disturbance is not the same disorder as the aphasia associated with vascular lesions. She identified three features that distinguish head injury from the stroke population that might underlie their different language outcomes. The first distinction is that, although aphasiology arose from investigations into the effects of open head injuries caused by fragment penetration in wartime on language, closed head injuries are the most common type of head trauma encountered in civilian life. As indicated previously, closed head injuries generally cause diffuse brain damage rather than focal lesions, as found most often in stroke cases. Consequently, the language disorders found in closed-head injury cases are likely to be the outcomes of more general and more pervasive memory and cognitive deficits (Holland, 1982). The language deficit in closed-head injury patients may, therefore, be superseded by other cognitive and memory deficits which need to be taken into account when treating these cases. Secondly Holland (1982) indicated that the difference in language outcome might be explained by the fact that as a result of the numerous cognitive and memory sequelae, closed-head injury rehabilitation is more interdisciplinary than post-stroke aphasia rehabilitation. Thirdly, Holland (1982) pointed out that there is an important demographic difference between the head injury and stroke populations, the occurrence of head injury being highest in the age group from 15 to 24 years of age (Kalsbeek et al., 1980) and stroke being more common in middle to old age. The lower mean age of head injury patients compared with stroke cases could be expected to be associated with a better prognosis for recovery of language function.

Despite the above difference between the post-stroke aphasic and head-injured populations, there are a number of similarities in the language disorder exhibited by these two groups (Holland, 1982). In particular anomia is a feature of the language disturbance in both groups, occurring with an equal frequency and showing a similar persistence and tenacity in each. However, a number of authors have suggested that there is a qualitative difference in the word-retrieval problems demonstrated by post-stroke aphasic and head-injured individuals and subsequently 'non-aphasic' naming errors have been described as a feature of head injury (Geschwind, 1967; Holland, 1982). These 'non-aphasic' misnamings are reported to occur in addition to the circumlocutions, paraphasias, etc. typical of aphasia of vascular origin and are characterized by either misnamings related to the patient's personal situation (e.g. the hospital they are in or their illness) or errors of confabulation. These errors tend to propagate and may be bizarre in nature. For example, patients may talk about driving their 'car' (wheelchair) down the 'road' (corridor) to the 'gate' (door). A similar phenomenon in one case of traumatic head injury was described by Weinstein and Kahn (1952) who classified it as part of an amnestic confabulatory syndrome. Other researchers, however, have found no evidence of such behaviour in the closed-head injured population (Heilman et al., 1971; Thomsen, 1975).

Another area of similarity between aphasia of vascular origin and the language deficit observed following closed head injury is that both types of patient initially demonstrate problems in auditory comprehension (Holland, 1982). Further, although in both cases comprehension abilities show an early improvement and spontaneous recovery is a feature of both disorders, problems in the comprehension of complex material usually persist well into the recovery phase. In addition, in both patient groups associated reading and writing deficits are usually present.

Holland (1982) believed that it is in the area of language pragmatics that post-stroke aphasia and head-injured communication contrast most vividly. Aphasic language can be considered a disorder of form while head-injured language is better considered a disorder of use. Milton, Prutting and Binder (1984) compared a group of five head-injured patients and five normal adults on the Western Aphasia Battery (Kertesz, 1980), the Communicative Abilities of Daily Living (Holland, 1980) and the Pragmatic Protocol (Prutting and Kirchner, 1983). No subject in either group scored below the aphasia cut-off score on the Western Aphasia Battery and the mean performance in the Communicative Abilities in Daily Living for the head-injured group was just above the cut-off score for normal functional communication. Error responses on the Communicative Abilities in Daily Living clustered around orientation explanations (e.g. giving directions from home to the doctor's office), visual attention and utilization of contextual cues (e.g. identifying that a person is smoking in a room where there is a 'No smoking' sign displayed), mathematical calculations and problem solving as well as divergent abstract reasoning (e.g. as used in interpretation of metaphors).

On the Pragmatic Protocol, Milton, Prutting and Binder (1984) reported a mean of 7.6 behaviours demonstrated by the head-injured group to be judged inappropriate while a mean of only 0.6 inappropriate behaviours was recorded for a group of normal subjects. The 10 pragmatic behaviours that were most frequently judged to be inappropriate in the head injured group were prosody, affect, topic selection, topic maintenance, turn-taking initiation, turn-taking pause time, turn-taking contingency, quantity/conciseness and fluency. The highest proportion of inappropriate pragmatic behaviours exhibited by the head injured adults was the illocutionary/perlocutionary act, suggesting that breakdown most often occurs in the way that head-injured adults function as discourse partners.

In summary, there is some disagreement in the literature as to the basis of the language impairment associated with head injury. Although a number of authors have classified the language disturbance following head injury as an aphasia, others believe that the associated language impairment is due to poor memory skills (post-traumatic amnesia) or confusion. Some researchers have developed terms such as 'sub-clinical' aphasia and 'non-aphasic misnaming' to refer to symptoms exhibited by head-injured patients but not seen in the classical syndromes of aphasia. One major difference between post-stroke aphasics and head-injured patients appears to be their pragmatic language abilities.

4.4.2 Language impairment and neurological status following head injury

Levin *et al.* (1981) investigated the long-term recovery of language function in 21 closed-head injured patients initially diagnosed as aphasic. Each subject was assessed between 6 months and 2 years post-trauma using the Multi-lingual Aphasia Examination, parts of the Neurosensory Centre Comprehensive Examination for Aphasics and the Wechsler Adult Intelligence Scale. Results were compared with the initial severity of the head injury, as determined by the Glasgow Coma Scale (Teasdale and Jennet, 1974) and computed tomography scan tests. Three general patterns of language recovery were observed by these workers. First, full language function was generally recovered if the head injury was initially mild with a haematoma in no more than one hemisphere, even when some ventricular expansion remained post-traumatically. Secondly, in other instances of mild closed head injury, specific expressive language deficits were retained for at least 6 months post-onset, characterized by anomia in the absence of any cognitive deficit or severe disability. Such cases were usually spared diffuse cerebral damage and similar cases have also been documented in several other long-term studies (Heilman *et al.*, 1971; Levin, Grossman and Kelly, 1976; Thomsen, 1975). Thirdly, more severe closed head injuries involving diffuse swelling and bilateral haematomas correlated with persistent expressive and receptive impairment (generalized linguistic disturbance) at least one-year post-injury. These latter subjects typically exhibited residual anomia, reduced word-association abilities, and deficient comprehension skills, accompanied by a global cognitive deficit for both verbal and visuo-spatial material. Repeat computed tomography scans at this stage showed dilatation of the ventricular system and minimal or circumscribed cortical atrophy. No signs of raised intra-cranial pressure, spasticity or progressive dementia were present.

4.4.3 Prognostic indicators for language function following head injury

Most linguistic recovery has been found to occur during the first 6 months following head trauma (Tabaddor *et al.*, 1984). According to Groher (1977), significant improvements in language and memory dysfunction subsequent to closed head injury are concentrated at 1 month post-trauma although some improvements continue for at least another 3 months. He noted that memory problems resolve faster than general language and orientation deficits. Sarno (1984) concluded that language problems secondary to closed head injury persist for at least one year if traumatic coma is experienced.

Researchers disagree, however, as to the most reliable method of predicting the occurrence, severity and outcome of language dysfunctions associated with head injury. It has been proposed that severe closed head injury (characterized by prolonged coma and involvement of the cerebral hemispheres and brainstem) is indicative of a high probability of aphasic disturbance (Levin *et al.*, 1976). Thomsen (1975), however, stressed that an extended period of unconsciousness is neither a necessary nor sufficient condition for aphasia following closed head injury, noting that language prognosis appeared to be closely related to the presence and degree of concomitant neuropsychological disorders. She also observed that recovery of linguistic competence is good though not complete in cases of severe closed head injury. A correlation was found by Groher (1977) between the degree of language and memory impairment and the extremes of coma duration following closed head injury. Although according to Sarno (1980, 1984) all closed-head injury patients experience some degree of linguistic impairment following coma, she discounted any relationship between the severity of aphasia and the length of coma, acknowledging only a minimal relationship between the severity of aphasia and anatomical discontinuity.

It has been reported that the behavioural measures of naming ability and word fluency can be used to reliably predict the severity of closed head injury and the occurrence of associated aphasias (Levin et al., 1976). Sarno (1980, 1984) also found that word-finding ability is the best predictor of the severity of linguistic disorders resulting from closed head injury. She noted that aphasic subjects invariably had the lowest scores on these sub-tests, while sub-clinical aphasics generally exhibited borderline performance. Levin et al. (1981) found that the severity of diffuse closed head injury (as indicated by the duration of coma) was significantly correlated with deficits evident on visual naming and language comprehension tasks. In a long-term study of linguistic recovery after closed head injury, these authors observed that head-injured patients who recovered to normal levels on all language tests generally had only mild diffuse brain injury while those who demonstrated a residual expressive and receptive impairment had sustained severe diffuse brain injury resulting in a global cognitive deficit. Those head-injured patients who presented with a persistent expressive language impairment (primarily in naming) had mild diffuse brain injury and, in 50% of cases, a focal left-hemisphere injury.

4.4.4 Speech disorders following head injury

A number of researchers have reported the occurrence of dysarthria following traumatic head injury (Groher, 1977; Najenson et al., 1978; Hagen et al., 1979; Sarno, 1980, 1984). Depending on the specific site of the lesion in either the central or peripheral nervous systems, a number of different types of dysarthria, including flaccid dysarthria, spastic dysarthria and mixed dysarthria may result from head injury. Flaccid dysarthria has been reported to follow head injury that damages either the cranial nerves which supply the muscles of the speech mechanism in their peripheral course or the cranial nerve nuclei in the brainstem. Spastic dysarthria, on the other hand, is associated with head injuries that cause diffuse upper motor neurone damage (Groher, 1983) (the speech characteristics of spastic and flaccid dysarthria are described in Chapter 8). As head injuries often simultaneously damage both the upper and lower motor neurones, as well as the other parts of the brain, many head-injured patients exhibit some type of mixed dysarthria such as a mixed spastic/flaccid dysarthria or a spastic/ataxic dysarthria (see Chapter 10).

When present following injury, dysarthria usually accompanies a language disturbance in the acute stage post-injury. Several authors, however, have noted that in such cases the dysarthric symptoms often persist after language function has returned to near normal levels (Najenson *et al.*, 1978, Sarno and Levin, 1985).

4.4.5 Mechanisms of recovery in head injury

In general the prognosis for recovery of speech–language function following head injury appears to be good. Despite this good prognosis, however, the mechanisms of recovery following head injury are still the subject of much controversy. The removal of haematomas and the associated relief of increased intra-cranial pressure together with the spontaneous recovery of physiological abnormalities account for at least some of the recovery observed during the first 1–3 weeks post-injury. During this period oedema may subside, blood flow to brain areas not irreversibly damaged may increase and increased intra-cranial pressure associated with haemorrhage or dynamic alterations of the circulation of the cerebrospinal fluid may return to normal. According to Heilman and Valenstein (1979), recovery continues at a maximal rate for up to 3 months and then slows. The mechanisms underlying recovery, however, are poorly understood. It is possible that recovery in the acute stage is the result of temporarily inactive but undamaged neural tissue resuming normal activity. This suggestion, however, is based on the assumption that damage to one part of the brain creates a state of shock by depriving other parts of the brain of their normal stimulation thereby causing them to become temporarily inactive. Longer term recovery (over weeks or months) may involve a process of functional re-organization of neural tissue while factors such as axonal re-growth, collateral sprouting and denervation hypersensitivity of the central nervous system may also contribute to the recovery of function (Fingers, 1978).

Although the severity of the head injury does have an influence, recovery from head injury appears to follow a set course (Jennet and Teasdale, 1981). Recovery from coma is characterized by opening of the eyes at times (evidence that the brainstem mechanisms concerned with wakefulness are recovering) and then uttering a few words. Some patients then go through a period of 'cerebral irritation' which is characterized by noisy disinhibited behaviour such as swearing, attempting to get out of bed and an aggressive attitude to others. During the subsequent phase the patient is quiet but confused about temporal, spatial and personal orientation. The end of post-traumatic amnesia marks a crucial stage of the recovery process. The duration of post-traumatic amnesia has been reported to be closely related to the ultimate degree of recovery and to the likelihood of cognitive sequelae (Brooks et al., 1980). Once the patient is out of post-traumatic amnesia, more subtle abnormalities of behaviour may be displayed. At this stage, although the patient may be able to return home, the family is warned that some erratic behaviour can still be expected. Neurological recovery is largely completed by 6 months post-injury and the patient therefore needs to be encouraged to adapt physically, mentally and socially to any deficit still evident at this time.

Groher (1977) investigated the language and memory abilities of closed-head injured patients after regaining consciousness to 4 months post-trauma. He reported that a significant improvement in both language and memory functioning most often occurred within the first month after regaining consciousness. In particular, verbal skills were the first to return and were superior to comprehension capabilities for the first month. After 4 months, expressive and receptive language skills were grossly functional for conversational purposes. Further, all memory tasks, with the exception of orientation skills, were within normal limits at this time. After 4 months, comprehension abilities were as good as verbal skills. Writing performance was just becoming functional after 4 months. At the conclusion of the study all patients were reported to be able to converse readily and could make their needs known. Despite this, signs of language impairment such as processing delays and word-finding difficulty were still evident, and although they scored well on most standardized measurements for aphasia, they still displayed problems of relevancy, inhibition of verbal output and with the sequential organization of ideas into logical outcomes. Although some of the head-injured patients examined by Groher (1977) were able to return to work, all reported problems in completing job tasks to their employer's satisfaction. In particular, the patients lacked judgement in the performance of their jobs, lacked initiative, forgot details easily and showed poor concentration. Groher (1983) has proposed that, although head-injured patients recover basic language skill within the first 6 months post-injury, they continue to show deficits in the analysis and synthesis of expressive and receptive language. In a study of 56 head-injured patients at a mean injury-test interval of 7 months, Sarno (1980) found that 7 patients had obvious aphasia but all others displayed only a 'sub-clinical aphasia' suggesting that although the prognosis for recovery of language abilities following head injury appears to be good, subtle language impairment may persist in the long term.

Jennet (1983) summarized the different physiological activities that operate at the different stages of recovery following head injury. He presumed that improvements in the first minutes after injury indicate the resolution of transitory dysfunction that may not have had any structural component. Recovery after several days post-injury, however, is more likely to be due to the resolution of temporary structural abnormalities such as oedema and vascular permeability. The mechanism that underlies recovery after months or years possibly involves the restoration of function in recovering damaged brain structures or the diversion of that function to other normally redundant undamaged areas of the brain. Jennet and Teasdale (1981) concluded that most recovery beyond the first month following trauma is due to the functional use of alternative or redundant neurological pathways.

It has often been suggested that neurobehavioural recovery following brain injury is better in children than in adults. The difference, for example, in the often observed rapid recovery rate of children with acquired aphasia compared to that of adult aphasics, was once interpreted as a demonstration of a better recovery potential in children (Lenneberg, 1967; Levin and Eisenberg, 1979b). It is not clear, however, how the basic mechanisms of injury might differ to produce the different recovery pattern in children as compared to adults. Strich (1969) suggested that the shearing strains produced by rotational acceleration in head trauma are less pronounced in smaller brains. If true this could result in a lesser amount of microscopic neuronal injury in young children than in adults following a comparable closed head injury. Alternatively it is conceivable that the reported better prognosis for head-injured children might be related to the different nature of their head injuries compared to those suffered by adults. Most head injuries suffered by children result from falls or low speed accidents and consequently many paediatric head injuries are associated with less severe rotational acceleration (Levin, Benton and Grossman, 1982). Most head injuries sustained by adults, on the other hand, result from high-speed motor-vehicle accidents which are, by their nature, likely to yield greater diffuse brain injury. Jamison and Kaye (1974) observed that persistent neurological deficits were present only in children injured in road traffic accidents.

Head injury represents an area in which the maxim of better recovery in children has been questioned in recent years (Levin, Ewing-Cobbs and Benton, 1984). Although the rate of spontaneous recovery in children following closed head injury is often striking, persistent long-term language disorders have been reported (Gaidolfi and Vignolo, 1980; Satz and Bullard-Bates, 1981; Jordan, Ozanne and Murdoch, 1988) and even when specific linguistic symptoms resolve, cognitive and academic difficulties often remain. Acquired language disorders in children are covered in greater detail in Chapter 11.

4.5 OTHER NEUROPSYCHOLOGICAL SEQUELAE OF HEAD INJURY

In addition to speech and language disorders, studies reported in the literature have shown that head-injured patients exhibit a variety of neuropsychological impairments. These include deficits in concentration, attention, memory, non-verbal problem solving, part/whole analysis and synthesis, conceptual organization, abstract thought and speed of processing. Hagen (1984) suggested that because these cognitive abilities are involved in language formulation and processing, then post-head injury language dysfunction would be influenced and perhaps even created by cognitive dysfunction. This view is supported by the observation that the prognosis for recovery of language function in head-injured patients is linked to the prognosis of cognitive ability (Brooks *et al.*, 1980; Hagen, 1984).

Some authors believe that the most consistent but unmeasurable feature of head injury is personality disorder. Relatives may note a reduction in drive, euphoric affect and a lack of social restraint and judgement. Rosenthal (1983) identified the emotional and behavioural changes that would appear to have their origins in the structural damage accompanying severe closed head injury. For example, decreased drive and initiative, dull or flat affect, disinhibition and apathy are known to be related to damage of the frontal lobes. Three types of secondary behavioural disturbances including denial, depression and the emotional dependence upon others have been identified as being the result of the stress of adapting to the trauma.

CHAPTER FIVE

Language disorders subsequent to right-hemisphere lesions

The extent to which language is represented in the right hemisphere has been a topic of controversy since the middle of last century. Although since the work of Broca (1861, 1865) the left hemisphere has been traditionally viewed as being dominant for language, in recent years there has been an accumulation of evidence suggesting that the right hemisphere of dextrals does play an important role in normal linguistic functioning (Searleman, 1977; Delis *et al.*, 1983).

5.1 LATERALIZATION OF LANGUAGE FUNCTION

Historically, hemispheric specialization for language was unanticipated until the reports of Broca (1861, 1865). According to Zangwill (1964), Broca was the first to ascribe speech dominance to the cerebral hemisphere contralateral to the preferred hand. Broca (1861) reported that the centre for speech was located in the third frontal convolution of the left cerebral hemisphere. In a later publication (Broca, 1965) he indicated that only the left hemisphere appeared to be involved in patients with impaired language function. Broca believed that the speech-dominant hemisphere in right-handed persons was the left hemisphere, while in left-handers the dominant hemisphere was the right hemisphere.

Broca's correlation of aphasia with focal brain damage in the left hemisphere was subsequently corroborated and the importance of the left hemisphere for language became widely accepted. Terms such as 'dominant' and 'non-dominant', 'major' and 'minor' were applied to the left and right cerebral hemispheres, respectively. Reports of aphasia following right hemisphere damage in left-handers added tentative support for Broca's idea (Zangwill, 1964). More recent reports, however, have provided convincing evidence that Broca's presumed one-to-one correlation between handedness and hemispheric dominance for language is overly simplistic (Schuell *et al.*, 1964; Zangwill, 1964; Penfield and Roberts, 1966; Rossi and Rosandini, 1967).

Currently it is believed that most people (approximately 96%) are left hemispheric dominant for language, which is related to handedness in the following way: approximately 93% of the population is right-handed and it is commonly estimated that 90–99% of all right-handers have their language functions predominantly subserved by the left hemisphere (Penfield and Roberts, 1959; Zangwill, 1960; Pratt and Warrington, 1972); the remaining 7% of the population are thought to be left-handed, with approximately 50–70% of these non-right-handers having their language functions localized primarily within the left hemisphere (Goodglass and Quadfasel, 1954; Zangwill, 1967; Hecaen and Sauguet, 1971; Warrington and Pratt, 1973). There is, therefore, a higher proportion of right-hemisphere language representation in non-right-handers than in right-handers (Strauss, Wada and Kosaka, 1983).

It has been suggested by some authors that most non-right-handers have considerable language function in both hemispheres (bilateral language dominance) (Luria, 1970; Gloning, 1977). The bilaterality of language function in non-right-handers is readily apparent from two observations – the greater frequency of aphasia following brain injury in non-right-handers (Gloning *et al.*, 1969) and their better recovery rate (Luria, 1970). Beaumont (1974) formulated a general model for cerebral organization which also accounted for the fact that language abilities are less clearly lateralized in non-right-handers by simply suggesting that more diffuse cerebral representation is a general characteristic of these individuals.

5.2 LINGUISTIC FUNCTIONS OF THE RIGHT HEMISPHERE

Research relating to the linguistic functions of the right hemisphere has involved a variety of experimental approaches which have utilized subjects of varying levels of neurological integrity including: neurologically normal subjects; hemispherectomized subjects; commissurotomized subjects; aphasic individuals with left hemisphere lesions; and patients with acquired right-hemisphere lesions.

In addition to the diversity of subjects, a variety of techniques have been used to study the language abilities of the right hemisphere. The techniques employed have included dichotic listening (Johnson, Sommers and Weidner, 1977; Pettit and Noll, 1979; Caldas and

Botelho, 1980), visual half-field tachistoscopic presentation (Moore and Weidner, 1974), intra-carotid amobarbital injection (Kinsbourne, 1971), regional cortical blood flow (Meyer et al., 1980) and electro-encephalography (Tikofsky, Kooi and Thomas, 1960). (The use of these techniques in various studies which have investigated the role of the right hemisphere in the recovery from aphasia is described later in this chapter.)

Studies involving neurologically normal subjects have contributed to both the current knowledge of language lateralization and the knowledge of the linguistic functions of the right hemisphere. In normal human subjects, while morbidity and trauma are presumed absent, it cannot be said that the cerebral hemispheres operate truly independently. Laterality differences, however, can be demonstrated in normal subjects and the findings generally complement those obtained from studies of clinical populations (Bradshaw and Nettleton, 1983). One of the richest sources of information on right-hemisphere linguistic function comes from studies of normal individuals who have suffered right-hemisphere damage secondary to stroke or trauma. In addition, researchers have also learned a great deal about language lateralization by studying the effects of various surgical procedures in the brain on language abilities. The procedures which have yielded the greatest amount of information are hemispherectomy and split-brain surgery.

5.2.1 Language symptoms of right-hemisphere damage

Although by using the more traditional aphasia test batteries it may be difficult to determine the presence of a language disturbance in a right-handed patient who has suffered either a vascular or traumatic lesion in the right cerebral hemisphere, the majority of speech patho-logists know from clinical experience that many right-handed patients with right-hemisphere damage do not communicate normally. Even though on the surface these patients appear to retain the basics of language, closer examination often reveals that they lack a complete understanding of the context of an utterance, the pre-suppositions entailed in the utterance or the tone of a conversational exchange.

The speech of persons with damaged right hemispheres has been described as excessive and rambling, inappropriate, confabulatory, irrelevant, literal and sometimes bizarre (Gardner, et al., 1975; Gardner, 1975; Collins, 1976). Some authors have highlighted that the comments made by patients with right-hemisphere damage are often off-colour and many exhibit inappropriate humour (Gardner et al., 1975). In addition, these patients have often been observed to focus on insignificant details in conversation or make tangential remarks and the usual range of intonation is frequently observed to be lacking (Gardner, 1975; Ross and Mesulam, 1979). Myers (1979, 1986) reported that the speech of these patients is characterized by difficulty in extracting critical pieces of information, in seeing the relationships among them and in reading conclusions or drawing inferences based on those relationships. In other words, right-hemisphere damaged patients appear to have a basic problem integrating and organizing information and an inability to form an overall gestalt of information given (Wapner, Hamby and Gardner, 1981; Myers, 1986).

It is well documented in the literature that severe disturbances in the linguistic components of language (i.e. phonology, syntax, lexical-semantic), are often seen in association with lesions of the left cerebral hemisphere. In contrast, lesions of the right hemisphere of dextrals rarely impairs the semantic or grammatical components of language. Consequently, the role of the right hemisphere in language has been considered rudimentary by some researchers. It must be remembered, however, that language is comprised of other components in addition to the forementioned linguistic components. Benson (1986) identified four major divisions of language: syntactic (relational) language; semantic (definitional) language; prosodic (vocal) language; and gestural (motor) language. While semantics and syntax belong to the linguistic facet of language, the latter two components, gestural and prosodic language, have been referred to as belonging to the extra- or para-linguistic facets of language. Evidence from research carried out in recent years suggests that although the traditional linguistic components of language are largely the concern of the dominant left hemisphere, the right hemisphere appears vital for the processing of the extra-linguistic aspects of language which contemporary students of language would include as part of pragmatics or the discourse function of language (Bates, 1976; Code, 1987; Sadock, 1974; Searle, 1969; Wapner, Hamby and Gardner, 1981). We will now examine in further detail the effects of right-hemisphere damage on both the linguistic and extra-linguistic components of language.

(a) Linguistic deficits following right-hemisphere damage

The findings of several studies have suggested that in dextrals the traditional linguistic components of language are lateralized primarily to the left cerebral hemisphere. In particular, of all the components of language, the syntactic component appears to be the most clearly lateralized, being almost totally an activity of the left hemisphere

(Zaidel, 1985; Zaidel and Peters, 1981). On the other hand, although the left hemisphere is also primarily responsible for the semantic component of language, the available evidence from split-brain and hemispherectomy studies (see below) suggests that the non-dominant right hemisphere also possesses some semantic linguistic functions. Lesser (1974) found that right-hemisphere damaged patients were not impaired in their use of syntax or phonological discrimination but had marked difficulty on a semantic test of language comprehension. Gainotti *et al.*, (1981) also concluded that right-hemisphere lesions consistently impair semantic-lexical discrimination but do not hamper phoneme discrimination.

Mild syntactic and semantic deficits were found by Hier and Kaplan (1980) to occur in only some right-hemisphere damaged patients. They did, however, report a high correlation between the occurrence of verbal deficits and hemianopia which led Millar and Whitaker (1983) to the conclusion that verbal deficits in these patients are likely to occur only in association with other deficits of the impaired right hemisphere. Myers (1986) argued that deficits in straightforward expressive and receptive tasks (e.g. naming, word discrimination, following simple commands, word and sentence reading and writing) should not be considered a major source of right-hemisphere communication impairment. In agreement with Millar and Whitaker (1983), she concluded that deficits evident on various language tests may be attributed to deficits in other processing functions of the right hemisphere. Further, Myers (1986) also highlighted that linguistic deficits, when evident in right-hemisphere damaged patients, tend to be mild. Millar and Whitaker (1983) indicated that there is good evidence that the right hemisphere contributes to many factors of human behaviour that play a role in communication. Consequently, they suggested that right-hemisphere damaged patients exhibit a cognitive affective disorder rather than a linguistic disorder.

Although linguistic tasks such as complex auditory comprehension (Hier and Kaplan, 1980), word fluency (Adamovich and Henderson, 1983) and naming (Gainotti *et al.*, 1981) have been demonstrated to be difficult for right-hemisphere damaged subjects, when they are present these deficits are mild and do not significantly affect communicative competence (Myers, 1984). Rather it is linguistic deficits at a higher level, where integration and organization of information is necessary, that disturbs the communicative abilities of patients with right-hemisphere damage. As these deficits exist at a complex level, researchers have not been able to delineate if these problems are purely linguistic or linguistic–cognitive in nature (Diggs and Basili, 1987).

The comprehension deficits exhibited by right-hemisphere damaged subjects has been described by Diggs and Basili (1987) as diminishing the ability of these patients to appreciate antonymic contrasts, interpret proverbs and figures of speech, solve incongruent statements and understand logico-grammatical and other linguistically complex sentences. They observed that comprehension problems in these patients increased with memory load and linguistic complexity. Hier and Kaplan (1980) also found the comprehension of logico-grammatical sentences and abstract interpretation of proverbs to be impaired in right-hemisphere damaged subjects.

Yet another feature of the comprehension abilities of right-hemisphere damaged patients reported in the literature is that they tend to interpret figurative language literally. The results of a study carried out by Foldi, Cicone and Gardner (1983) in which right-hemisphere damaged patients were required to interpret both direct and indirect speech indicate that they interpret indirect speech literally, without taking contextual information into account. Similarly, Myers and Linebaugh (1981) observed that right-hemisphere damaged adults tend to relate more readily to denovative rather than connotative aspects of language, a tendency which could possibly prevent them from grasping intended or implied meaning embedded in an utterance. Myers and Linebaugh (1981) also reported that their subjects with right-hemisphere damage responded literally to idiomatic expressions more frequently than control subjects and had difficulty determining the correct context of utterances. Further, their right-hemisphere damaged subjects tended to interpret an idiom analytically, breaking it down into its constituent elements rather than processing it as a whole using context to aid interpretation.

Several authors have investigated the ability of right-hemisphere damaged adults to interpret and process narratives (Wapner, Hamby and Gardner, 1981; Gardner et al., 1983). These studies have shown that, as in the case of idioms, right-hemisphere damaged patients tend to use a logical analytical style to infer directly information leading to literal and abnormal interpretation of emotional, moral and humorous content. When asked to retell the narrative, these patients were generally able to recall isolated details but had difficulties inhibiting tangential and confabulatory responses. They were reported to constantly violate the story boundary allowing personal experience and opinions to intrude. The right-hemisphere damaged patients also displayed problems in ordering and integrating specific information and were unable to judge the appropriateness of facts, situations and characterizations to the story. It has been suggested by Wapner, Hamby and Gardner (1981) that these results indicate a poor integration of language elements and a diminished appreciation of narrative form. They have proposed that these deficits in right-hemisphere damaged adults are due to an underlying problem acquiring a sense of the overall gestalt of linguistic entities. If Wapner and co-workers are correct, this problem would lead to a disturbance of the basic schema of language and would make the appreciation of the relationships between language elements difficult. They summarized the effect of these deficits by suggesting that without this organizing principle, right-hemisphere damaged adults are consigned to an undirected rambling discourse and are unable to judge which details matter and what overall points they yield. This suggestion is supported by the findings of studies which have shown that patients with right-hemisphere damage provide less information using an equal or greater number of words when compared to normal subjects (Myers, 1979; Rivers and Love, 1980; Diggs and Basili, 1987).

Overall, right-hemisphere damaged patients primarily have difficulty in the interpretation, integration and organization of linguistic elements and context. They fail to appraise information holistically and to form an overall structure as a reference base to aid their interpretation and production of complex language.

(b) Extra-linguistic deficits following right-hemisphere damage

Although in the past, most studies have examined the effects of right-hemisphere damage on the more traditional linguistic components of language, an increasing amount of research in recent years has focused on the contribution of the right hemisphere to the pragmatic, prosodic and emotional aspects of communication. In fact, Myers (1986) considered the most important part of the communication deficit exhibited by right-hemisphere damaged patients to be the impairment in these so-called 'extra-linguistic' components of language.

Myers (1986) distinguished between two different categories of extra-linguistic deficits: affective–prosodic disturbances and pragmatic disorders. Affective–prosodic disturbances relate to the comprehension and expression of emotional tone (affect) conveyed through mechanisms such as facial expression, gesture (body language) and prosodic features of oral speech (intonation contour, volume, rate and rhythm). Pragmatics refers to the use of language in linguistic and situational context. According to Code (1987, p. 88) 'pragmatics represents an interface between language and other aspects of behaviour and is concerned not with linguistic entities themselves but with the total behavioural–social context in which communication takes place and the actual intentions behind an utterance or message'.

Reports in the literature have indicated that patients with right-hemisphere lesions have a disturbance in the affective (emotional) component of prosody (Tucker, Watson and Heilman, 1977; Ross and Mesulam, 1979). In particular, right-hemisphere damaged patients have been described as having 'flat-affect' and monotonous speech (Ross and Mesulam, 1979). Subjects sustaining right-hemisphere damage have also been reported to be impaired in their ability to evaluate emotional situations presented through non-verbal means, particularly those presented through facial expressions (Benowitz *et al.*, 1983). Affective gestural impairment was also reported by Lundgren, Moya and Benowitz (1983) who found impaired ability in right-hemisphere damaged patients to perceive affect from facial expression and to interpret body language. An inability to express emotion through either vocal melody or gesture, a condition referred to by Ross (1981) as 'aprosodia', has been reported to be associated with damage to the right posterior inferior frontal area (the analogue of Broca's area in the left hemisphere). Further, Ross (1981) also suggested that the emotional quality of speech cannot be comprehended if there is damage in the posterior temporal–parietal area of the right hemisphere. These findings have lead some authors to conclude that the right hemisphere is dominant for processing affective–prosody (Ross and Mesulam, 1979; Ross, 1981; Emmorey, 1987).

Based on his observations of the effects of right-hemisphere damage on prosody, Ross (1981) proposed a theory that the functional anatomical organization of affective prosody in the right hemisphere mirrors that of propositional language in the left hemisphere. He proposed a method of assessment and classification of prosodic disorders (aprosodias) similar to that used for aphasic language disorders based on prosodic comprehension, repetition and expression. Ross hypothesized that after damage to the anterior right hemisphere, prosodic expression and repetition would be impaired but prosodic comprehension would remain intact and that damage to the posterior right hemisphere would result in impaired prosodic comprehension and repetition with fluent expression. Ross (1981) also forecast the existence of transcortical aprosodic syndromes in which prosodic repetition is basically intact.

Ross's theory of aprosodia extends beyond his results and was based on case histories of ten patients only. Within his limited subject group Ross was unable to demonstrate all of the aprosodias that he proposed. Although his own research does not provide conclusive support for Ross's theory, the basic premise of differential localization in affective prosody in the right-hemisphere gains support from the work of Roberts, Kinsella and Wales (1982). In agreement with the differential anterior–posterior damage profile advanced by Ross (1981), these latter authors found that patients with posterior right-hemisphere damage performed significantly worse on prosodic discrimination tasks than patients with anterior right-hemisphere damage.

It could be argued that these reported affective and prosodic deficits described in the research relating to the effects of right-hemisphere damage on language might be related to the global problem that right-hemisphere damaged adults have in integrating and processing information. The inability to discriminate and produce affective and prosodic information may be due to inefficient and inaccurate integration of affect and prosody with language leading to deficits in the interpretation and production of messages.

Clinically, one other important finding of research in this area is that after right-hemisphere damage, affect may not match mood (Ross, 1985). Hence, 'flat-affect' displayed by right-hemisphere damaged patients need not reflect an underlying depressed mood but rather a specific affective prosodic deficit. Roberts, Kinsella and Wales (1982), after comparing performance on tasks discriminating linguistic prosody and affective prosody, found right-hemisphere damaged adults to have differential difficulty with certain emotional tones of voice. Their results suggested that it is not emotion itself that is difficult for patients with right-hemisphere lesions, but rather the prosodic element of emotional (affective) speech. These findings were supported to some extent by Kinsella (1986) who demonstrated similar deficits of prosodic discrimination in right-hemisphere damaged subjects when required to judge sentence type by prosodic cues alone.

Pragmatic deficits associated with damage to the right hemisphere are interrelated with the deficiencies in affective–prosodic and linguistic–cognitive processing describe above (Myers, 1986). Clinical observation of the conversational abilities of right-hemisphere damaged patients shows that these individuals are unable to appreciate the context and tone of a conversation or the presuppositions entailed. Their discourse often focuses on insignificant and tangential details and includes inappropriate humour and comments, giving their language an excessive and rambling nature. Myers (1986, p. 456), suggested that the reason why right-hemisphere damaged patients are unable to use contextual cues results from their difficulty in 'evaluating the significance of sensory input, in associating it with prior knowledge and integrating multiple features of experience into a meaningful pattern or context'. Further, she suggested that the inability of these patients to use contextual cues occurring in conjunction with an essentially intact linguistic system is the reason why patients with right-hemisphere damage react to literal rather than metaphorical, humorous or idiomatic speech forms and why they confabulate, miss the point and include tangential detail in their conversation. Myers (1986) has suggested that it is these two deficits (i.e. the tendency to interpret words and events on a literal, superficial basis and the failure to establish adequate organizational framework) that constitute the two major pragmatic deficits in patients with right-hemisphere damage.

The inability of right-hemisphere damaged patients to interpret contextual information and organize their response was well illustrated in a study carried out by Myers (1979). Eight right-hemisphere damaged patients and eight non-brain-damaged matched control patients were asked to describe the 'Cookie Theft' picture from the Boston Diagnostic Aphasia Examination (Goodglass and Kaplan, 1972). Using the list of concepts normally used in describing the picture, Myers classified the concepts as either literal (e.g. reaching up; has finger to mouth) or interpretive (e.g. asking for cookie; saying 'shhh') and then compared the frequency of these two types of concepts in both the right-hemisphere damaged and control groups. The results indicated that patients with right-hemisphere damage used significantly fewer interpretive concepts compared to the control group. Myers (1979) also noted that the right-hemisphere damaged group produced a large number of irrelevant comments. She concluded that right-hemisphere damaged patients have difficulty integrating information both on a perceptual and on a more formal level and that this deficit is manifest in their verbal output as irrelevant and often excessive information and a literal treatment of questions and events.

Prutting and Kirchner (1987) analysed the pragmatic abilities of right-hemisphere damaged adults in conversation. They reported that these patients were similar to aphasic adults when comparing mean percentage of appropriate pragmatic behaviours (84% compared with 82%). However, the right-hemisphere damaged adults manifested a distinctive cluster of inappropriate behaviours. The pragmatic behaviours most frequently judged to be inappropriate in this group included eye gaze, prosody, adjacency, contingency and quantity and conciseness. Adjacency and contingency are turn-taking behaviours that relate to the ability to continue conversation and share a topic.

The observation that left-hemisphere damaged aphasic patients communicate better than they talk (Holland, 1977) and right-hemisphere damage patients perhaps talk better than they communicate has led to the hypothesis that the right hemisphere is responsible for simultaneously organizing and integrating different elements of conversation while the left hemisphere is critical for literal language (phonology, syntax and low-level semantics) (Foldi, Cicone and Gardner, 1983).

5.2.2 Hemispherectomy

The surgical removal of a total cerebral hemisphere (hemispherectomy) has occasionally been performed as treatment for various diseases of the brain (Basser, 1962; Smith and Burklund, 1966; Smith and Sugar, 1975). The most striking aspect of the literature on hemispherectomy, as pointed out by Searleman (1977), concerns the enormous differences between children and adults, particularly in relation to post-operative speech and language functioning. The development of essentially normal language skills following the removal of either hemisphere in children, reflects what Zangwill (1960) called the 'equipotentiality' of the two hemispheres to perform language functions during the early developmental years so that a shift of language competency from one hemisphere to the other is easily accomplished at an early age. The results from adult hemispherectomy subjects attest to the specialization of the two cerebral hemispheres.

Most reported hemispherectomies have been performed in children with intractable epilepsy secondary to a severely damaged hemisphere. Where the operation has involved the removal of the right hemisphere, as most cases have, aphasia, at least in the classic sense, has not been recorded. Relatively normal right-hemisphere language has also been reported following removal of the left cerebral cortex in children (Dennis and Whitaker, 1977). Basser (1962) claimed that hemispherectomy in a child up to the age of puberty would not result in a permanent aphasia.

Several authors have hypothesized that both cerebral hemispheres possess equal potential for language until at least 4 or 5 years of age (Osgood and Miron, 1963; Lennenberg, 1967). Further, it is believed by several researchers that the non-dominant hemisphere retains an ability to assume responsibility for language through the first 10 or 12 years of life and possibly beyond (Smith, 1966; Lenneberg, 1967; Cummings *et al.*, 1979). For this reason, although children who have begun to develop language may exhibit a language disturbance following cerebral damage, the resulting language disorder is usually only temporary and mild. It appears, therefore, that cerebral dominance for language develops with age and consequently the younger the child at the onset of acquired aphasia, the less complete the dominance for language in one hemisphere and the better the other hemisphere can assume language function.

In contrast to the minimal effects in children of hemispherectomies on speech and language function, hemispherectomy in late childhood or adult life has been reported to have a serious effect on cognition as well as speech and language function (Whitaker and Ojemann, 1977). Although hemispherectomies involving the removal of the dominant left hemisphere in adults are performed infrequently, they produce quite a different language outcome to the equivalent operation in children (Basser, 1962; Smith, 1966; Smith and Sugar, 1975). Removal of the left hemisphere in this situation, usually for the treatment of glioma, results in a permanent aphasia of a type similar to that seen in cases of global aphasia (Basser, 1962; Smith, 1966). Right hemispherectomy in adults, on the other hand, is usually characterized by impaired visuo-spatial functioning and intact speech and language (Basser, 1962; Searleman, 1977). Where there is evidence that the left hemisphere has been diseased since childhood, later removal of that hemisphere has also been reported to result in little or no language impairment, suggesting that the right hemisphere is able to assume responsibility for language function in these cases (Smith and Sugar, 1975).

5.2.3 Commissurotomy

The corpus callosum is the major commissure connecting the two cerebral hemispheres. Occasionally, the corpus callosum is severed in a surgical operation (split-brain surgery or commissurotomy) primarily to control the spread of intractable epilepsy from one hemisphere to the other (Searleman, 1977). A variety of techniques have been developed to investigate the language abilities of the isolated right hemisphere following section of the corpus callosum (Gazzaniga, 1970; Zaidel, 1976; Sperry and Gazzaniga, 1967). Overall, these techniques have demonstrated that the isolated right hemisphere is capable of considerable language function. Sperry and his co-workers found in a series of studies in commissurotomized patients that each hemisphere, when structurally isolated from the other, functions autonomously (Gazzaniga and Sperry, 1967; Sperry and Gazzaniga, 1967). For example, Gazzaniga and Sperry (1967) found that linguistic information (words) presented visually to the right hemisphere could not be communicated in writing or speech, but had to be expressed through non-verbal responses. If, for instance, the word spoon was flashed to the subject's left visual field, the subject would subsequently be able to point to a picture of a spoon with the left hand, although would not be able to name it. Through the use of non-verbal responses such as this, Sperry and Gazzaniga (1967) were able to show that the right (non-dominant) hemisphere is not 'word-blind' or 'word-deaf' as first suggested by Geschwind (1965a,b) who proposed that the right hemisphere is completely incapable of utilizing or comprehending verbal stimuli. Nebes and Sperry (1971) further showed that the right hemisphere is not limited to just matching single words to objects but rather that it can pick out from a group of hidden objects the one that is verbally described.

The work of Sperry and colleagues with commissurotomized patients suggests that the comprehension of both spoken and written language is represented in both the left and right cerebral hemispheres, although the degree of representation in the right is unknown. Levy (1970) was able to demonstrate that the right hemisphere can comprehend verbal stimuli and carry out spoken commands. In a further series of experiments with commissurotomized patients, Levy, Trevarthen and Sperry (1972) demonstrated what appears to be an upper limit for right-hemisphere linguistic abilities. They found that although the right hemisphere can usually comprehend verbal stimuli, it is totally unable to make phonological transformations. For example, the right hemisphere may comprehend the meaning of the words 'ache' and 'lake', but it would never know that they rhyme (Levy, 1974). Zaidel (1983) claimed, in direct contrast to the earlier findings of Gazzaniga and Hillyard (1971), that the right hemisphere of commissurotomized patients possesses considerable syntactic capabilities.

Zaidel (1976), from his study of both hemispherectomy and commissurotomy subjects, noted that his data, in conjunction with the data of other authors, suggest a complex model of development of language laterality of the brain. He felt that some, but not all, auditory language functions continue to develop in the right hemisphere past what is generally regarded as the critical period of language acquisition. From the results of a series of further studies in this area, Zaidel (1978a,b) concluded that the right hemisphere represents the limited linguistic competence that can be acquired by a more general-purpose, non-linguistic cognitive apparatus through repeated exposure to experience and through the formation of associations. He attributed the following characteristics to the right hemisphere: no speech; little writing; surprisingly rich auditory lexicons; substantial visual vocabularies; a good grasp of pictorial semantics in terms of everyday social experience; poor ability to evoke the sound image of a word; a rudimentary level of syntax; practically no ability to read sentences; no grapheme-to-phoneme correspondence rules; and better auditory comprehension than comprehension of written words or phrases. Gazzaniga *et al.* (1984) also reported that the right hemisphere of split-brain subjects has a rich lexicon, although not as rich as that of the left hemisphere, as well as a limited control of syntax.

Based on a review of split-brain subjects, Gazzaniga (1983) concluded that language and speech in the right hemisphere can exist at either a rudimentary or a sophisticated level. He determined that, among the small sub-set of his subjects who possessed a well-developed level of language in the right hemisphere, in almost every case it was attributable to brain pathology occurring prior to com-missurotomy. Levy (1983) critically reviewed Gazzaniga's (1983) paper and while agreeing that the normal right hemisphere is non-linguistic, contested some of Gazzaniga's other findings. Levy claimed, for example, that the results provided by Gazzaniga did not imply that early brain damage produces re-organization and bilateralization of language. Further, Zaidel (1983) opposed the claim of Gazzaniga (1983) that when right-hemisphere language does occur, it does not show a consistent and unique form of organization but varies widely as a function of the extent of pre-surgical left hemisphere damage. Zaidel (1983) felt that this latter claim implied that right-hemisphere language carried no special theoretical linguistic significance.

Overall, therefore, the results from split-brain studies suggest that the right hemisphere has a considerable capacity to comprehend verbal stimuli. Gazzaniga (1970), however, has warned that it is often possible to over-estimate the linguistic abilities of the right hemi-sphere of split-brain subjects, due to what he termed 'cross-cueing' strategies. In the process of cross-cueing, one hemisphere may learn how to transmit information extra-callosally to the other by initiating gestures, orienting responses or even verbalizations that can be picked up and acted upon by the other hemisphere. In addition to cross-cueing, other methodological and theoretical problems also arise when interpreting the results of human commissurotomy studies and when generalizing these results to the normal population. It must be remembered that the surgery is rather traumatic, involving manipu-lation of the brain as well as the disruption of its blood supply, factors which could possibly affect the patient's later performance on particu-lar linguistic tasks. Commissurotomy subjects may also be prone to bilateral language representation as a result of early cerebral damage from epileptic seizures. Functional re-organization of the brain may occur in patients who have had long-standing epilepsy and a higher percentage of bilateral language representation in these subjects could account for the relatively high degree of linguistic skills observed in several of the reported studies (Searleman, 1977). One additional problem is the possibility of inter-hemispheric transmission by other mid-brain commissures. Finally, the disconnected right hemisphere is not subject to the mutual facilitation and inhibition of the normal brain, with the result that its functioning must in some sense inevitably be abnormal. These shortcomings need to be taken into account when interpreting the results obtained from studies involving commissurotomized subjects.

As a consequence of the above problems reported to be associated with commissurotomy and due to the undesirability of dividing the optic chiasma in humans, a number of techniques have been developed by various researchers to present language material to the right hemisphere only of both commissurotomized and normal subjects. These techniques include methods such as the use of a scleral contact lens system (Zaidel, 1976; 1978a,b), tachistoscopy (Zurif and Bryden, 1969) and a key-tapping method (Tsunoda, 1975). In general, these studies have tended to confirm the findings of the reported split-brain investigations. Overall, the results of these latter investigations suggest that the right hemisphere of commissurotomized subjects is capable of some language function. In particular, the right hemisphere appears to be able to recognize (comprehend) some printed and spoken words. Its ability to express language material, however, is extremely limited.

5.2.4 Anatomical differences between the left and right hemispheres

A number of researchers have provided evidence that the speech and language centres of the left hemisphere are anatomically better developed than equivalent regions in the right hemisphere. Geschwind and Levitsky (1968) reported that the planum temporale (that part of the auditory association cortex posterior to Heschl's gyrus on the superior–lateral surface of the temporal lobe (Brodmann areas 22 and 42) and which includes the speech reception area of Wernicke) was larger on the left in 65% and larger on the right in only 11% of the 100 brains they studied. Their findings have since been confirmed by other researchers (Wada, Clarke and Hamm, 1975; Rubens, Mahowald and Hutton, 1976).

In general, studies of the morphological asymmetries between the two cerebral hemispheres have indicated the presence of a complex and interrelated set of differences in the anatomy of the posterior Sylvian region. These include differences in the position of the Sylvian fissure and size of the parietal operculum as well as the development of the planum temporale. Although it is possible that these differences in the structure of the posterior Sylvian region of the two hemispheres may in some way provide a neuroanatomical substrate for left hemisphere dominance for speech–language (Geschwind and Levitsky, 1968; Le May and Culebras, 1972; Wada, Clarke and Hamm, 1975; Galaburda *et al.*, 1978), most authors have advised caution in applying a functional importance to these asymmetries. Despite this cautionary advice, however, a number of factors highlight the possible link between hemisphere asymmetry and cerebral dominance for language. First, the posterior Sylvian region of the dominant hemisphere coincides with Wernicke's area and is known to be involved in language processing. Secondly, the greater development of the planum temporale in the left hemisphere has been demonstrated in infant and foetal brains (Witelson and Pallie, 1973; Wada, Clarke and Hamm, 1975) as well as adults, thereby suggesting that physical asymmetry may precede functional asymmetry. Thirdly, the asymmetry of the parietal opercula has been found to be absent in the majority of left-handers (Le May and Culebras, 1972). Finally, Ratcliff *et al.* (1980) reported asymmetries of the posterior Sylvian branches of the middle cerebral artery in favour of the dominant hemisphere in 59 patients for whom lateralization was known from Wada testing. The asymmetry, however, was significantly reduced in patients with atypical cerebral dominance.

5.3 OTHER NEUROPSYCHOLOGICAL SEQUELAE OF RIGHT-HEMISPHERE DAMAGE

In addition to the language disorders outlined above, damage to the right hemisphere is also associated with a number of other neuropsychological deficits. The primary information-processing functions of the right hemisphere are those of attention and perception. Consequently, the deficits that occur following right-hemisphere damage fall primarily in the areas of neglect, denial, visual and spatial perceptual disorders as well as constructional disturbances.

Unilateral neglect is a condition in which the patient fails to attend and respond to stimuli on one side of the body. It occurs more commonly and with greater severity following right-hemisphere lesions than left-hemisphere lesions. Once considered to be a unitary disorder, the neglect syndrome has been classified by Myers (1986) into two distinct types: parietal and frontal neglect. The symptoms of parietal neglect range from a failure to respond to simultaneous bilateral stimulation to a failure to report any incoming stimuli from an area to the left of the patient's midline. When specifically directed to left-sided stimuli, patients may report what is there, but on their own, without constant reminders, they will attend to the right side only. The symptoms of frontal neglect include impaired performance on cancellation and scanning tasks as well as on tests of block design, copying and spontaneous drawing, indicating impaired constructional abilities.

Other neuropsychological disorders that are the observed effect of impaired attention and perception following right-hemisphere damage include impaired facial recognition (prosopagnosia) and impaired geographical and topological orientation. Prosopagnosia is generally accepted as a deficit in visual integration and along with topological disorientation, can lead to the wrong conclusion that the patient is suffering from general confusion. Myers (1986) described topological disorientation as being the failure to assimilate visual cues. Patients with this problem have difficulty reading a map, remembering familiar routes and learning new ones. Re-duplicative paramnesia has also been reported in association with lesions of the right hemisphere. This disorientation is specific to location such that patients may know they are in a specific ward in a hospital but may be convinced that the hospital itself is far away in another city.

Lesions in either the left or right parietal lobes are often associated with constructional disturbances. As in the case of neglect, right-hemisphere lesions tend to be associated with more frequent and severe constructional deficits than left-hemisphere lesions.

5.4 THE ROLE OF THE RIGHT HEMISPHERE IN RECOVERY FROM APHASIA

As early as the middle of last century, Broca (1861) postulated that the minor hemisphere possessed an inherent ability to mediate speech following major hemisphere damage. Since that time a number of researchers have implicated the right hemisphere in the recovery of language function following a left-hemisphere lesion. Luria (1963) offered three possible explanations for the recovery of language function in these cases: (1) the injury is of a transitory nature; (2) re-organization within the left hemisphere allows compensatory functioning of adjacent undamaged structures; or (3) there is a complete transfer of language to corresponding areas in the right hemisphere ('dominance shift'). Klingman and Sussman (1983) further suggested that the observed recovery may be a product of the right hemisphere working in conjunction with the damaged left hemisphere to mediate language function.

A variety of experimental techniques have been used to assess the relative involvement of the left and right hemispheres in the recovery from aphasia. The techniques employed have included dichotic listening (Sparks, Goodglass and Nickel, 1970; Johnson, Sommers and Weidner, 1977; Pettit and Noll, 1979; Caldas and Botelho, 1980), visual half-field tachistoscopic presentation (Moore and Weidner, 1974), intra-carotid amobarbital injection (Kinsbourne, 1971), regional cortical blood flow (Meyer *et al.*, 1980) and electroencephalography (Tikofsky *et al.*, 1960).

The most commonly used method to study hemispheric language lateralization in aphasia has been the dichotic listening test. Studies using this method have generally demonstrated either an emerging left-ear preference during the course of language recovery (Johnson, Sommers and Weidner, 1977; Pettit and Noll, 1979; Caldas and Botelho, 1980), or no ear advantage (Schulhoff and Goodglass, 1969). The emergence of a left-ear preference parallel to language recovery has been taken to suggest that the right hemisphere may participate in some way in language processing. However, development of a left-ear preference does not appear to occur uniformly across different aphasic groups. Caldas and Botelho (1980) administered three different dichotic listening tests to a group of aphasic subjects at various stages of the recovery process. Interestingly, the non-fluent aphasics demonstrated a tendency to increase left-ear preference, while fluent aphasics in direct contrast, demonstrated an increase of right-ear preference in a verbal dichotic test. Caldas and Botelho (1980) suggested that these findings indicate a compensatory function of the right hemisphere in cases of non-fluent aphasia (due to a pre-rolandic lesion), but not in cases of fluent aphasia of post-rolandic origin.

The importance placed upon such results obtained from dichotic listening studies has been questioned (Berlin and McNeil, 1976; Linebaugh, 1978). Many researchers have failed to recognize the vulnerability of the dichotic listening technique to both the acoustic and phonetic variables of the dichotic (simultaneous) stimuli. Acoustic parameters such as voice onset timing, signal intensity, frequency range and length of stimuli have been shown to alter the right-ear advantage which, it is claimed, is typically displayed by normal subjects (Berlin and McNeil, 1976). A significant phonetic variable has also been discovered in that it has been noted in dichotic listening experiments, performed with consonant–vowel syllables, that voiceless stop consonants are more intelligible than voiced consonants (Lowe *et al.*, 1970; Berlin *et al.*, 1973). This effect is so powerful that, if voiceless consonants were presented to the left ear and voiced consonants to the right ear as a dichotic listening task, the results could be expected to show a left-ear advantage (Berlin and McNeil, 1976). Ideally, dichotic stimuli should be truly simultaneous, with maximum phonetic and acoustic competition. Unfortunately, however, the necessary methodological processes for accurate stimulus matching have not been widely used to date since they depend on computerized simulated speech.

Interpretation of the left-ear advantage as an indication of right-hemisphere involvement in language functioning has been criticized to the extent that the dichotic paradigm is considered merely to illustrate a 'lesion effect' (Linebaugh, 1978). That is, the lesion of the left auditory cortex causes degradation of the contralateral right-ear signal. The existence of a lesion effect is supported by the fact that the presence of pathology in the left hemisphere has been demonstrated to variably produce extinction of stimuli presented to the left ear, to the right ear or to neither after dichotic stimulation (Sparks, Goodglass and Nickel, 1970). The observed variability can be attributed to the anatomical locus of the pathology (Benson, 1979).

A right-hemisphere superiority in visual linguistic decoding has been demonstrated in aphasic subjects with the technique of visual half-field tachistoscopic presentation (Moore and Weidner, 1974). Normal subjects typically exhibit a left-hemisphere superiority. The reversal of hemisphere dominance for the decoding of visual language in aphasics supports the 'dominance shift' hypothesis in the recovery of language following left hemisphere pathology. However, several studies reviewed by Searleman (1977) suggested that both the dichotic and tachistoscopic techniques may be measuring factors unrelated to language lateralization and may vary considerably in terms of reliability. In addition, both techniques only measure low-level speech perception and comprehension abilities (Searleman, 1977; Klingman and Sussman, 1983). Research in both audition and vision has provided evidence that hemispheric asymmetries are found primarily at later higher-order stages of information processing than those stages investigated by dichotic and tachistoscopic paradigms (Berlin and McNeil, 1976; Madden and Nebes, 1980; Bradshaw and Nettleton, 1983).

The technique of intra-carotid amobarbital injection, developed by Wada and Rasmussen (1960), has been used to investigate the source of residual aphasic speech. Kinsbourne (1971) reported three right-handed aphasic subjects in which speech production continued during anaesthetization of the left hemisphere. In two of those cases, a subsequent contralateral injection arrested speech production demonstrating that dominance for residual speech production had shifted to the right hemisphere. Unfortunately, other aphasic subjects tested failed to support the 'hemispheric shift' hypothesis, leading Kinsbourne to conclude that the right hemisphere is not always the source of aphasic speech.

Regional cortical blood-flow measurements, obtained during the period of aphasic language recovery from left-hemisphere infarction, have yielded interesting results. A failure of regional cerebral blood flow to increase in Broca's area during behavioural activation in eight right-handed aphasic subjects who had demonstrated good speech recovery within 3 months post-onset of aphasia was reported by Meyer *et al.* (1980). In contrast, these subjects showed a significant increase in regional cerebral blood flow in the corresponding posterior frontal and peri-sylvian regions of the right hemisphere, thereby indicating right-hemisphere participation in the recovery from aphasia. Unfortunately, regional cerebral blood-flow measurements cannot be performed without some risk to the subject, which poses limitations for its use as an experimental technique (Kertesz, 1979).

Further evidence to support a role for the right hemisphere in language recovery in aphasia has come from electroencephalographic studies. Tikofsky *et al.* (1960) established a relationship between electroencephalogram readings and recovery of speech in aphasic subjects. They found that abnormal electroencephalogram recordings over the right hemisphere significantly lessened the prognosis for speech recovery, whereas abnormal readings over the left hemisphere had no observable relationship to speech recovery.

Klingman and Sussman (1983) adapted the linguistic–manual time-sharing paradigm initially developed by Kinsbourne and Cook (1971) as a speech output-based lateralization measure. The paradigm involves the concurrent performance of a linguistic and uni-manual task (e.g. separate left or right index-finger tapping). When normal right-handed subjects perform simultaneous right index-finger tapping and verbalization tasks (both programmed by the left hemisphere), disruption in finger-tapping rate is evident in comparison to tapping in silence. Left finger tapping and verbalization (programmed by the right and left hemispheres, respectively) produces non-significant disruption in manual performance compared to the control condition. Performance impairments of the right hand indicate left-hemisphere lateralization of speech production in normal subjects (Kinsbourne and Cook, 1971). Theoretically, the basis for the laterality effect, revealed by the concurrent performance paradigm, relies upon the notion of 'functional cerebral space' (Kinsbourne and Hicks, 1978). Hypothetically, the cerebral programmes which control the two performances share the same intra-hemispheric functional space, such that concurrent task performance over-taxes the capacity of the processing centre and one task must necessarily suffer.

The linguistic–manual paradigm was administered by Klingman and Sussman (1983) to eight non-fluent aphasics and a normal matched control group. As expected, the normal subjects only displayed disruption in right index-finger tapping rates during concurrent linguistic and manual tasks. Collectively, the aphasic subjects displayed symmetrical manual disruption indicative of bilateral language representation. Three aphasic subjects revealed consistent right hemisphere lateralization.

Perhaps the most salient evidence for right hemisphere mediation in aphasic language recovery has come from the accumulated case reports documenting the effect of right-hemisphere lesions in recovered aphasic patients subsequent to original left-hemisphere pathology (Russell and Espir, 1961; Levine and Mohr, 1979; Cambier *et al.*, 1983). These studies have tended to show that the language abilities of recovered aphasics are impaired following damage to the right hemisphere to a greater extent than would normally be expected subsequent to a right-hemisphere lesion, thereby suggesting that there has been a shift of language function in these patients, at least in part, from the left to the right hemisphere. Although there appears to be a considerable corpus of evidence available to support a role for the right hemisphere in the processes of recovery from aphasia, the exact nature of this role remains uncertain.

CHAPTER SIX

Language disturbances in dementia

Dementia is an acquired clinical syndrome in which there is a persistent impairment in intellectual function as a consequence of brain dysfunction (Cummings and Benson, 1983). According to Cummings, Benson and LoVerme (1980) at least three of the following areas of mental activity are disrupted: language, memory, visuo-spatial skills, emotion or personality and cognition (abstraction, calculation, judgement, etc.). Although most definitions of dementia indicate that language impairment may be present, recent research suggests disturbed language function is present at all stages of some dementia syndromes (Irigaray, 1973; Obler and Albert, 1981; Bayles, 1982).

Dementia is a disorder which affects the adult and geriatric population, being most common in people over the age of 65 years. Up to 20% of persons in this age group have been reported to have some degree of dementia (Reifler, Larson and Hanley, 1982). Currently the size of the geriatric population in most Western countries is increasing, both in terms of absolute numbers and as a percentage of the overall population. Statistics provided by the Australian Bureau of Statistics (1982) indicate that by the year 2001, 11.7% of the Australian population will be over 65 years of age. Estimates taken in the USA suggest that persons over 65 years of age will comprise between 17 and 20% of that country's population by the year 2030, this figure representing approximately 51 million persons (Plum, 1979; Wells, 1981).

If, as suggested in the literature, language impairment is a common feature of dementia, the rapidly increasing incidence of the disorder has important implications for the future case loads of those speech–language pathologists involved in the treatment of adult neurologically disordered clients. As we move towards the twenty-first century we can expect that demented persons will form an ever increasing proportion of the clients requiring speech-pathology services. During the past decade, a greater level of interest in dementia and its effects on language function has been reflected in a sudden rise in the number of studies reported in the literature devoted to investigation of the language abilities of demented patients. It is now evident that different patterns of language impairment may accompany the different types of dementia. For this reason, prior to reviewing the findings of studies which have investigated the language abilities of demented patients, the major dementia-producing diseases will be discussed in terms of their characteristics and underlying neuropathologies.

6.1 TYPES OF DEMENTIA

Dementia can be the product of a number of different diseases and depending on the aetiology, the impairment in mental abilities can be either reversible or irreversible. The major types of dementia are listed in Table 6.1.

Dementia is classically associated with degenerative disorders of the cerebral cortex as occurs in conditions such as Alzheimer's disease and Pick's disease. More recently, it has also been recognized that dementia is also seen in neurological disorders that involve predominantly sub-cortical structures such as the basal ganglia, thalamus and brainstem (e.g. Parkinson's disease and Huntington's chorea). A number of important differences recognized clinically serve to distinguish cortical dementia from sub-cortical dementia. First, sub-cortical dementing illnesses affect the motor system and are therefore associated with movement disorders such as the rigidity and bradykinesia of Parkinson's disease, whereas motor impairments are not evident in cortical dementias until late in their clinical course. Secondly, cortical dementias are characterized by the presence of a language disorder, agnosia and alexia, features typically absent in sub-cortical dementia. Further, the progressive intellectual and memory impairment in cortical disorders may be more severe and progress more rapidly than that of the sub-cortical syndromes. Patients with sub-cortical dementia are described as apathetic and often depressed (Cummings and Benson, 1984), whereas patients with cortical disorders often lack insight and tend not to be depressed.

The dementia associated with hydrocephalus is mainly of the sub-cortical type. In many of the other types of dementia, however, features of both cortical and sub-cortical dementia are present. Examples of these latter mixed-dementia types include: multi-infarct dementia; metabolic and toxic encephalopathies; infectious disorders (e.g. Jakob–Creutzfeldt's disease); and neoplastic dementia.

Table 6.1 Major causes of dementia

Cortical dementias
  Alzheimer's disease
  Pick's disease
Extra pyramidal syndromes (sub-cortical dementias)
  Parkinson's disease
  Hepato-lenticular degeneration (Wilson's disease)
  Huntington's chorea
  Progressive supranuclear palsy
  Idiopathic basal ganglia calcification

Hydrocephalus

Metabolic and toxic encephalopathies
  Nutritional deficiencies, e.g. vitamin B₁ deficiency (Wernicke-Korsakoff
                                            syndrome)
  Endocrine disturbances, e.g. hypothyroidism, hyperthyroidism, Cushing's
                                            disease, Addison's disease
  Drug toxicity
  Heavy-metal exposure
  Alcohol abuse

Cardiovascular disease
  Multi-infarct dementia, e.g. Binswanger disease

Infectious diseases
  Slow virus dementia, e.g. Jakob–Creutzfeldt disease
  Neurosyphilis
  Herpes simplex encephalitis

Miscellaneous dementia syndromes
  Neoplastic
  Post-traumatic
  Post-anoxic

Currently, cortical dementias are untreatable, whereas sub-cortical dementias or dementia syndromes with mixed cortical and sub-cortical features are either reversible or at least partially treatable. Consequently, as the presence or absence of a language disturbance is an important criterion for differentiating between cortical and sub-cortical types of dementia, language testing may play an important role in indicating the presence or otherwise of a treatable syndrome in a demented patient presenting at a speech-pathology clinic. Bayles, Tomoeda and Caffrey (1982) emphasized that understanding the pattern of linguistic dissolution associated with each dementing illness, and their relation to morphological changes in the brain, is essential to making accurate diagnoses of dementia type.

The dementia-producing diseases most commonly seen by a speech pathologist include: Alzheimer's disease, Pick's disease; multi-infarct dementia; dementia associated with extra-pyramidal syndromes; and Korsakoff's disease.

6.1.1 Characteristics of Alzheimer's disease

Alzheimer's disease is the single most frequently diagnosed cause of dementia (Terry and Katzman, 1983) accounting for approximately one-half of all cases of irreversible dementia (Reifler, Larson and Hanley, 1982). For this reason, it has received the greatest amount of attention in the literature relating to language disturbances in dementia.

Although originally described as pre-senile dementia, Alzheimer's disease is now considered to be essentially the same as senile dementia of the Alzheimer-type. The neuropathological findings and the behavioural alterations in these diseases are essentially indistinguishable (Haase, 1977) and the *Diagnostic and Statistical Manual of Mental Disorders* (3rd edn) (American Psychiatric Association, 1980) eliminated the age of onset as a criterion for the diagnosis of Alzheimer's disease. The term 'Alzheimer's disease' is now used for all age groups.

Diagnosis of Alzheimer's disease is usually made between 70 and 79 years of age and the condition is associated with a marked decrease in life expectancy. According to the *Diagnostic and Statistical Manual of Mental Disorders*, the symptomatology of Alzheimer's disease includes: intellectual dysfunctions sufficient to interfere with social behaviour, absence of the characteristics of delirium, memory impairment, indication of brain damage and at least one of either personality change, impairment in abstract thinking, poor judgement or other disturbance of higher cortical function (e.g. aphasia, apraxia, agnosia or constructional difficulty).

Pathological changes reported to occur in the brains of Alzheimer patients include both morphological and neurochemical alterations. The brains of Alzheimer patients have been found to be underweight (usually < 1 kg) (Corsellis, 1976; Tomlinson, 1977) and characterized by diffuse ventricular dilatation, atrophy and neuronal loss in both grey and white matter. The morphological changes, however, are not spread diffusely throughout the cerebral cortex. On gross examination the areas of brain damage, although widespread, can be seen to primarily involve the temporal and frontal lobes, the hippocampus, para-hippocampus and hippocampal gyrus (Tomlinson, 1977). At the same time, areas of cortex subserving primary motor, somato-sensory and visual functions are relatively spared (Jervis, 1937; McMenemy, 1940). Motor abnormalities do not usually appear until the late stages of the disorder and sub-cortical structures do not become involved until after the dementia is well established (Cummings and Benson, 1983).

A number of characteristic histological changes are evident in the neurones of the cerebral cortex in Alzheimer's disease. These changes include neurofibrillary tangles, senile (neuritic) plaques and granulo-vacuolar degeneration. Neurofibrillary tangles are the most character-istic morphological change in Alzheimer's disease and occur when neurofibrils within the neurone become twisted and contorted into a helical shape within the neuronal cytoplasm (Jervis, 1971). When neurofibrillary tangles are present, the normal function of the neuro-fibril in facilitating transport of intracellular elements in the neurone is disrupted. Senile plaques are small spherical areas of tissue degener-ation consisting of granular deposits and remnants of neuronal processes. The plaques have a three-layered structure with an outer zone of degenerating neuritic processes, a middle zone of swollen axons and dendrites and a central amyloid core (Kidd, 1964). Synapses are markedly reduced throughout the plaque (Krigman, Feldman and Bensch, 1965). Granulo-vacuolar degeneration refers to the accumulation within the neurones of granular debris and fluid-filled spaces (vacuoles).

Neuropathologies such as senile plaques, neurofibrillary tangles and granulo-vacuolar degeneration can only be definitely identified via direct histological examination of the brain tissue. Consequently, confirmation of Alzheimer's disease in most cases is currently dependent on autopsy, with the occasional rare case being verified by biopsy (Shore, Overman and Wyatt, 1983; Martin et al., 1985b). Such limitations emphasize the need to make ante-mortem diagnoses on the basis of some other identifiable symptom complex. Berg (1985) reported that the term 'Dementia of the Alzheimer's type' is used to refer to the disorder diagnosed by clinical criteria, but not histo-logically verified as Alzheimer's disease. However, the terms 'dementias of the Alzheimer's type' and 'Alzheimer's disease' are used interchangeably by different authors.

As indicated above, the histological changes in Alzheimer's disease are not spread diffusely throughout the cerebral cortex but rather occur in specific topographic patterns. Granulo-vacuolar degeneration, for instance, occurs almost exclusively in the hippocampus (Corsellis, 1976). The neurofibrillary tangles and senile plaques are most evident in the temporal–parietal–occipital junction area with the frontal lobe being affected to a lesser degree. Significant involvement also occurs in the temporal–limbic cortex and the posterior cingulate gyrus. In the limbic cortex the hippocampus, entorhinal area (the anterior part of the para-hippocampal gyrus) and the amygdala are affected. Additional histological changes evident in the affected cortical areas in Alzheimer's disease include a loss of neurones, accumulation of neuronal lipofuscin and astrocyte hyperplasia (Corsellis, 1976; Terry *et al.*, 1981; Dowson, 1982). Although the cerebral cortex is the primary region affected in Alzheimer's disease, histological changes including senile plaques and neurofibrillary tangles have also been reported in sub-cortical structures such as the thalamus, hypothalamus and mammillary bodies (McDuff and Sumi, 1985).

The importance of the histological changes outlined above is that they interfere with nerve impulse transmission at the cellular level and consequently the changes in cognitive function observed in Alzheimer's disease can largely be related to the predominant brain areas affected by the histological alterations. For instance, the memory impairment seen in Alzheimer's disease is believed to be associated with the damage to the limbic system, especially the hippo-campus. Although the causes of the morphological changes are unknown, in recent years it has been suggested that they may be the by-products of a malfunction in the cholinergic system (a neuronal network that transmits nerve impulses through acetycholine) (White-house *et al.*, 1981; Gottfries *et al.*, 1983). Levels of enzymes involved in the manufacture of acetylcholine have been found to be reduced by 80% in Alzheimer's disease patients (Davies, 1983). The distribution of this biochemical reduction closely parallels the topography of the histological alterations, the reduction in enzyme activity being most marked in those brain areas with the greatest concentration of senile plaques, neurofibillary tangles and granulo-vacuolar degeneration.

It is possible that the cholinergic deficit in Alzheimer's disease is related to histological changes in the nucleus basalis of Meynert located beneath the globus pallidus (Coyle, Price and de Long, 1983). In the region of 70% of the cholinergic activity in the cortex appears to reside in axon terminals with cell bodies located in this nucleus. In Alzheimer's disease, extensive reduction of cholinergic neurones occurs within the nucleus basalis along with histological changes, suggesting that the lack of cortical cholinergic input and the develop-ment of senile plaques, neurofibrillary tangles and granulo-vacuolar degeneration are related.

In addition to the cholinergic hypothesis, other alternative causes of Alzheimer's disease proposed in the literature include aluminium intoxication (Crapper, Krishnan and Quittkart, 1976), disordered immune function (Behan and Behan, 1979; Mitler *et al.*, 1981), viral infection (de Boni and Crapper, 1978) and dysfunction of cellular microtubules (Heston *et al.*, 1981). A hereditary factor has also been suggested. Larrson, Sjörgren and Jacobson (1963) reported that first-order relatives of persons suffering from Alzheimer's disease have a four times greater chance of having the disease than individuals without such familial relationship. Further evidence for a genetic factor comes from the finding that individuals with Down's syndrome develop the neuropathological changes typical of Alzheimer's disease if they survive to middle life (Reisberg, 1981). Like Down's syndrome, therefore, Alzheimer's disease may be the consequence of chromosomal abnormality (Reisberg, 1981). An increased incidence of Down's syndrome has been found in families of persons with Alzheimer's disease. Alternatively, the disease may represent an exaggeration of normal ageing, since the associated histological changes are also found in brains from normal elderly individuals, but to a lesser extent.

6.1.2 Characteristics of Pick's disease

Pick's disease is a rare primary degenerative dementia that clinically resembles Alzheimer's disease. The aetiology is unknown and the condition is mostly regarded as an idiopathic degenerative process affecting the neuronal cell body. Onset is usually between the ages of 40 and 60 years (Haase, 1977). Death usually occurs within 2–15 years (Slaby and Wyatt, 1974) and women are affected more frequently than men. Pick's disease is characterized by progressive intellectual deterioration; however, memory is less impaired, especially in the early stages, than in Alzheimer's disease. The relative sparing of memory may be due to the fact that the hippocampal formation is less affected in Pick's patients than in persons with Alzheimer's disease. Language abnormalities and a wide variety of personality changes are among the earliest intellectual alterations to occur in Pick's disease (Wechsler *et al.*, 1982). The changes in personality may include irritability, depression, socially inappropriate activity, hypersexuality, loss of personal propriety and other behavioural disinhibitions. Abnormalities of motor and sensory systems are conspicuously absent during most of the course of the disease, but as it advances, a parkinsonian-type of extra-pyramidal syndrome often develops and pyramidal system abnormalities may also appear. In the late stages of Pick's disease, intellectual deterioration may affect all areas of intellectual function and terminally these patients are mute and immobile with both urinary and faecal incontinence.

Pathologically, the brains of patients with Pick's disease are reduced in size. The atrophy, as evident on computed tomography scans and at autopsy, selectively involves either the frontal or temporal lobes, or both (Corsellis, 1976). In particular, the anterior and medial temporal areas and the orbito-frontal cortex show the most severe atrophic changes while the parietal lobe, the pre-central gyrus (primary motor cortex) and the posterior one-third of the superior temporal gyrus tend to be spared (Corsellis, 1976; Cummings and Duchen, 1981).

In most cases of Pick's disease, two characteristic histo-pathological features are evident in the neurones within the affected areas of the cerebral cortex. These features include intracytoplasmic Pick bodies and 'inflated' neurones. Pick bodies are dense intracellular structures about the same size as the nucleus and which occur in the neuronal cytoplasm. These bodies are characteristic of Pick's disease in that they do not occur in normal ageing and are rare in diseases other than Pick's disease. Inflated neurones are enlarged nerve cells in which the nuclei are displaced to one side even though no body is evident in the cytoplasm to cause the displacement. Other histo-pathological changes including loss of neurones, astrocyte hyperplasia (gliosis) and microglial cell proliferation are also evident in the affected cortical areas in Pick's disease.

Neurone loss and astrocytic gliosis also occur to a variable extent in the basal ganglia, thalamus and sub-thalamic nucleus. Pick bodies, however, tend to be confined to the cerebral cortex and are not found in sub-cortical structures (Corsellis, 1976; Cummings and Duchen, 1981). There is no apparent selective involvement of one transmitter system in Pick's disease as occurs in the cholinergic system of Alzheimer patients.

6.1.3 Characteristics of multi-infarct dementia

Vascular dementia results from multiple infarctions in the cerebral tissue as a consequence of multiple vessel occlusions. The specific location of the involved vessels as well as the total amount of infarcted tissue determines the characteristics of the associated dementia (Tomlinson, Blessed and Roth, 1970) so that the mental and behavioural changes resulting from multi-infarct dementia vary from one case to another. This type of dementia accounts for approximately 14–20% of dementia cases (Tomlinson, 1977).

Although hypertension is the most common aetiological factor, multi-infarct dementia may be caused by a large number of different disease entities (e.g. atherosclerosis, diabetes, Binswanger disease, cardiac disease, etc.). Pathologically, the infarcts usually involve sub-cortical structures such as the basal ganglia, internal capsule and thalamus. However, in some cases, cortical infarcts are also present in which case language abnormalities may be evident.

Features identified by Hachinski *et al.* (1975) and which serve to differentiate multi-infarct dementia from other dementias include: abrupt onset, stepwise deterioration in intellectual function; fluctuating course; nocturnal confusion; relative preservation of personality; depression; somatic complaints; emotional lability; a history of hypertension; a history of strokes; evidence of associated atherosclerosis; and focal neurological signs and symptoms. If a number of these features are present, then multi-infarct dementia should be suspected. Most of the patients exhibit pseudobulbar-like bilateral abnormalities including rigidity, spasticity, hyper-reflexia, gait abnormalities and dysarthria. Because the exact nature of the cognitive and language deficits depends on the site, location and extent of the cerebral infarctions, speech–language deficits in the form of an aphasia, apraxia of speech and/or dysarthria may form components of the overall symptom complex in multi-infarct dementia.

6.1.4 Characteristics of dementia in extra-pyramidal syndromes

Extra-pyramidal syndromes are disorders in which the major pathological changes occur in the basal ganglia, thalamus and upper brain-stem while the cerebral cortex is relatively spared. The principal extra-pyramidal syndromes in which dementia may be present include Parkinson's disease, Huntington's chorea, progressive supra-nuclear palsy, Wilson's disease and idiopathic basal ganglia calcification. The characteristics and neuropathology of each of these disorders is described in Chapter 9 in relation to hypokinetic and hyperkinetic dysarthrias.

The dementia syndromes associated with the various extra-pyramidal disorders have similar characteristics and are collectively referred to as 'sub-cortical dementia'. The features of sub-cortical dementia include: mental slowness; inertia and lack of initiative; memory impairment that profits from structure and cues; slowness of cognitive processing and disturbances of mood (including depression).

Sub-cortical dementias differ from cortical dementias in two important ways. First (with the possible exception of some parkinsonian patients), sub-cortical dementias in general lack the language disturbance, agnosia, apraxia and alexia that is characteristic of the cortical dementia syndromes such as seen in Alzheimer's and Pick's disease. Secondly, the sub-cortical dementias are different in that each of the extra-pyramidal syndromes is associated with a prominent movement disorder (e.g. bradykinesia, tremor, rigidity, chorea, myoclonus, dystonia, etc.) as part of their clinical symptomatology. Consequently, the majority of patients with extra-pyramidal disorders exhibit a dysarthria (see Chapter 9).

A cortical dementia may possibly occur in some parkinsonian patients. Alzheimer-like degenerative changes have been observed in patients with Parkinson's disease and are thought to be related to the presence of dementia (Wisniewski, Terry and Hirano, 1970). Whether these histo-pathological changes are a consequence of the same pathological process that causes the concomitant sub-cortical changes in these individuals or the result of co-occurring Alzheimer's disease is unknown. Some authors have suggested that there may be two forms of idiopathic Parkinson's disease, one a sub-cortical disorder associated with motor disturbances and the other a disorder in which motor dysfunction and dementia occur in association with both sub-cortical and cortical histo-pathological changes (Lieberman *et al.*, 1979; Boller *et al.*, 1980).

6.1.5 Characteristics of Korsakoff's syndrome

Korsakoff's syndrome (formerly known as 'Korsakoff's psychosis') has been defined as an isolated loss of recent memory in a fully awake, alert and co-operative patient who has normal immediate recall, remote memory and manipulation of old knowledge (Ross, 1980a). The condition may be encountered in a variety of disorders (e.g. closed head injury, vascular infarction of the left or both medial temporal lobe(s), herpes simplex, encephalitis, anoxia, tuberculous meningitis, etc.) and may be either permanent or transient (Ross, 1980a). Traditionally, however, it is linked to alcoholism and Wernicke's encephalopathy.

Wernicke's encephalopathy and Korsakoff's syndrome represent two clinical aspects of the same pathological process that occurs in response to chronic alcohol abuse. Wernicke's encephalopathy represents the acute stage of this process and Korsakoff's syndrome the chronic residual mental deficit that usually occurs in the late stages of Wernicke's encephalopathy. Excessive alcohol intake over a long period of time causes changes in the stomach that interferes with the absorption of vitamins, especially thiamine (vitamin B_1). The resultant deficiency of vitamin B_1 causes damage to several regions of the brain including the brainstem, thalamus, mammillary bodies (which receive strong hippocampal input), hypothalamus and the frontal and associative areas of the neocortex. These morphological changes are associated with a triad of symptoms which characterize Wernicke's encephalopathy. These include mental changes, paralysis of eye movements and an ataxic gait.

The most striking mental change in Korsakoff's syndrome is an amnesia for recent events (recent memory disturbance) coupled with an inability to acquire new memories. Although lesions of the mammillary bodies were long believed to be the anatomical basis of the distinctive memory deficit, the results of more recent clinico-pathological studies have suggested that lesions of the medial thalamus, especially those involving the dorso-medial nucleus and the medial pulvinar (see Chapter 3) correlate best with the amnestic symptoms (Haller, 1980). With thiamine supplementation and a balanced diet, patients with Wernicke–Korsakoff syndrome may become more alert but the memory problems persist.

It should be noted that in addition to Wernicke–Korsakoff syndrome, there is now evidence to suggest that an alcoholic dementia also exists. In contrast to Wernicke–Korsakoff syndrome in which difficulties in learning new information represent the predominant mental disability (Cutting, 1978), alcoholic dementia is characterized by neuropsychological deficits which include poor memory, poor abstraction, impaired planning and poor constructions (Ron, 1977; Lishman, 1981).

6.2 LANGUAGE DISORDERS IN CORTICAL DEMENTIAS

6.2.1 Language disorders in Alzheimer's disease

Although, according to the *Diagnostic and Statistical Manual of Mental Disorders* (3rd edn) (American Psychiatric Association, 1980), the recognized diagnostic criteria do not specify that the presence of a language impairment is mandatory in order to make a diagnosis of Alzheimer's disease, the results of a number of studies have indicated that a language deficit, in a more or less severe form, is present at all stages of the disorder (Irigaray, 1973; Obler and Albert, 1981; Appell, Kertesz and Fisman, 1982, Bayles, 1982; Shore, Overman and Wyatt, 1983; Cummings *et al.*, 1985; Murdoch *et al.*, 1987). In fact, a language disorder was noted by Alzheimer (1907) as part of the clinical findings in the first description of Alzheimer's disease. Based on their performance on the Western Aphasia Battery (Kertesz, 1980), Appell, Kertesz and Fisman (1982) found all 25 subjects in their sample of Alzheimer patients to be aphasic. Both Cummings *et al.* (1985) and Murdoch *et al.* (1987) concluded that the presence of a language impairment is an important diagnostic criterion of Alzheimer's disease. As discussed above, the lack of language impairment in association with a profound dementia is considered by some authors to be indicative of sub-cortical dementias (Obler and Albert, 1981; Fisher *et al.*, 1983).

Patients with Alzheimer's disease exhibit a progressive decline in communicative ability with different language abilities being affected at different stages of the disorder. A number of authors have suggested that the patterns of language impairment evidenced at certain stages correspond to specific aphasia syndromes. During the early stages of dementia, speech output is fluent, well articulated and syntactically preserved and auditory comprehension abilities for conversational material and reading aloud are intact. Hier, Hagenlocker and Shindler (1985) reported that the language deficit exhibited by early-stage Alzheimer patients shows considerable similarity to anomic or semantic aphasia with respect to mean length of utterance, number of subordinate clauses, anomic index, empty word and conciseness index on a picture description task. A number of other authors have also reported anomic aphasia as a prominent feature of early-stage Alzheimer's disease (Constantinidis, Richards and de Ajuriaguerra, 1978; Kertesz, 1979; Sandson, Obler and Albert, 1987). In order to attain the correct name of an object on a naming task, early-stage Alzheimer patients may circumlocute in a manner similar to healthy elderly persons. However, they require more time on average than normal elderly people to retrieve the target word and are less able to utilize phonemic cues to facilitate retrieval (Obler and Albert, 1981). Although verbal paraphasias may be present in the spontaneous speech of Alzheimer patients, in the early stage of the disorder these, where they occur, tend to be self-corrected.

During the mid-stage of Alzheimer's disease, language becomes increasingly paraphasic. Moderately affected Alzheimer patients demonstrate an increasing number of uncorrected verbal and literal paraphasic errors in discourse and in the responses to naming tasks. Neologisms also become more frequent and auditory comprehension ranges from mild to moderately impaired. Overall the language abilities in the mid-stage resembles either a transcortical sensory aphasia (in the absence of a repetition deficit) or a Wernicke's aphasia (in the presence of a repetition deficit). Murdoch *et al.* (1987) found that the language of patients with moderate to moderately-severe dementia of the Alzheimer type resembled that of transcortical sensory aphasia. Cummings *et al.* (1985) also reported that the language disorder in Alzheimer's disease throughout most of its course resembles a transcortical sensory aphasia. However, these latter authors reported that as the disease progresses, repetition abilities deteriorate and the language disorder then resembles a Wernicke's aphasia.

In the late stage of Alzheimer's disease, breakdown in pragmatic function is the primary alteration in language abilities. Sandson *et al.* (1987) described late-stage Alzheimer patients as non-fluent, echolalic, palilalic and perseverative. At the terminal or end stage, the patient may be mute or restricted to echolalia or palilalia and auditory comprehension for spoken language is severely impaired. The severe language impairment at this time is reminiscent of global aphasia.

Patients with Alzheimer's disease do not show an equal degree of impairment across the different types of linguistic knowledge. The semantic and pragmatic language systems appear to be more impaired in dementia than do syntactic and phonological systems (Irigaray, 1973; de Ajuriaguerra and Tissot, 1975; Whitaker, 1976; Schwartz, Marin and Saffran, 1979; Bayles and Boone, 1982; Murdoch *et al.*, 1987) possibly as a consequence of their greater reliance on conscious processing (Bayles, 1985).

(a) Pragmatic language abilities in Alzheimer's disease

The term 'pragmatic language abilities' refers to an individual's ability to use language effectively and represents a speaker's ability to judge contextual effects, perceive and express emotions and use conversational conventions. Pragmatics appears to be the area of linguistic knowledge most dependent upon cognition, which may explain why pragmatic deficits are more apparent than phonologic or syntactic difficulties in Alzheimer patients (Bayles *et al.*, 1982). Although pragmatics is a difficult aspect of language to quantify, a number of researchers have investigated language functions thought to have a pragmatic focus, such as testing the ability of Alzheimer patients to describe, their ability to provide a narrative and their ability to relate context and meaning. Hier *et al.* (1985) suggested that the most obvious linguistic deficit in Alzheimer patients is their failure to use language to convey information. Irigaray (1967) and Obler and Albert (1981) reported that Alzheimer patients exhibit a difficulty with sentence construction and narrative (description) tasks. Such difficulty is manifested by a paucity of commands, questions, indirect requests, second-person pronouns and terms inferring speaker awareness of the truth value of a statement (e.g. perhaps). These findings reflect a breakdown in the use of language as a tool for communication, information conveyance, director of action, and concept and proposition generator (Appell, Kertesz and Fisman, 1982).

As a test of their functional communicative abilities, Murdoch *et al.* (1988) compared the performance of a group of Alzheimer subjects on the Communicative Abilities in Daily Living (CADL) (Holland, 1980) with that of an appropriately constituted control group. They found that the Alzheimer patients were impaired in all areas of functional communication compared to the control group except humour/metaphor/absurdity with the poorest performance being observed in those categories of the CADL more highly dependent on cognition.

(b) Syntactic language abilities in Alzheimer's disease

On the basis of a large scale study, Irigaray (1973) concluded that phonologic and syntactic linguistic codes are comparatively preserved in Alzheimer's disease compared with lexical semantic abilities. Although his observation has been supported by many researchers (including Whitaker, 1976; Schwartz, Marin and Saffran, 1979; Appell, Kertesz and Fisman, 1982), it is now evident that Irigaray's findings may not represent the true extent of the alterations in syntactic abilities that occur in dementia. De Ajuriaguerra and Tissot (1975) concluded that the intellectual regression which occurs in dementia has implications for language, extending beyond the lexical semantic domain. Schwartz, Marin and Saffran (1979) further suggested that it is not uncommon to find reports of syntactic disturbances in the late stages of deteriorating dementia. To the extent that syntax maps logical relations which are beyond the limited competence of the dementia patient, syntactic ability will also be affected.

Similarly, Constantinidis, Richards and de Ajuriaguerra (1978) reported that close examination reveals, in many instances, that syntax does not escape the general disorganization of language production present in Alzheimer patients. Breakdown may occur in the use of phrase markers and grammatical agreement, and sentences and phrases are often left unfinished (aposiopoisis). Critchley (1984) noted that aposiopoisis is a conversation trait manifested by the memory deficits of the aged and, hence, is not unique to patients with Alzheimer's disease. In agreement, Hier, Hagenlocker and Shindler (1985) found no significant difference between the number of sentence fragments provided by Alzheimer and matched control subjects. Furthermore, Obler and Albert (1981) reported that the ability of patients to repeat long and short syntactic sequences depended upon the frequency of occurrence of the semantic (lexical) components included in these sequences.

(c) Semantic language abilities in Alzheimer's disease

Performance on various naming tasks is frequently used to determine the status of a patient's semantic language abilities. Although naming impairment is one of the most commonly reported language deficits associated with Alzheimer's disease (Rochford, 1971; Bayles, 1982; Bayles and Boone, 1982; Bayles and Tomoeda, 1983; Martin and Fedio, 1983; Seltzer and Sherwin, 1983; Kirshner, Webb and Kelly, 1984; Blackburn and Tyrer, 1985; Chui et al., 1985; Hier, Hagenlocker and Shindler, 1985), the cause of the naming deficit is still disputed. Some investigators hypothesize that misnaming in Alzheimer's disease results from misperception of the stimuli, while others attribute failure to linguistic factors such as the erosion of referential boundaries of semantic classes (Schwartz, Marin and Saffran, 1979; Bayles and Tomoeda, 1983). Bayles et al. (1982) and de Ajuriaguerra and Tissot (1975) suggested that a decrease in functional vocabulary and word-finding difficulties are the cause of the poor performance of Alzheimer patients on naming tasks.

The finding that naming is facilitated in Alzheimer patients when the function of an object is demonstrated led to the hypothesis that dementia patients are perceptually impaired (Lawson and Barker, 1968). In agreement with this hypothesis, it was found by Rochford (1971) that the types of naming errors produced by demented patients are consistent with an impairment of visual recognition and relatively unimpaired verbal ability. The naming performance of the demented patients was described by Rochford as verbally fluent, but perceptually off course.

Several studies, however, have failed to support the suggestion that visual perceptual problems may be responsible for the naming difficulty exhibited (Schwartz, Marin and Saffran, 1979; Bayles, 1982; Martin and Fedio, 1983; Kirshner, Webb and Kelly, 1984; Smith, Murdoch and Chenery, 1989). Bayles and Tomoeda (1983) found that, when misnaming occurred, it was most likely to be semantically associated to the target (e.g. 'truck' for 'bus'), many of the error responses provided by Alzheimer patients being both semantically and visually similar to the stimulus (for example, a watch is semantically and visually similar to a clock). These authors suggested that the proportion of error responses related only visually to the target was smaller than would be expected if perceptual impairment caused misnaming, as Rochford (1971) and Lawson and Barker (1968) suggested.

Smith, Murdoch and Chenery (1989) investigated the lexical semantic abilities of a group of Alzheimer subjects using both visual and tactile naming tasks. The results of their study support a semantic network disruption rather than a visual perceptual deficit as the basis of the naming disturbance observed in Alzheimer patients. Based on the pattern of error responses provided in the naming tasks, they concluded that patients with Alzheimer's disease are able to identify the semantic class to which the target in the naming task belongs, but cannot provide the lexeme corresponding to the correct individual class member.

The ability of Alzheimer patients to demonstrate, through gesture, the recognition of objects they cannot name, is further evidence against the interpretation of a perceptual impairment as the basis of the naming impairment (Schwartz, Marin and Saffran, 1979). Further, the observation that naming errors become more semantically unrelated as the dementia worsens does not support Rochford's interpretation. Rather, it seems reasonable to predict that, if the visual signal was degraded, as would occur in a perceptual disorder, the naming errors would be more random and only related semantically to the stimulus by chance.

Martin et al. (1985b) believe that Alzheimer patients display either a loss of, or an inability to utilize, those attributes which serve to distinguish semantically related words. Similarly, Grober et al. (1985) suggested that the saliency of essential attributes may be reduced in Alzheimer's disease so that the more important, referent-defining attributes are considered by Alzheimer patients to be no more important than other less essential attributes. The reduced importance given to the essential attributes in turn changes the organization of semantic information from a set of ordered attributes (ordered with respect to importance of the concept) to a set of more equally weighted attributes, such that the patient may not appreciate the relative importance of the attribute in delineating meaning or for lexical mapping. Such a change in the semantic organization can consequently affect the encoding of new information into both the semantic memory and episodic memory (Grober et al., 1985). Further, any cognitive process into which the disorganized knowledge enters could be profoundly affected. Hence, it is possible that the naming difficulties exhibited by Alzheimer patients may represent semantic deficits causing memory defects rather than the reverse condition.

The loss of semantic information in Alzheimer's disease may represent an extreme case where the weights assigned to specific attributes have become so reduced that they are at a level below that needed for correct identification. Disorganization of the patient's semantic knowledge may result in deficits of lexical access (Bayles and Tomoeda, 1983; Martin and Fedio, 1983; Martin et al., 1985a). Such difficulties may be manifested as circumlocutions, semantic paraphasias or the use of indefinite anaphoric reference (the use of empty words, use of vague high order (superordinate) or generic words as replacements for words identifying specific meanings or entities) (de Ajuriaguerra and Tissot, 1975; Obler and Albert, 1981). Jargon, extremely laconic speech or mutism may also present as manifestations of more profound lexical disturbances (Obler and Albert, 1981).

Schwartz, Marin and Saffran (1979) suggested that, as a result of semantic re-organization in dementia, names no longer specify a unique referent (or class of referents) but rather, a population of referents representing a more global extension of the word's application. Similarly, Martin and Fedio (1983) found that patients with Alzheimer's disease often substitute either a more general, high order (superordinate category) name, or the name of an object from the same semantic category. That is, Alzheimer patients tend to produce semantic field errors involving either a hierarchical relationship or a linear, within-category relationship.

Schwartz, Marin and Saffran (1979) suggested that the lexical loss in Alzheimer's disease is characterized by a systematic and progressive loss of semantic attributes that define referents – the more specific distinguishing attributes being lost before the more general ones. Warrington (1975) provided similar reasoning. Bayles and Tomoeda (1983) concluded that the referential boundaries of items in the mental lexicon appear to erode concurrently with the demise in the patient's ability to abstract and generalize. These latter authors further noted that the more abstract sets of semantic features and intellect appear to deteriorate concurrently and they argued that linguistic impairment, rather than perceptual impairment, better explains the majority of naming errors produced by demented patients.

6.2.2 Language disorders in Pick's disease

Diagnosis of Pick's disease can only be made with certainty following histological examination of the brain tissue at autopsy. Consequently, few reports in the literature have appeared which document the language abilities of patients with Pick's disease. Although in a number of ways Pick's disease and Alzheimer's disease are similar, it appears that the two conditions do differ in the pattern of their associated language impairment.

In some cases of Pick's disease, a language abnormality is the earliest disturbance in intellectual function to appear (Wechsler et al., 1982). Certainly a disturbance in language function appears early in the course of the disease and is a prominent feature of the condition by mid-stage (Cummings and Benson, 1983). Speech, in the initial stage of Pick's disease has been described as slow and deliberate with verbal paraphasias (Holland et al., 1985). By mid-stage, typical language disturbances include anomia, auditory agnosia, excessive use of verbal stereotypes, circumlocutions and echolalia (Cummings and Duchen, 1981; Cummings, 1982; Wechsler *et al.*, 1982; Holland *et al.*, 1985). Speech output at this time is non-fluent (Morris *et al.* 1984; Holland *et al.*, 1985). As the condition progresses, impaired auditory comprehension abilities also become evident and in the terminal stage of the disorder, many patients become completely mute (Gustafson, Hagberg and Ingvar, 1978; Cummings and Duchen, 1981).

Although the language disturbance observed in patients with Pick's disease is similar to that evidenced in Alzheimer patients in terms of their word-finding difficulties and naming impairments, the speech output of the two groups differs markedly in relation to fluency. Non-fluent speech is a feature of Pick's disease at a relatively early stage. In patients with Alzheimer's disease, however, it is not until the late stage of the disorder that speech becomes non-fluent. These similarities and differences in speech–language abilities are easily explained by reference to the neuroanatomical involvement in each condition. As described earlier, in Pick's disease the brain areas which show the greatest histological changes are the pre-frontal areas of the cortex and the superior temporal gyrus. Particularly in the early stages, there is relative sparing of the occipital and parietal region. Consequently, Pick patients exhibit impaired speech output in the form of non-fluent speech while their auditory comprehension abilities remain relatively intact until the later stages. In contrast, the cortical areas most affected in early- to mid-stage Alzheimer's disease include the temporal and parieto-temporal association areas. Damage in the frontal areas tends to be less severe and temporal lobe involvement is usually greater than parietal (Kemper, 1984). Therefore, the neuropathological changes that occur in Alzheimer's disease in particular involve the posterior language area and posterior border-zone region. The transcortical sensory-like and/or Wernicke's-like aphasia observed in Alzheimer patients by a number of authors (Cummings *et al.*, 1985; Murdoch *et al.*, 1987), therefore, could be expected. Loss of fluent speech in late-stage Alzheimer's disease corresponds to further involvement of the frontal lobe at that time. At the terminal stage of the disorder, extensive atrophy to all cortical association areas parallels the cortical damage found in global aphasia (i.e. a large perisylvian lesion involving the frontal, temporal and parietal lobes).

6.3 RELATIONSHIP BETWEEN THE LANGUAGE OF DEMENTIA AND APHASIA

The use of the term 'aphasia' to describe the language deficit occurring in dementia has been questioned by a number of authors (Critchley, 1964; Bayles, 1984). To these authors, aphasia implies the presence of a language deficit in association with a focal brain lesion, the loss in ability to interpret and formulate language being dispro-portionate to impairment in other cognitive functions. Darley (1982) suggested that the term 'aphasia' be applied only in those cases where the aetiology is known and where the patient has a focal and not a diffuse lesion. Further, Bayles *et al.* (1982) believed the term should be used for conditions with an abrupt onset and for non-progressive diseases. Clearly, these criteria would be inconsistent with the use of the term 'aphasia' to describe the language disorder in dementia. It is interesting to ask, however, that if the term was applied with rigid adherence to these criteria, how then would one refer to the language disability manifested, for example, by head injury patients with multiple contusions or by patients with neural abscess or progressive cerebral tumours that have spread to involve the patient's language area?

Although not universally accepted, most definitions of aphasia suggest that language impairment can exist in the absence of other cognitive dysfunction. Kitselman (1981) suggested that the primary difference between language impairment in aphasia and dementia is the degree to which the language impairment occurs in isolation. For most people then, language disturbance in dementia cannot be defined as aphasia because in dementia the language disturbance is embedded within a variety of other cognitive deficits including memory impairment. In aphasia, however, the language disturbance is the primary problem.

Support for considering language disorders in dementia as aphasia syndromes comes from a number of sources. As noted by Joynt (1984), the first descriptions of Alzheimer's disease (1907) and Pick's disease (1892) emphasized the presence of a language disturbance as part of the clinical findings. Since these early reports, a number of other studies have also documented the occurrence of a language deficit in association with dementia (Irigaray, 1973; Obler and Albert, 1981; Appell, Kertesz and Fisman, 1982; Bayles, 1982, 1985; Shore, Overman and Wyatt, 1983; Cummings *et al.*, 1985; Murdoch *et al.*, 1987). Nicolosi, Harryman and Krescheck (1983, p. 11), defined aphasia as 'the inability to speak or to comprehend words arranged in phrases'. Hence they did not suggest any restriction of usage based on characteristics such as onset or lesion site. Based on their definition, it would be appropriate to refer to the language difficulties of demented patients observed in the above studies as aphasia. It is obvious, there-fore, that much of the controversy regarding the relationship between language disturbance in aphasia and dementia is related to ter-minology.

Many authors have compared language deficits in dementia to

aphasia syndromes. Based on the lack of a direct parallel between the degree of dementia and the extent of the language disturbance, Seltzer and Sherwin (1983) reported that the language dysfunction demonstrated by their demented patients reflected a specific language disorder (aphasic disturbance) rather than a non-specific feature of general intellectual deterioration. During the early stages of dementia, it has been reported that the language deficit exhibited by patients with dementia of the Alzheimer type shows considerable similarity to anomic or semantic aphasia (Hier, Hagenlocker and Shindler, 1985). In the later stages of dementia, expressive language and comprehension abilities decline and there is a progressive increase in the emptiness of speech. On a picture description task, the number of prepositional phrases and conciseness of expression decrease while the anomia and number of empty words provided increase. Many researchers have likened these language abilities to Wernicke's aphasia based on the combined presence of comprehension deficit and incoherent but fluent output (Whitaker, 1976; Albert, 1980; Obler and Albert, 1981).

In contrast to these authors, Hier, Hegenlocker and Shindler (1985) and Whitaker (1976) suggested that, if demented patients' repetition abilities were considered in these comparisons, their language abilities would more closely parallel the characteristics of a transcortical (sensory or mixed) rather than Wernicke's aphasia. Therefore, the importance of considering all available data when attempting to categorize, label or compare patients is emphasized. It is suggested that many such comparisons based on minimal behavioural information complicate and distort the relationship between aphasia and dementia syndromes. Finally, in the terminal stages of dementia, researchers have reported a poverty of output, jargon and comprehension abilities reminiscent of global aphasia.

Schwartz, Marin and Saffran (1979) drew attention to the similarities between the overall language pattern exhibited by Alzheimer patients (preserved syntactic and phonologic capacities in the face of marked semantic loss) and transcortical aphasia patients' language abilities, especially those with the mixed sensori-motor type of transcortical aphasia. Similarly, Cummings *et al.* (1985) reported that Alzheimer patients had fluent paraphasic output and impaired auditory comprehension with relative preservation of the ability to repeat, which are characteristics of the transcortical sensory aphasic. They noted that Alzheimer patients performed poorly when executing overlearned speech sequences – which is also evident in many transcortical sensory aphasics – but, in contrast to transcortical sensory aphasics, showed less paraphasia and echolalia. Consequently, Cummings *et al.* (1985) concluded, in agreement with above researchers, that the language abilities of Alzheimer patients in the later stages more closely resembles the characteristics of Wernicke's aphasia. Whitaker (1976) noted a similar relationship between these two syndromes. However, Schwartz, Marin and Saffran (1979) point out that the nature of the semantic loss in transcortical sensory aphasia has not been clearly defined and, therefore, they suggest that Alzheimer patients demonstrate a unique pattern of language alteration which is similar, but not identical to transcortical sensory aphasia. That is, the comparison does not necessarily infer similarity of underlying primary deficits, only of clinical manifestations.

The relationship of the language deficits of early and late stages of Alzheimer's disease to specific aphasic syndromes, however, is not as neat as it may first appear. Early Alzheimer patients have been shown to use fewer total words and prepositional phrases than the anomic aphasic, being more comparable to Wernicke's aphasics in this respect (Hier, Hagenlocker and Shindler, 1985). Similarly, late Alzheimer patients have shown values for mean length of utterance, total words and subordinate clauses which are within the range of the anomic aphasic (Hier, Hagenlocker and Shindler, 1985).

The comparison is further complicated by the clinical observation of significant increase in fluent aphasia (compared with Broca's non-fluent aphasia) in older adults. Obler *et al.* (1978) demonstrated that different aphasia types distribute differently across older adulthood. Broca's aphasia is more prevalent among patients in their early 50s. (mean age 51 years) while Wernicke's aphasia is most prevalent in patients in their early 60s (mean age 63 years). Hence, the similarity between demented patients' language abilities and Wernicke's aphasia may be influenced by the normal ageing effects of these patients' language independent of, or in addition to, the aphasia or dementia syndrome.

CHAPTER SEVEN

Neurological disturbances associated with aphasia

A number of neurological disturbances are commonly encountered in patients with aphasia. These associated neurological disturbances include apraxia, agnosia, alexia, agraphia and Gerstmann syndrome. Although each of these disturbances may occur in isolation, they are most often observed as part of an aphasic disturbance. Where they do occur as part of an aphasia, the associated neurological disorders complicate the clinical features of the aphasia and therefore have important implications for the evaluation and treatment of aphasic patients.

7.1 APRAXIA

Apraxia is a disorder of learned movement in which the patient is unable to carry out, at will, a complex or skilled movement that they were previously able to perform, where that inability is not due to paralysis of the muscles, ataxia, sensory loss, comprehension deficits or inattention to commands (Geschwind, 1975). Although both developmental and acquired varieties of apraxia have been identified, this chapter will describe only the acquired forms of the disorder.

Three major types of apraxia have been identified in the literature. These include ideomotor apraxia, ideational apraxia and limb-kinetic apraxia. In addition, a number of special types of apraxia are also recognized by some authors, including constructional apraxia, dressing apraxia, verbal apraxia etc. It has been suggested by some authors (e.g. Benson, 1979), however, that many of these special types of apraxia represent either fixed motor or visuo-spatial disturbances and, therefore, cannot strictly be defined as apraxias.

7.1.1 Ideomotor apraxia

Ideomotor apraxia is the most common type of apraxia. It is essentially a disorder in which language alone is unable to initiate and direct correctly the performance of certain learned motor tasks that the patient is otherwise able to perform under different circumstances and different sensory inputs. In other words, the patient is unable to carry out motor acts to verbal command but can carry out the same motor tasks at a reflex or automatic level. For example, the individual involved may be unable to carry out various facial expressions, such as a smile or frown, in response to a verbal command and yet show normal facial expressions in spontaneous situations.

The disorder in movement in ideomotor apraxia cannot be accounted for by muscular weakness, disturbances in tone or posture, incoordination, sensory loss or incomprehension of the verbal command. Further, the disorder may be isolated to particular muscle groups such as the muscles in the bucco-facial region (oral apraxia) or respiratory muscles, or it can involve the limbs (limb apraxia) either bilaterally or the non-dominant (left) limb unilaterally, depending on the location of the lesion.

Oral apraxia (bucco-facial apraxia) was first described by Jackson (1932) in a patient with aphasia who could not protrude his tongue when asked to, but managed quite well to recover with his tongue a breadcrumb which remained stuck to his lip. Depending on the degree of severity, oral apraxia can interfere with a variety of voluntary non-speech movements of the tongue, pharynx and cheeks. For example, patients may be unable to move their tongues toward the corners of the mouth, the nose, the chin, etc. when requested. They may also have trouble moving the tongue to make clicking sounds or the apico-dental 'tsk-tsk'. Some patients with oral apraxia have problems whistling when requested or mimicking laughter, etc. Oral apraxia is commonly associated with Broca's aphasia.

According to Benson (1979), ideomotor apraxia can be demonstrated in as many as 40% of aphasic patients if correctly tested. Apraxia of both bucco-facial and limb activities is seen frequently in Broca's aphasia, transcortical motor aphasia and conduction aphasia. Although clinically the presence of ideomotor apraxia in patients with aphasia is often overlooked, it is important that speech–language pathologists recognize that this finding has important implications for the evaluation and treatment of their aphasic patients. Of very real

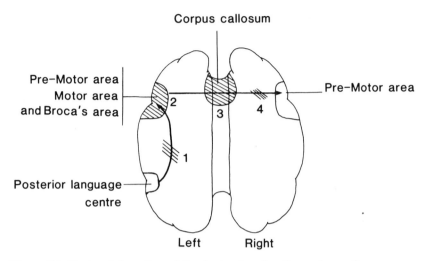

Figure 7.1 Horizontal section of the brain showing the major pathways thought to be involved in the performance of motor acts to strictly verbal commands. 1. Lesion involving the arcuate fasciculus – oral and bilateral limb apraxia. 2. Lesion involving the pre-motor area, motor area and Broca's area – oral and left limb apraxia. 3. Lesion involving the anterior portion of the corpus callosum – left limb apraxia. 4. Lesion in the sub-cortical white matter of the right hemisphere – left limb apraxia.

importance is the interference that apraxia produces in the testing of comprehension, particularly if comprehension is tested only through motor activities. For instance, ideomotor apraxia is recognized as another factor which may contribute to a patient's failure of the Token Test. On the other hand, a comprehension defect must be ruled out before an apraxia can be diagnosed.

Ideomotor apraxia is caused by an interruption of pathways between the centre for formulation of a motor act and the motor areas necessary for its execution. These pathways are depicted schematically in Figure 7.1. Briefly, the verbal command is initially comprehended and interpreted in the posterior language centre and a response formulated. Information is then passed to the pre-motor cortex of the left hemisphere via the association fibres, including the arcuate fasciculus. In the pre-motor cortex, a programme is developed which determines the sequence of muscle contractions required to execute the command. Once developed, information regarding the programme is conveyed to the primary motor cortex of the left hemisphere which in turn initiates the necessary nerve impulses that pass via the descending motor pathways to the muscles that execute the command. At the same time that information is passed to the primary motor cortex of the dominant hemisphere, the same information must also be directed to the motor areas of the non-dominant hemisphere so that the muscles on the left side of the body (e.g. muscles of left hand) can perform the skilled movement commanded if required. This transfer of information from the pre-motor cortex of the dominant hemisphere to the motor areas of the non-dominant hemisphere occurs through the anterior part of the corpus callosum.

Ideomotor apraxia can result from damage at any point along the pathways depicted in Figure 7.1. Depending on the point of disruption, however, the nature of the apraxic disturbance will vary. In the case of a lesion that involves the left inferior parietal lobe, particularly in the area of the supra-marginal gyrus, it is likely that the arcuate fasciculus will be involved and a conduction aphasia will result (see Chapter 2). In addition, however, in most cases the patient will also have an ideomotor apraxia that involves all four limbs and the bucco-facial muscles, since neither the right or left pyramidal motor system is able to receive information from the posterior language area. Depending on the extent of the lesion, in some cases the limbs are not affected or are only minimally involved.

A lesion involving the anterior language centre and primary motor strip of the left hemisphere, as may occur in cases of Broca's aphasia, precludes apraxia testing in the right limbs due to the presence of a concomitant right hemiplegia. In this instance, the patient exhibits a bucco-facial and left-limb apraxia, since the right motor system is unable to receive information from the posterior language centre. On the other hand, a lesion in the anterior corpus callosum disconnects the right pre-motor and pyramidal motor cortices from the left hemisphere, thereby rendering the patient apraxic in the left limbs (callosal apraxia). Since the nuclei of the cranial nerves (with the exception of the nucleus supplying the lower part of the face) receive bilateral upper motor neurone innervation (see Chapter 8), the patient in this case should not exhibit bucco-facial apraxia. Lesions located in the sub-cortical white matter of the right hemisphere which disconnect the fibres which have passed through the corpus callosum may also cause left-limb apraxia.

The motor performance of patients with ideomotor apraxia usually improves when they are asked to imitate the clinician's movements. Even so, the improvement in motor performance is usually only partial. In cases where the movement involves the use of an object (e.g. hammering), motor performances also usually improves if the patient is given the actual object to use. In these circumstances, the apraxic patient is able to by-pass their language-motor disconnections through intact visual and somesthetic connections to the pyramidal motor cortices.

Patients with ideomotor apraxia do not present with any particular difficulties in performing spontaneous movements in everyday life. Consequently, the disorder does not produce any symptoms as such. Generally, the defective bucco-facial and limb movements are only disclosed by specific testing.

7.1.2 Ideational apraxia

Ideational apraxia is a disorder of gestural behaviour which involves the loss of ability to formulate the ideational plan for the execution of the components of a complex act (Liepmann, 1908). In this disorder, the individual component movements of a complex act cannot be synthesized into a purposeful plan, even though the simple isolated component movements may be performed normally. In some cases the individual component movements may be performed in a faulty sequence or portions only of the complex act may be performed. Occasionally the patient with ideational apraxia may perform an action similar to, but not the same as, the one required. For example, patients affected with such an apraxia are incapable of carrying out a task such as making a cup of coffee, lighting a candle with a match, etc. One example described by Miller (1986) was that of a woman who, when attempting to light her gas stove, first lit the match, then blew it out and then turned on the gas which she left going. According to Miller she was only prevented from either gassing herself or causing an explosion by the intervention of a neighbour. On another occasion the same woman, again while lighting the gas stove, turned on the gas, then filled the kettle and finally struck a match causing a minor explosion. Clearly in this case, although the woman was capable of performing each of the partial acts comprising the overall action, their sequencing and planification were deranged.

Unlike patients with ideomotor dyspraxia, where the disorder is not evident in everyday life, ideational apraxics stand out as being abnormal even in everyday activities. Clearly, due to the poor sequencing of movements in a complex act, there is often a great potential for serious injury to be inflicted (e.g. the gas may be turned on prior to filling the kettle when making a cup of coffee; machinery may be started before safety procedures are checked, etc). Patients with ideational apraxia are able to imitate an action in that, in such situations, the plan of the action is provided to them from outside.

Ideational apraxia is always manifested bilaterally and rarely occurs as an isolated phenomenon. Clinically it is most often observed as a manifestation of diffuse brain disorders (e.g. cerebral arteriosclerosis) although in some cases it has been reported secondary to focal brain lesions (Poeck and Lehmkuhl, 1980). Ideational apraxia may co-occur with other apraxias (especially ideomotor apraxia) making it difficult to distinguish clinically and is most often seen by speech–language pathologists as part of either a severe aphasia or dementia syndrome.

7.1.3 Limb-kinetic apraxia

Limb-kinetic apraxia (also called kinetic apraxia) was first described by Liepmann (1908) and is characterized by an inability to execute fine acquired motor movements. The disturbance is often confined to one limb and in most cases affects the finer movements of the distal portions of one upper extremity. Furthermore, the impairment is present in automatic as well as volitional acts. Movements affected include those required for tasks such as writing, doing up buttons, sewing, playing musical instruments, opening a safety pin, etc.

Commonly seen as part of a Broca's aphasia, limb-kinetic apraxia is associated with lesions in the pre-motor area of the cerebral cortex. Consequently, it manifests unilaterally on the side contralateral to the corresponding lesion. It may co-occur with an ideomotor apraxia.

Liepmann (1908) suggested that limb-kinetic apraxia results from the loss of the so-called kinetic memories for a single limb. More recently, however, several authors have excluded it from the true apraxias (Kerschensteiner, Poeck and Lehmkuhl, 1975). Most contemporary authors regard it as a pyramidal movement disorder rather than an apraxia.

7.1.4 Constructional apraxia

In this type of apraxia, first described by Kleist (1922), the patient is unable to form a construction in space. Miller (1986, p. 56), defined constructional apraxia as 'a disorder of planned movements for any kind of task involving the structuring or arranging of objects, parts of objects, or lines in two and three dimensional space'. Patients with this disorder, therefore, are unable to copy simple geometric figures by drawing or by the arrangement of blocks, matches, etc. In the more severe cases, the patients are not even able to draw such simple geometric figures as squares, circles and crosses. In general, patients are not aware of their inability to perceive spatial relationships.

In the majority of cases, constructional apraxia is associated with lesions in either of the parietal lobes, although constructional deficits have occasionally been described secondary to frontal-lobe lesions. As an isolated disorder, it is more common following lesions of the non-dominant (right) hemisphere, in which case the disturbances associated with this condition also tend to be more severe. For instance, individuals with lesions in the right parietal lobe tend to produce drawings which, although elaborate, tend to be done hastily and without care. In addition, the presence of a model is of no help and disorientation on the page is marked and neglect of the left-hand side of the page is regularly observed. Further, the axes of the figures tend to be grossly disturbed.

In contrast, in those cases where the lesion involves the dominant (left) hemisphere, the individual involved tends to produce drawings which are carefully done but poor in content. With the aid of a model, however, the diagram may be more elaborate. In addition the axes of the figures are more truly represented than in the drawings of patients with right parietal lesions. Neglect of the right-hand side of the page occurs without exception. Frequently there may be (associated with a constructional apraxia due to a lesion of the dominant hemisphere, in addition to an apraxia), all or some of the elements of a Gerstmann syndrome (see below).

Constructional apraxia is of importance to the speech–language pathologist in that, when it co-occurs with an aphasia, it has important implications for the assessment and clinical management of the aphasic patient. For instance, the presence of a constructional apraxia affects the performance of the aphasic individual on tests used by speech–language pathologists and related professionals and which require use of the patient's constructional abilities. For example, performance of the patient on the supplementary non-language tests of the Boston Diagnostic Aphasia Examination (Goodglass and Kaplan, 1983), which include construction of three-dimensional block designs, tests of drawing to command and stick constructions, could reasonably be expected to be influenced by the presence of a constructional apraxia.

7.1.5 Dressing apraxia

Dressing apraxia is a disturbance in which patients cannot dress because they are unable to relate the parts of the garment to the parts of their bodies. This disturbance is seen most frequently in association with lesions in the parietal lobe of the non-dominant (right) hemisphere although it has also been reported to follow left parietal lesions. Dressing apraxia is often accompanied by a constructional apraxia and may co-occur with aphasia. In some cases the dressing apraxia is further complicated by the presence of a hemiplegia, hemianaesthesia, hemineglect and visual-field defects.

7.2 APRAXIA OF SPEECH

7.2.1 Clinical characteristics of apraxia of speech

Darley (1982, p. 10) described apraxia of speech (verbal apraxia) as 'a disorder in which the patient has trouble speaking because of a cerebral lesion that prevents his executing voluntarily and on command the complex motor activities involved in speaking, despite the fact that muscle strength is undiminished'.

As indicated in Chapter 1, apraxia of speech is a disorder of motor speech programming manifested primarily by errors in articulation and secondarily by what are thought by many researchers to be compensatory alterations of prosody (e.g. pauses, slow speech rate, equalization of stress, etc.). Articulation errors, therefore, are the primary features of this motor speech disorder. As they speak, patients with apraxia of speech struggle to correctly position their articulators. As they struggle, affected individuals appear to visibly and audibly grope to achieve the correct individual articulatory postures and sequence of articulatory postures to produce sounds and words. The resulting articulation, however, is frequently off-target. As patients with apraxia of speech are aware of their articulatory mistakes, however, they usually attempt to correct them. Often these attempted corrections are also erroneous, but importantly are not always the same as the original error in articulation. In fact, on a series of trials, the articulatory errors exhibited by individuals with apraxia of speech are highly variable. For example the initial consonant v may, at different times, be produced by the same individual as v, z, p, f, r, b, h and w. Although on occasions patients with apraxia of speech are able to produce all phonemes correctly, at times the substitution of incorrect phonemes (literal paraphasias) as indicated, occurs frequently and inconsistently. It should be noted that, by using broad phonetic transcriptions to record what was heard with the naked ear, the majority of researchers who have reported the perceptual features of apraxia of speech have suggested that articulatory errors seen in these cases involve substitutions rather than the distortion of individual phonemes as occurs in dysarthria (see Chapters 8, 9 and 10) (Johns and Darley, 1970; Trost and Canter, 1974; LaPointe and Johns, 1975). More recent studies using acoustic analysis, however, have disputed this finding (Wertz, LaPointe and Rosenbek, 1984). Further, the number of articulatory errors exhibited by patients with apraxia of speech is greater during repetition than during conversational speech (Rosenbek, Kent and LaPointe, 1984). Consequently, apraxia of speech is most clearly illustrated when the patient is asked to repeat spoken language.

The number of articulatory errors produced by individuals with apraxia of speech also increases as the complexity of the articulatory exercise increases. Fewer errors, for instance, occur in the production of vowels than during the production of single consonants, with the greatest number of articulatory errors occurring during the production of consonant clusters (Trost and Canter, 1974; LaPointe and Johns, 1975; Burns and Canter, 1977; Dunlop and Marquardt, 1977; Wertz, LaPointe and Rosenbek, 1984). Repetition of a single consonant such as *puh, tuh* or *kuh* is ordinarily more easily accomplished by these patients than the repetition of the sequence *puh-tuh-kuh* (Rosenbek, Wertz and Darley, 1973). Further, initial consonant phonemes tend to be misarticulated more often than final consonant phonemes (Johns and Darley, 1970; Trost and Canter, 1974).

Some authors have reported that palatal and dental phonemes are significantly more susceptible to misarticulation than other phonemes in patients with apraxia of speech (LaPointe and Johns, 1975). Also phonemes occurring with a relatively high frequency in the English language tend to be more accurately articulated than phonemes which occur less often (Trost and Canter, 1974; Wertz, LaPointe and Rosenbek, 1984). The number of articulatory errors produced by these patients has also been found to increase as the length of the word increases. For example, as these patients produce a series of words with an increasing number of syllables (e.g. door, doorknob, doorkeeper, dormitory) more errors occur during production of the longer words (Johns and Darley, 1970).

In addition to the impairment in articulation, the prosodic features of the speech of patients with apraxia of speech are also disturbed. The prosodic deficits occur as a result of the fact that as these patients speak, they slow down their rate of speech, space their words and syllables more evenly and stress each of them more equally in an attempt to avoid articulatory errors (Darley, Aronson and Brown, 1975a). The non-fluent speech output observed in these patients is primarily caused by the pausing and hesitating that occurs as the person gropes for articulatory placement and makes repeated efforts to produce words correctly.

Patients with apraxia of speech display a marked discrepancy between their relatively good performance at automatic and reactive speech productions and their relatively poor volitional–purposive speech performance. These patients have been reported to sound normal when producing words and phrases that are well known to them through either practice or usage (Schuell, Jenkins and Jimenez-Pabon, 1964). For instance, on those occasions where these patients are either speaking off-hand or reciting an over-learned expression or reacting to a sudden stimulus, they may produce words without articulatory inaccuracy or groping (e.g. when counting, swearing, repeating rhymes and jingles, etc. and uttering greetings and farewells). In contrast, these same patients exhibit effortful and off-target groping in spontaneous speech when they are required to select a particular target word. Deal (1974) reported that patients with apraxia of speech exhibit both a consistency effect and an adaptation effect when required to read the same material repeatedly. Although during the performance of such a task these patients tend to produce their articulation errors in the same place from trial to trial, they also tend to make fewer errors on successive readings.

7.2.2 Differentiation of apraxia of speech from aphasia

Although in a small number of cases apraxia of speech can occur in a relatively pure form, in most instances it is observed clinically as part of an aphasia syndrome, particularly either a Broca's or conduction aphasia. Apraxia of speech, when it occurs without aphasia, is a uni-modality disorder, affecting speech out of proportion to other language modalities (Halpern, Darley and Brown, 1973). Tests of performance in various language modalities (i.e. reading, writing and speaking) reveal that in these patients, speaking performance is significantly worse than performance in listening, reading and writing. On the other hand, in aphasic patients a more widespread disturbance of language function is seen, disturbances being evident in reading and writing as well as speaking. It was his observation of disproportionate impairment of speaking in his two patients that led Broca (1861) to call this disorder 'aphemia' in an attempt to distinguish it from what he called 'verbal amnesia' in which there was considered to be a general impairment of language function.

According to Mohr (1976), although apraxia of speech has a number of features in common with Broca's aphasia, it is associated with a lesser degree of comprehension loss (there may be none demonstrable) and agrammatism. In addition, as indicated in Chapter 2, Mohr et al. (1975) have shown that apraxia of speech is associated with lesions restricted to Broca's area, whereas Broca's aphasia is produced by more extensive lesions which usually involve the territory of the upper division of the middle cerebral artery.

Patients with apraxia of speech retain the ability to process meaningful linguistic units and although they may have difficulty in articulating a particular word, it can be easily demonstrated that the articulation problem is not related to a word-finding difficulty. For instance, the patient may be able to write the word that they have difficulty articulating. In addition, they are able to choose the word in question from among a group of words when given a choice. Halpern *et al.* (1973) found that patients with apraxia of speech performed significantly better than aphasics on auditory retention and naming tasks. These same authors also reported that apraxic patients were better than aphasics on writing to dictation and syntax and fluency tasks. Apraxia of speech, therefore, is an independent and separate disorder from aphasia (Bay, 1964; Darley, 1975) being a motor speech disorder rather than a language disorder.

In addition to occurring concurrently with an aphasic syndrome, apraxia of speech may also occur in association with an oral apraxia (i.e. an apraxia involving the muscles of the bucco-facial area in non-speech movements). Although apraxia of speech may occur independently of oral apraxia, when an oral apraxia is demonstrated in patients, an apraxia of speech is also usually present.

7.2.3 Differentiation of apraxia of speech from dysarthria

Apraxia of speech and dysarthria are both motor speech disorders but each represents a breakdown at a different level of speech production. Apraxics, on neurological examination, show no significant evidence of slowness, weakness incoordination, paralysis or alteration of tone of the muscles of the speech mechanism that can account for the associated speech disturbance. Dysarthric patients, on the other hand, dependent on the type of dysarthria present, may exhibit either hyper- or hypo-tonus of the speech muscles, ataxia, a restricted range of movement, etc. (see Chapters 8, 9 and 10). Whereas dysarthric patients show variable disturbances in all of the basic motor processes which underlie speech production including respiration, phonation, resonation, articulation and prosody, in apraxia of speech the continuing impairment is specifically articulatory, with prosodic alterations at times occurring as compensatory phenomena.

One further difference between apraxia of speech and dysarthria lies in the nature of the articulatory errors seen in each disorder. In dysarthria the articulatory errors are characteristically errors of simplification (e.g. distortions or omissions) whereas in apraxia of speech the articulatory errors largely take the form of complications of speech (e.g. substitution of one phoneme for another, additions of phonemes, repetition of phonemes and prolongations of phonemes).

7.3 ALEXIA AND AGRAPHIA

Alexia and agraphia represent two separate disorders of written language. In general, disorders encountered in written language parallel those of the oral modalities in brain-damaged patients (Assal, Buttet and Jolivet, 1981). Dissociations between oral and written language disorder, however, have been reported and reading and writing disturbances, although usually encountered as part of an aphasia syndrome, can occur as isolated language deficits (Hier and Mohr, 1977). When it occurs in isolation, a reading disturbance caused by brain damage is referred to as an 'alexia'. On the other hand, an isolated inability to write, as a result of brain damage, is called an 'agraphia'.

Four kinds of alexia are commonly described in the literature including: alexia without agraphia; alexia with agraphia; frontal alexia; and deep alexia.

7.3.1 Alexia without agraphia

Alexia without agraphia is a rare type of alexia which also goes by the names of occipital alexia, pure alexia, agnosic alexia and pure word blindness. The disorder is characterized in most cases by a severe reading disorder contrasted with preserved writing abilities. Patients with this disorder are unable to read aloud or cannot comprehend what they read but retain the ability to name seen objects, write spontaneously and copy written material slowly and slavishly despite being unable to read what they have written. These patients are better able to read letters than words, and letters in isolation better than letter strings. In some cases, the affected individual is able to understand commonly written words such as the name of the city where they live, their own name and commonly used language symbols. A colour-naming difficulty involving both an inability to name seen colours or match colours to a given name in the presence of an ability to sort and match colours normally and to use colour names in conversation is a variable finding in patients with alexia without agraphia.

Dejerine (1892) proposed that alexia without agraphia is caused by an inability to transfer visually perceived language stimuli from the primary visual areas to an area in the left hemisphere involved in the interpretation of visual language. This latter area is thought to be located in the angular gyrus of the left cerebral hemisphere. The underlying brain lesion in cases of alexia without agraphia usually involves the left medial occipital lobe (including the left primary visual cortex – Brodmann area 17) and either the splenium of the corpus callosum or the sub-cortical white matter adjacent and inferior to the splenium (Dejerine, 1891; Vignolo, 1983) (see Figure 7.2). The usual pathology associated with the condition is a cerebrovascular accident

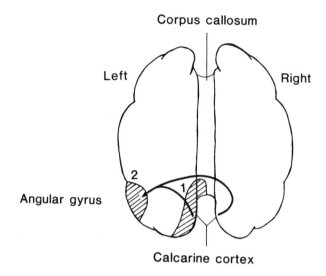

Figure 7.2 Horizontal section of the brain showing the major pathways connecting the primary visual cortex to the angular gyrus. 1. Lesion involving the left primary visual cortex and splenium of the corpus callosum – alexia without agraphia. 2. Lesion involving the angular gyrus of the dominant hemisphere – alexia with agraphia.

involving occlusion of the posterior cerebral artery; however, the syndrome has also been described in association with tumours and arterio-venous malformations (Benson, 1979).

The lesion, as shown in Figure 7.2 disconnects the right posterior occipital lobe from communication with the left hemisphere via the splenium. In addition, the damage to the left primary visual cortex produces a right hemianopsia (defective vision or blindness in half of the visual field). The left visual cortex monitors events in the right visual field and hence, when damaged, as in patients with alexia without agraphia, is associated with the loss of the right visual field in each eye. As the right primary visual cortex is not involved in the lesion, however, the patients are able to see in their left visual field. Due to the involvement of the splenium, however, visual information perceived in the right primary visual cortex cannot be passed to the area in the left cerebral hemisphere involved in the interpretation of visual language (i.e. the angular gyrus). Consequently, even though patients are able to see written words with their left visual field, they are unable to comprehend the meaning of the words due to the disconnection between the right visual cortex and the left angular gyrus. Since the lesion does not involve the central or peri-central speech areas, no aphasic defects in spoken language occur and hence the patient with alexia without agraphia is able to spell words and recognize verbally-spelled words. These patients can also recognize individual letters when they are drawn in the palm or when they are palpated from embossed blocks. Object naming abilities are also preserved and, oddly enough, most of these patients are able to read arabic numbers (Brown, 1974). However, these patients are unable to read sentences and passages they are able to write spontaneously unless they have been memorized.

Although the course of alexia without agraphia is variable, many patients show a slow and persistent improvement. Reading ability, however, rarely returns to normal.

7.3.2 Alexia with agraphia

Alexia with agraphia, also called 'parietal alexia' or 'parietal–temporal alexia' was first described by Dejerine (1891). His case manifested an almost total inability to read, limited writing ability, a mild aphasia and a calculation disturbance (acalculia). The loss of the ability to read is almost invariably severe in cases of alexia with agraphia, both the ability to read out loud and the ability to comprehend written language being disturbed. In addition, the reading disturbance often includes words, letters, numbers and musical notes, although the degree of impairment in the latter two categories varies from case to case (Benson, 1979). Unlike the majority of patients with alexia without agraphia, individuals with alexia with agraphia are unable to comprehend words spelled aloud (Brown, 1974). Likewise they are unable to understand symbols such as letters and numbers presented via tactile means.

As suggested by the name of the condition, a writing disturbance is an important component of alexia with agraphia. In some cases it appears that writing is less disturbed than reading while in others the reading and writing disorders are roughly equivalent in terms of severity. Usually, however, the writing disturbance is not so severe that the patient is unable to produce written letters. The writing is, however, invariably paragraphic, consisting of recognizable letters combined into unrecognizable words. Although patients with alexia with agraphia can produce written letters spontaneously, unlike persons with alexia without agraphia they are unable to write to dictation or at least produce a paragraphic output in this situation. Also in contrast to patients with alexia without agraphia, these patients show a significant disturbance in their ability to copy written material. While patients with alexia without agraphia copy slowly and slavishly but correctly, persons with alexia with agraphia usually make mistakes in reproducing the actual form of the letters.

Dejerine (1891) localized the neuropathology in alexia with agraphia to the angular gyrus of the dominant (left) cerebral hemisphere. Although a variety of aetiologies, including cerebrovascular accidents, neoplastic disease and head trauma, may produce this syndrome, reports in the literature suggest that involvement of the dominant angular gyrus is a constant finding (Benson and Geschwind, 1969). The most common aetiology is a cerebrovascular accident which, in the majority of cases, involves occlusion of the angular branch of the left middle cerebral artery (Benson, 1979).

As the lesion associated with alexia with agraphia involves the posterior language centre, this syndrome is usually but not always accompanied by some form of aphasia. Often the associated aphasic disturbance takes the form of an anomic aphasia; however, on occasions a Wernicke's aphasia is present in these cases. In addition to aphasia, a number of patients with alexia with agraphia exhibit either a full or partial Gerstmann syndrome (see below). A visual-field defect in the form of either a hemianopsia or quadrantanopsia, although not an invariable finding, is present in some cases.

The prognosis for recovery from alexia with agraphia varies depending on the underlying aetiology and size of the brain lesion. In some cases a relatively rapid recovery of some or all reading skills occurs. The majority of these patients, however, show only a partial recovery.

7.3.3 Frontal alexia

Benson (1977) described a third kind of alexia which is present in individuals with frontal-lobe pathology. Called 'frontal alexia', this syndrome differs in a number of ways from the two classical alexias first documented by Dejerine which were described above. According to Benson (1979), the majority of patients with frontal alexia are able to comprehend some written material. In particular, these patients can usually read aloud and comprehend some individual words, especially nouns or action verbs, within a written sentence. As an example, Benson (1979) indicated that patients with frontal alexia in some cases are able to decipher newspaper headlines but are unable to understand the sentences that make up the article.

Although patients with frontal alexia are able to read some words, they have difficulty in reading (naming) individual letters within the words they can read. In other words, in contrast to patients with alexia without agraphia, these patients exhibit a literal alexia rather than a verbal alexia. Further, individuals with frontal alexia are unable to recognize most words spelled aloud, the few words they can recognize being primarily substantive words.

Frontal alexia is usually accompanied by a non-fluent aphasia usually in the form of a Broca's aphasia. A concomitant right hemiplegia or hemiparesis is also usually present and sensory changes and visual-field deficits are evidenced by some individuals with this disorder. Frontal alexics almost always demonstrate a severe writing disturbance. Their writing in most cases consists of poorly formed letters and spelling errors are common. Their ability to copy written material is also impaired.

The lesion underlying frontal alexia involves the frontal lobe of the dominant (left) hemisphere. Benson (1979, p. 116) describes the lesion as involving 'the posterior portion of the inferior frontal gyrus with extension into the subcortical tissues in the anterior insula'. A cerebrovascular accident usually forms the underlying aetiology in most cases of frontal alexia.

7.3.4 Deep alexia

The concept of deep alexia (also called 'phonemic alexia') has arisen from research carried out since the late 1960s that has investigated alexia from either a linguistic or psycholinguistic point of view (Marshall and Newcombe, 1966, 1973). Deep alexia is characterized by the presence of semantic errors when the patient is reading aloud. The three major symptoms include: the presence of semantic substitution errors (e.g. 'child' being read as 'girl'; 'wed' as 'marry'; 'listen' as 'quiet', etc.); the presence of derivational errors (e.g. 'direction' being read as 'directing'); and the omission or misreading of grammatical morphemes during reading aloud (e.g. 'an' being read instead of 'you').

It appears that patients with deep alexia, when reading aloud, pass directly to the semantic value of a word, deriving a semantic impression from its printed form without appreciating the sound of the word in question (Kaplan and Goodglass, 1981). Consequently, the patients may produce a spoken word that is not exactly the equivalent of the written word leading to semantic substitution errors. Likewise, as there is no semantic referent for grammatical morphemes, these cannot be dealt with by this information-processing route thereby leading to their omission, or at least, misreading. According to Kaplan and Goodglass (1981) the two problems central to deep dyslexia are that patients with this disorder are first unable to use graphophonemic recoding and secondly are unable to access the phonology of a word directly. In other words, these patients are no longer able to use a phonological reading system.

Components of a deep dyslexia may be observed in most types of aphasias. Most frequently, however, it is seen in association with a Broca's aphasia with agrammatism (Kaplan and Goodglass, 1981).

7.3.5 Agraphia

The various types of agraphia are largely delineated by their associations with other neuropsychological disorders. Consequently, the types of agraphia described in the literature include aphasic agraphia (agraphia with Broca's, Wernicke's and conduction aphasia, etc.), agraphia with alexia, constructional agraphia (agraphia with visuospatial disability), apraxic agraphia (agraphia associated with limb apraxia) and pure agraphia (agraphia in the absence of other neuropsychological deficits).

Pure agraphia, involving the selective impairment of written communication in isolation of other language disorders, is a condition that is only rarely described in the literature. In fact, the rarity of pure agraphia has led some authors to express doubts regarding the existence of this disorder as an autonomous entity (Kreindler and Fradis, 1968). Basso, Taborelli and Vignolo (1978) reported that of 500 brain-damaged patients that they examined, only 2 cases exhibited a pure agraphia. One other case of pure agraphia was reported by Rosati and de Bastiani (1979) and a further case by Miceli, Silveri and Caramazza (1985). Chedru and Geschwind (1972) described the occurrence of pure agraphia in association with confusional states (e.g. dementia) and suggested that this mechanism may underlie the majority of cases of isolated agraphia reported in the literature to that time. Some reports of pure agraphia occurring secondary to focal brain lesions have also been reported in the literature (Basso, Taborelli and Vignolo, 1978; Rosati and de Bastiani, 1979; Miceli, Silveri and Caramazza, 1985).

Pure agraphia has been described in patients with lesions in multiple cerebral sites, including the second frontal convolution (Sinico, 1926; Hecaen and Albert, 1978; Kaplan and Goodglass, 1981), the superior parietal lobule (Basso, Taborelli and Vignolo, 1978), the posterior peri-sylvian region (Rosati and de Bastiani, 1979) and the region of the left caudate and internal capsule (Laine and Mattila, 1981).

Rather than as an isolated disorder, agraphia most commonly occurs as part of an aphasia syndrome (aphasic agraphia). A writing disturbance in one form or another occurs in each of the major clinically recognizable types of aphasia. (The nature of the writing disorder seen in each aphasia type is described in Chapter 2.) Consequently the aetiology of agraphia is similar to that of aphasia. In general, any brain disorder that affects the dominant hemisphere has the potential to cause agraphia, the only prerequisite being that the patient is able to write prior to becoming brain damaged. Therefore, cerebrovascular disorders of the brain, head trauma, tumours, encephalitis and atrophic brain disorders can all cause agraphia.

7.4 AGNOSIA

Agnosia refers to an inability to recognize familiar objects perceived by the senses due to brain damage. It is important to note that the disorder of recognition in this condition does not result from sensory loss (e.g. visual-field defects; hemianaesthesia, etc), mental deterioration, disorders of consciousness or attention or a non-familiarity with the object. Prior to a diagnosis of agnosia being made, therefore, it is necessary that the presence of these factors be ruled out. Agnosia usually affects recognition through only one sensory modality so that an object that cannot be recognized by one sensory route (e.g. vision) can be recognized by a different sensory system (e.g. touch).

The occurrence of agnosia is thought to be associated with lesions that involve the sensory association areas of the cerebral cortex. The sensory pathways connecting the sensory receptors to the cerebral cortex themselves remain intact as does the primary receptive area for the particular sensory modality involved. According to Geschwind (1965a,b), the various agnosias represent isolated naming disturbances resulting from lesions that disconnect the perceptual recognition areas of the right or both hemispheres from the speech–language areas of the left hemisphere.

The three major types of agnosia are delineated according to the sensory modality involved. These include visual agnosia, auditory agnosia and tactile agnosia.

7.4.1 Visual agnosia

Visual agnosia is a rare condition characterized by an inability to visually recognize, describe or name objects. In addition, patients are unable to locate visually an object in their immediate environment when that object is named. Patients with visual agnosia are also unable to choose the correct name of a seen object if provided with a list of alternative names. Visual agnosics, however, are able to identify objects through other sensory modalities, a feature which distinguishes the naming deficit in this condition from that observed in anomic aphasia. In anomic aphasia, the naming of objects is affected in all sensory modalities. Also, in contrast with anomic aphasics, patients with visual agnosia not only fail to recognize seen objects, they are also unable to describe what the object is used for. Anomic aphasics, on the other hand, although unable to name objects, can often describe the use of the object.

Visual agnosias may be specific for objects (visual object agnosia), colours (colour agnosia), faces (prosopagnosia), geometric figures and pictures (agnosia for images) plus a number of other factors. In general, the lesions associated with visual agnosia tend to be extensive and in most reported cases involve the occipital lobes bilaterally with the temporal and/or parietal lobe also being involved in some patients. Reports of visual agnosia occurring subsequent to unilateral lesions in either the right or left hemisphere have also appeared in the literature.

7.4.2 Auditory agnosia

In this condition, although hearing is not impaired, the patient is unable to recognize or distinguish various sounds which are heard. For instance, provided no visual or tactile stimuli are provided, the patient cannot recognize familiar noises such as the jingling of coins or keys. In some cases this inability may also extend to spoken speech.

Two different types of auditory agnosia which may occur independent of one another are currently recognized. First, an auditory agnosia for non-linguistic sounds (also called 'psychic deafness') has been described which involves an inability to recognize non-linguistic auditory stimuli such as animal noises, bells ringing, etc. Secondly, in some cases the auditory agnosia may selectively involve a failure to recognize linguistic stimuli. This latter condition has been referred to as auditory verbal agnosia or pure word deafness (sub-cortical word deafness). Pure word deafness is a rare disorder in which, although patients cannot comprehend verbal language and therefore cannot repeat words or write to dictation, their spontaneous speech, writing and reading abilities are not disrupted.

Although it is somewhat controversial, auditory agnosia for non-linguistic stimuli is thought to be caused by lesions which involve the auditory association cortex of both hemispheres but which spare Heschl's gyrus. Both bilateral and unilateral lesions have been described in association with the occurrence of pure word deafness.

Where the disorder is the result of bilateral lesions, the damage has in most cases been described as being located in the mid-portion of the superior temporal gyri of both hemispheres. According to Geschwind (1965a,b), where the disorder results from a unilateral lesion, it is necessary that the lesion involve the sub-cortical area of the temporal lobe so as to prevent the posterior language area from receiving input from the primary auditory receptive areas of either hemisphere.

7.4.3 Tactile agnosia

A syndrome analogous to auditory and visual agnosia, which involved a failure to recognize objects by touch in the presence of a preserved ability to name objects on the basis of either auditory or visual stimuli, was reported by Beauvois *et al.* (1978). They referred to this syndrome as 'bilateral tactile aphasia'. Although sensation in the hands was otherwise normal in the case reported by them, the patient was unable to recognize objects placed in his hands unless he was allowed to see or hear them. The lesion was presumed to involve both parietal lobes.

7.4.4 Special forms of agnosia

Two special forms of agnosia that may co-occur with language disorders include autotopagnosia and anosognosia. Autotopagnosia refers to the impaired recognition of body parts. Patients with this condition may deny that some body part such as the arm belongs to them. Autotopagnosia is thought to be caused by lesions in the posterior-inferior portion of the parietal lobe.

Anosognosia involves a lack of awareness of disease or denial of disease. A patient with this disorder may for instance deny that they have a hemiplegia or may deny that they are blind (Anton's disease). The condition is associated with lesions of the parietal lobe in the region of the supramarginal gyrus.

7.5 GERSTMANN SYNDROME

First described by Gerstmann (1931) this syndrome is characterized by four primary symptoms: a disability in calculation (acalculia); finger agnosia (an inability to recognize one's own fingers or the fingers of others); a right–left disorientation; and agraphia. Gerstmann syndrome is usually associated with a focal lesion of the dominant (left) cerebral hemisphere in the region of the angular gyrus and may occur independently or co-occur with aphasia. A partial Gerstmann syndrome is often seen in association with alexia with agraphia.

In that a full Gerstmann syndrome occurs only rarely in brain-damaged individuals, a number of authors including Benton (1977) and Critchely (1966) have questioned the existence of this syndrome. Benson (1979) suggested that Gerstmann syndrome is only part of a larger disorder called 'angular gyrus syndrome' which includes, in addition to the symptoms of Gerstmann syndrome, anomic aphasia and alexia with agraphia.

CHAPTER EIGHT

Dysarthria associated with upper and lower motor neurone lesions

Darley, Aronson and Brown (1975a, p. 2) have defined dysarthria as 'a collective name for a group of related speech disorders that are due to disturbances in muscular control of the speech mechanism resulting from impairment of any of the basic motor processes involved in the execution of speech'. According to this definition, the term 'dysarthria' is restricted to those speech disorders which have a neurogenic origin (i.e. those speech disorders that result from damage to the central or peripheral nervous system), and does not include those speech disorders associated with either somatic structural defects (e.g. cleft palate, congenitally enlarged pharynx, congenitally short palate and malocclusion) or psychological disorders (e.g. psychogenic aphonia).

Speech is a complex behaviour that requires the co-ordinated contraction of a large number of muscles for its production. Contraction of the muscles of the speech mechanism is controlled by nerve impulses which originate in the motor areas of the cerebral cortex and then pass to the muscles by way of the motor pathways. Overall the control of muscular activity can be considered as if the nervous system involved a series of levels of functional activity.

The lowest level of motor control is provided by the neurones which connect the central nervous system to the skeletal muscle fibres. These neurones, referred to as lower motor neurones, have their cell bodies located in either nuclei in the brainstem (in which case their axons run in the cranial nerves having a motor function) or the anterior horns of grey matter of the spinal cord (in which case their axons run in the various spinal nerves). The lower motor neurones form the only route by which nerve impulses can travel from the central nervous system to cause contraction of the skeletal muscle fibres and, for this reason, are also known as the final common pathway.

Table 8.1 Clinically recognized types of dysarthria together with their lesion sites

Dysarthria type	Lesion site
Flaccid dysarthria	Lower motor neurones
Spastic dysarthria	Upper motor neurones
Hypokinetic dysarthria	Basal ganglia and associated brainstem nuclei
Hyperkinetic dysarthria	Basal ganglia and associated brainstem nuclei
Ataxic dysarthria	Cerebellum and/or its connections
Mixed dysarthria, e.g.	
Mixed flaccid – spastic dysarthria	Both lower and upper motor neurones (e.g. amyotrophic lateral sclerosis)
Mixed ataxic – spastic – flaccid dysarthria	Cerebellum/cerebellar connections, upper motor neurones and lower motor neurones (e.g. Wilson's disease)

The motor areas of the cerebral cortex responsible for the initiation of voluntary muscle activity constitute the highest level of motor control. These areas can dominate the lower motor neurones arising from the brainstem and spinal cord, either directly via the pyramidal pathways, or indirectly via the extra-pyramidal pathways. The neurones which make up the pyramidal and extra-pyramidal pathways are collectively referred to as upper motor neurones. The extra-pyramidal pathways include a number of different tracts of nerve fibres and involve a multiplicity of connections which render them open to influence from many sub-cortical structures, particularly the basal ganglia. Co-ordination of muscular contraction is a function of the cerebellum.

The type of dysarthria that results from damage to the neuromuscular system depends very much upon where in the neuromuscular system that damage is located. Parts of the neuromuscular system that can be affected include the lower motor neurones, upper motor neurones, extra-pyramidal system, cerebellum and neuromuscular junction as well as the muscles of the speech mechanism themselves. Damage to each of these sites is associated with a particular type of dysarthria (see Table 8.1).

Prior to looking in detail at the individual characteristics of the different types of dysarthria and attempting to explain the occurrence of the various deviant speech dimensions seen in association with each, it is necessary that the reader has an understanding, not only of he neuropathology underlying each condition, but also a knowledge of the neuroanatomy of the motor pathways.

8.1 FLACCID DYSARTHRIA (LOWER MOTOR NEURONE DYSARTHRIA)

Lower motor neurones form the ultimate pathway through which nerve impulses are conveyed from the central nervous system to the skeletal muscles, including the muscles of the speech mechanism. The cell bodies of the lower motor neurones are located in either the anterior horns of the spinal cord or in the motor nuclei of the cranial nerves in the brainstem. From this location, the axons of the lower motor neurones pass via the various spinal and motor cranial nerves of the peripheral nervous system to the voluntary muscles. Lesions of the motor cranial nerves and spinal nerves represent lower motor neurone lesions and interrupt the conduction of nerve impulses from the central nervous system to the muscles. As a consequence, voluntary control of the affected muscles is lost. At the same time, because the nerve impulses necessary for the maintenance of muscle tone are also lost, the muscles involved become flaccid (hypotonic).

In addition to loss of muscle tone, lower motor neurone lesions are characterized by muscle weakness, a loss or reduction of muscle reflexes, atrophy of the muscle involved and fasciculations (spontaneous twitches of individual muscle bundles – fascicles). All or some of these characteristics may be evidenced in the muscles of the speech mechanism in a patient with flaccid dysarthria. In particular, however, the degree of muscle atrophy may show some variability depending upon the nature of the underlying neurological disorder and fasciculations are not manifest in all of the diseases that can cause damage to lower motor neurones.

Damage to either the lower motor neurones (including those that innervate the respiratory musculature and/or those that run in the cranial nerves to innervate the speech musculature) and/or the muscles of the speech mechanism results in speech changes collectively referred to as flaccid dysarthria, although the term 'peripheral dysarthria' has been used by some authors (Edwards, 1984). The actual name, 'flaccid dysarthria', is of course derived from the major symptom of lower motor neurone damage, flaccid paralysis. The speech characteristics of each patient with flaccid dysarthria, however, vary depending upon which particular nerves are affected and the relative degree of weakness resulting from the damage. The actual

Table 8.2 Lower motor neurones associated with flaccid dysarthria

Speech process	Muscle	Site of cell body	Nerves through which axons pass
Respiration	Diaphragm	3rd–5th cervical segments of spinal cord	Phrenic nerves
	Intercostal and abdominal	1st–12th thoracic and 1st lumbar segments of the spinal cord	Intercostal nerves. 6th thoracic to 1st lumbar spinal nerves
Phonation	Laryngeal muscles	Nucleus ambiguus in medulla oblongata	Vagus nerves (X)
Articulation	Pterygoids, masseter, temporalis, etc.	Motor nucleus of trigeminal in pons	Trigeminal nerves (V)
	Facial expression, e.g. orbicularis oris	Facial nucleus in pons	Facial nerves (VII)
	Tongue muscles	Hypoglossal nucleus in medulla oblongata	Hypoglossal nerves (XII)
Resonation	Levator veli palatini	Nucleus ambiguus in medulla oblongata	Vagus nerves (X)
	Tensor veli palatini	Motor nucleus of trigeminal in pons	Trigeminal nerves (V)

lower motor neurones which, if damaged, may be associated with flaccid dysarthria are listed in Table 8.2.

8.1.1 Innervation of the speech mechanism

With the exception of the muscles of respiration, the muscles of the speech mechanism are innervated by the motor cranial nerves which arise from the bulbar region (pons and medulla oblongata) of the brainstem. These nerves include cranial nerves V, VII, IX, X, XI and XII.

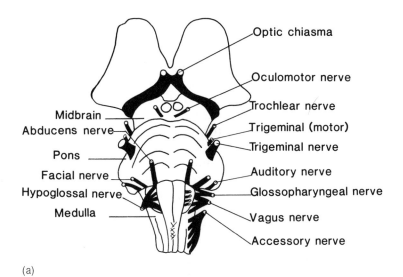

(a)

(b)

Figure 8.1 (a) Inferior view of the brainstem showing the origins of the cranial nerves and (b) lateral view of the brainstem showing the origins of the cranial nerves.

(a) Trigeminal nerves (V)

The trigeminal nerves emerge from the lateral sides of the pons and are the largest of the cranial nerves (see Figure 8.1). Each trigeminal nerve is composed of three branches – the ophthalmic branch, the maxillary branch and the mandibular branch. Of the three branches, the ophthalmic and maxillary are both purely sensory, while the mandibular is mixed sensory and motor. A large ganglion, the Gasserian ganglion, which is homologous to the dorsal root ganglion of the spinal nerve, is located at the point where the trigeminal divides into three branches.

The ophthalmic branch exits the skull through the superior orbital fissure and provides sensation from the cornea, ciliary body, iris, lacrimal gland, conjunctiva, nasal mucous membrane and the skin of the eyelid, eyebrow, forehead and nose. The maxillary branch leaves the skull through the foramen rotundum and supplies sensory fibres to the skin of the cheek, lower eyelid, side of the nose and upper jaw, teeth of the upper jaw and mucous membrane of the mouth and maxillary sinus.

The mandibular branch unites with the motor root immediately after it has exited from the cranial cavity via the foramen ovale. The motor root arises from the motor nucleus of the trigeminal in the pons. Because the trigeminal nerve is mainly sensory, the motor root is much smaller than the sensory portion. Sensory fibres in the mandibular branch provide sensation from the skin of the lower jaw and the temporal region. In the mouth they supply the lower teeth and gums and the mucous membrane covering the anterior two-thirds of the tongue. The motor fibres of the mandibular branch innervate the muscles of mastication which include the temporalis, masseter and medial and lateral pterygoid muscles. In addition, the motor fibres also supply the mylohyoid, anterior belly of the digastric, the tensor veli palatini and the tensor tympani of the middle ear.

The functioning of the motor portion of the trigeminal nerve can be tested clinically by observing the movements of the mandible. Normally, when the mouth is opened widely, the mandible is depressed in the midline. In unilateral trigeminal lesions, however, the mandible deviates towards the paralysed side due to the unopposed contraction of the pterygoid muscles on the active side (i.e. the side opposite to the lesion) when the mouth is opened. As a further test of trigeminal function, the masseter and temporalis muscles should be palpated while patients clench their teeth. In patients with unilateral lesions, it will be noted that the muscles of mastication on the same side as the lesion will either fail to contract or contract only weakly. Where bilateral trigeminal lesions are present, the muscles of mastication on both sides will undergo flaccid paralysis.

(b) Facial nerve (VII)

Each facial nerve emerges from the lateral aspect of the brainstem at the lower border of the pons, in the ponto-medullary sulcus, in the form of two distinct bundles of fibres of unequal size. The larger more medial bundle arises from the facial nucleus of the pons and carries motor fibres to the muscles of facial expression. The smaller, more lateral bundle, carries autonomic fibres and is known as the nervus intermedius. The two roots run together for a short distance in the posterior cranial fossa to enter the internal auditory meatus in the petrous temporal bone along with the VIIIth nerve (auditory nerve). Within the temporal bone the facial nerve passes through the facial canal and eventually emerges from the skull at the stylomastoid foramen. From here the motor fibres are distributed to the muscles of facial expression including the occipito-frontalis, orbicularis oris and buccinator. Other muscles supplied by the facial nerve include the stylohyoid and the posterior belly of the digastric. Within the facial canal, a small number of motor fibres are given off to supply the stapedius muscle in the middle ear.

The autonomic fibres pass into two fine branches of the facial nerve which emerge independently from the temporal bone. One of these is the chorda tympani, which exits the skull via the petro-tympanic fissure to join the lingual nerve, a branch of the mandibular division of the trigeminal nerve. The lingual nerve delivers the fibres of the chorda tympani to the sub-mandibular ganglion. Here they synapse with post-ganglionic neurones which pass to the sub-mandibular and sub-lingual salivary glands. The chorda tympani also conveys taste sensation from the anterior two-thirds of the tongue. The second small branch which carries autonomic fibres supplies the lacrimal gland in the orbit and is known as the greater petrosal nerve.

The motor portion of the facial nerve is tested by observing the patient's face, both at rest and during the performance of a variety of facial expressions such as pursing the lips, smiling, corrugating the forehead, blowing out the cheeks, showing the teeth and closing the eyes against resistance. Normally, all facial movements should be equal bilaterally. Unilateral facial nerve lesions cause weakness or paralysis of the half of the face on the same side as the lesion. At rest, the face of patients with unilateral flaccid paralysis of the muscles of facial expression appears to be asymmetrical. The mouth on the affected side droops below that on the unaffected side and saliva may constantly drool from the corner. In addition, due to loss of muscle tone in the orbicularis oris muscle, the lower eyelid may droop caus-ing the palpebral fissure on the affected side to be somewhat wider than on the normal side. When the patient smiles, the mouth is retracted on the active side but not on the affected side. Likewise when asked to frown, the frontalis muscle on the contralateral side will corrugate the forehead, but, on the side ipsilateral to the lesion, no corrugation will occur.

In bilateral facial nerve paralysis, as might occur in Mobius syndrome, saliva may drool from both corners of the mouth. The seal produced by compression of the lips may be so weak that patients cannot puff out their cheeks and the lips may be slightly parted at rest.

(c) Glossopharyngeal nerve (IX)

Each glossopharyngeal nerve arises from the medulla oblongata as a series of rootlets at the upper end of the post-olivary sulcus. The IXth nerve leaves the cranial cavity via the jugular foramen along with the vagus and accessory nerves.

The glossopharyngeal nerve contains both sensory and motor as well as autonomic fibres. The motor fibres arise from the nucleus ambiguus and innervate the stylopharyngeus muscle. The sensory fibres provide sensation from the pharynx, the posterior one-third of the tongue, the fauces, tonsils and soft palate. They also carry the sense of taste from the posterior one-third of the tongue.

The autonomic fibes within the IXth nerve pass to the otic ganglion where they synapse with post-ganglionic neurones which in turn regulate secretion from the parotid salivary gland.

(d) Vagus nerve (X)

Each vagus nerve arises from the lateral surface of the medulla oblongata by numerous rootlets which lie immediately inferior to those which give rise to the glossopharyngeal nerve. It then leaves the cranial cavity via the jugular foramen.

The vagus nerve contains sensory, motor and autonomic fibres and is the only cranial nerve to venture beyond the confines of the head and neck, supplying structures within the thorax and the upper parts of the abdominal cavity.

After emerging from the jugular foramen, the vagus receives additional motor fibres from the cranial portion of the accessory nerve. The motor fibres of the vagus arise from the nucleus ambiguus and in combination with those from the accessory nerve, supply the muscles of the pharynx, larynx and the levator veli palatini and musculus uvulae of the soft palate. The first branch of the vagus nerve important for speech is the pharyngeal nerve which supplies the levator muscles of the soft palate. As the vagus descends in the neck it gives off a second branch, the superior laryngeal nerve, which supplies the crico-thyroid muscle (the chief tensor muscle of the vocal cords). At a lower level in the neck, a third branch is given off, the recurrent laryngeal nerve, which supplies all of the intrinsic muscles of the larynx except for the crico-thyroid and is, therefore, responsible for regulating adduction of the vocal cords for phonation and abduction of the vocal cords for unvoiced phonemes and inspiration.

Prior to entering the larynx, the left recurrent laryngeal nerve descends into the thorax, loops under the aortic arch and then ascends along the lateral aspects of the trachea to enter the larynx from below and behind the left crico-thyroid joint. The right recurrent laryngeal nerve enters the larynx at the equivalent point on the right side but descends in the neck only as far as the right sub-clavian artery before commencing its ascent to the larynx. Looping of the left recurrent laryngeal nerve under the aortic arch makes it vulnerable to compression by intra-thoracic masses (e.g. lung tumours) and aortic arch aneurysms.

The autonomic component of the vagus supplies organs in the thorax and abdomen including the heart, lungs, major airways and blood vessels and the upper part of the gastrointestinal system.

Functioning of the vagus nerve can be easily checked clinically by noting (1) the quality of the patient's voice, (2) their ability to swallow and (3) the position and movements of the soft palate at rest and during phonation. Unilateral vagus nerve lesions cause paralysis of the ipsilateral vocal cord leading to dysphonia. The paralysed cord can be neither abducted or adducted. By asking patients to open their mouth and say 'ah', movements of the soft palate can be observed. Normally the uvula and soft palate rise in the midline during phonation. However, unilateral lesions of the vagus nerve cause the palate to deviate to the contralateral side (the side opposite to the lesion) during phonation. In addition, the distance between the soft palate and the posterior pharyngeal wall is less on the paralysed side and the arch of the palate at rest will droop on the side of the lesion.

In bilateral lesions of the vagus nerves, both sides of the soft palate and both vocal cords may be paralysed. Both sides of the soft palate rest at a lower level than normal, although their symmetry at rest may appear normal to inexperienced clinicians. However, despite the apparent symmetry, there is less space under the arches of the soft palate and the curvature is flatter. The extent of movement on phonation is reduced and in severe cases , the palate may not rise at all. When observed by either direct or indirect laryngoscopy, abduction and adduction of both vocal cords is severely impaired.

(e) Accessory nerve (XI)

There are two parts to each accessory nerve – a cranial portion which arises from the nucleus ambiguus in the medulla oblongata and a spinal portion which arises from the first five segments of the cervical region of the spinal cord. The cranial accessory emerges from the lateral part of the medulla oblongata in the form of four to five rootlets immediately below those that form the vagus nerve. Prior to leaving the cranial cavity via the jugular foramen, the cranial accessory is joined by the spinal accessory to form the accessory nerve. The spinal accessory fibres arise from the anterior horns of the first five cervical segments of the spinal cord. These fibres emerge from the lateral parts of the spinal cord and unite to form a single nerve trunk which ascends alongside the spinal cord and enters the skull through the foramen magnum to join the cranial accessory.

After exiting through the skull, the cranial accessory leaves the spinal accessory and joins the vagus nerve and is distributed by that nerve to provide motor supply to the muscles of the pharynx, larynx, musculus uvulae and levator veli palatini muscles. The spinal accessory, on the other hand, provides the motor supply to the trapezius muscle and the upper portion of the sternocleidomastoid muscle.

Disorders of the cranial accessory are recognized clinically as disorders of the vagus nerve while disorders of the spinal accessory are evident in atrophy and paralysis of the trapezius and sternocleidomastoid muscles.

(f) Hypoglossal nerve (XII)

Each hypoglossal nerve arises from motor cells in the hypoglossal nucleus and emerges from the medulla oblongata as a series of rootlets in the groove that separates the pyramid and olive. The nerves leave the cranial cavity via the hypoglossal canal which lies in the margin of the foramen magnum.

The hypoglossal nerves provide the motor supply to the muscles of the tongue. Tongue muscles can be divided into two groups – the intrinsic muscles, which lie entirely within the substance of the tongue and are responsible for changes in its shape, and the extrinsic muscles. The latter muscles are attached at one end to structures outside the tongue and are responsible for moving the tongue within the mouth. The hypoglossal nerves innervate all of the tongue muscles with the exception of the palato-glossus. Other muscles in the region of the neck also supplied by the hypoglossal nerves include the sternohyoid, sternothyroid, inferior belly of the omohyoid and the geniohyoid muscles.

Functioning of the hypoglossal nerves can be tested by observing the tongue at rest and during movement. Unilateral hypoglossal nerve damage is associated with atrophy and fasciculations in the ipsilateral side of the tongue. When observed in the mouth the tongue on the side of the lesion may appear smaller and the surface corrugated, indicative of atrophy. Fasciculation of the tongue may in some cases be the earliest sign of lower motor neurone disease. When the patients are asked to protrude their tongue, it will deviate to the paralysed side. Another test for weakness of the tongue is to have patients press their tongue against their cheek while the examiner presses against the bulging cheek with the hand.

In bilateral hypoglossal involvement, both sides of the tongue may be atrophied and show fasciculations. Although protrusion occurs in the midline, the degree of protrusion may be severely limited by weakness and in the more severe cases the patient may not be able to extend the tongue far beyond the lower teeth. Elevation of the tip and body to contact the alveolar ridge or hard palate may be difficult or impossible.

8.1.2 Neurological disorders associated with lower motor neurone lesions

Flaccid paralysis of the muscles supplied by nerves arising from the bulbar region of the brainstem is commonly called bulbar palsy. Diseases which cause bulbar palsy may affect either the cell body of the lower motor neurone or the axon of the lower motor neurone as it courses through the peripheral nerve. A variety of neurological diseases can cause damage to the lower motor neurones that innervate the muscles involved in speech production. Viral infections, tumours, cerebrovascular accidents, progressive degeneration and congenital conditions may impair the cranial nerve nuclei or anterior horn cells of the spinal cord. Traumatic head injuries, tumours, cardiovascular defects (aneurysms), bony prominences and toxins or infections that produce neuritis may affect the spinal and cranial nerves once they exit from the central nervous system.

Depending whether they affect the nerve in its peripheral course or involve the nerve cell bodies in either the cranial nerve nuclei or anterior horns of the spinal cord, disorders of lower motor neurones

Table 8.3 Disorders of lower motor neurones causing dysarthria

Site of lesion	Disorder	Aetiology	Signs and symptoms
Peripheral nerves (especially cranial nerves V, VII, IX, X, XI and XII)	Polyneuritis	Inflamation of a number of nerves. Acute type – may follow viral infections, e.g. glandular fever. Chronic type – may be associated with diabetes mellitus and alcohol abuse	Sensory and lower motor neurone changes usually begin in the distal portion of the limbs and spread to involve other regions including the face, tongue, soft palate, pharynx and larynx. The muscles of respiration may also be involved. Bilateral facial paralysis may occur in idiopathic polyneuritis (Guillain–Barré syndrome)
	Compression of and damage to cranial nerves	Neoplasm, e.g. acoustic neuroma causing compression of the VIIth nerve. Aneurysm, e.g. compression of the left recurrent laryngeal nerve by an aortic arch aneurysm. Trauma, e.g. damage to the recurrent laryngeal nerve during thyroidectomy	Localized lower motor neurone signs dependent on the particular nerves involved
	Idiopathic facial paralysis (Bell's palsy)	Pathogenesis unknown in most cases but may be related to inflammatory lesions in the stylomastoid foramen. Approximately 80% of cases recover	Abrupt onset of unilateral facial paralysis

Cranial nerve nuclei and/or anterior horns of spinal cord	Brainstem cerebrovascular accidents		
		Lateral medullary syndrome (Wallenberg's syndrome) – caused by occlusion of the posterior-inferior cerebellar artery, vertebral artery or lateral medullary artery	Damage of the nucleus ambiguus (origin of the IXth, Xth and cranial portion of the XIth nerve) leads to dysphagia, hoarseness and paralysis of the soft palate on the side of the lesion. Impaired sensation over the face, vertigo and nausea are also present
		Medial medullary syndrome – caused by occlusion of the anterior spinal or vertebral arteries	Damage to the hypoglossal nucleus leads to unilateral paralysis and atrophy of the tongue. A crossed hemiparesis (sparing the face) and sensory changes are also present
		Lateral pontine syndrome (Foville's syndrome) – caused by occlusion of the anterior-inferior cerebellar artery or circumferential artery	Damage to the facial nucleus causes flaccid paralysis of the facial muscles on the side of the lesion. Other symptoms may include deafness, ataxic gait, vertigo, nausea and sensory changes
		Medial pontine syndrome (Millard–Gubler syndrome – caused by occlusion of the paramedian branch of the basilar artery	Symptoms include facial paralysis on the side of lesion, diplopia, crossed hemiparesis and impaired touch and position sense

Table 8.3 cont'd

Progressive bulbar palsy	A type of motor neurone disease in which there is progressive degeneration of the motor cells in some cranial nerve nuclei	Progressive weakness and atrophy of the muscles of the speech mechanism
Poliomyelitis	Viral infection which affects the motor nuclei of the cranial nerves and the anterior horn cells of the spinal cord	Paralysis and wasting of affected muscles will lower motor neurone signs. Paralysis may be widespread or localized and can effect the speech muscles, limb muscles and muscles of respiration
Neoplasm	Brainstem tumours – these are more common in children than adults	Tumour may progressively involve the cranial nerve nuclei causing gradual weakness and flaccid paralysis of the muscles of the speech mechanism
Syringobulbia	Slowly progressive cystic degeneration in the lower brainstem in the region of the 4th ventricle. Congenital disorder with onset of symptoms usually in early adult life	As the cystic cavity develops there may be progressive involvement of the cranial nerve nuclei leading to the lower motor neurone signs in the muscles of the speech mechanism
Mobius syndrome (congenital facial diplegia)	Congenital hyperplasia of the VIth and VIIth cranial nerve nuclei	Bilateral facial palsy (VII) and bilateral abducens palsy (VI)

which cause flaccid dysarthria can be divided into two groups. The major disorders of lower motor neurones which can cause dysarthria are listed in Table 8.3.

In addition to lesions in the lower motor neurones themselves, flaccid dysarthria can also be associated with either impaired nerve impulse transmission across the neuromuscular junction (e.g. myasthenia gravis) or disorders which involve the muscles of the speech mechanism themselves (e.g. muscular dystrophy and poly-myositis).

8.1.3 Clinical characteristics of flaccid dysarthria

Darley, Aronson and Brown (1969 a,b) found that the combination of speech characteristics that best distinguished flaccid dysarthria from other types of dysarthria were marked hypernasality often coupled with nasal emission of air, continuous breathiness in the voice and audible inspiration. Other prominent speech characteristics reported by these workers included imprecise consonants, monopitch, harsh voice quality, short phrases and monoloudness. The ten main aspects of flaccid dysarthria listed by Enderby (1986) in rank order of frequency of occurrence included poor lip seal, abnormality of lips at rest, abnormality of spread of lips, dribbling, reduced elevation of tongue, abnormality of tongue at rest, poor alternating movements of tongue, reduced phonation time, poor intelligibility of repetition and poor intelligibility of description.

As indicated earlier, the particular speech characteristics exhibited by a patient with flaccid dysarthria depend upon which nerves and muscles are affected.

(a) Phrenic and intercostal nerve lesions

The muscles of respiration are important for the motor production of speech in that the exhaled breath provides the power source for speech. It follows, therefore, that interruption of the nerve supply to the respiratory muscles would interfere with normal speech pro-duction.

Lesions involving either the phrenic or intercostal nerves may lead to respiratory hypofunction in the form of a reduced tidal volume and vital capacity and impaired control of expiration. Respiratory hypo-function may in turn affect the patient's speech resulting in speech abnormalities such as short phrases due to more rapid exhaustion of breath during speech and possibly to a reduction in pitch and loudness due to limited expiratory flow volume (Darby, 1981; Darley *et al.*, 1975a).

(b) Vagus nerve lesions

The vagus nerves supply the muscles of the larynx and the levator muscles of the soft palate. Consequently, lesions of the vagus can affect either the phonatory or resonatory aspects of speech production or both, the speech abnormality exhibited by the patient varying according to the location of the lesion along the nerve pathway. Lesions which involve the nucleus ambiguus in the brainstem (as occurs in lateral medullary syndrome) or the vagus nerve near to the brainstem (e.g. in the region of the jugular foramen) cause paralysis of all muscles that are supplied by the vagus. In such cases, the vocal cord on the affected side is paralysed in a slightly abducted position leading to flaccid dysphonia characterized by moderate breathiness, harshness and reduced volume. Additional voice characteristics that may also be present include diplophonia, short phrases and inhalatory stridor. Further, the soft palate on the same side is also paralysed causing the presence of hypernasality in the patient's speech. If the lesion is bilateral, the vocal cords on both sides are paralysed and can be neither abducted or adducted and elevation of the soft palate is also impaired bilaterally, causing more severe breathiness and hypernasality. The major clinical signs of bilateral flaccid vocal cord paralysis include: breathy voice (reflecting incomplete adduction of the vocal cords that results in excessive air escape); audible inhalation (inspiratory stridor – reflecting inadequate abduction of the vocal cords during inspiration); and abnormally short phrases during contextual speech (possibly as a consequence of excessive air loss during speech as a result of inefficient laryngeal valving). Other signs seen in some patients include monotony of pitch and monotony of loudness.

Bilateral weakness of the soft palate is associated with hypernasality, audible nasal emission, reduced sharpness of consonant production (as a consequence of reduced intra-oral pressure due to nasal escape) and short phrases (reflecting premature exhaustion of expiratory air supply as a result of nasal escape).

Lateral medullary syndrome is one neurological disorder in which the origin of the vagus nerve in the brainstem can be affected, thereby leading to impaired phonation and resonation. Lateral medullary syndrome is caused by a cerebrovascular accident involving occlusion of the posterior inferior cerebellar artery, vertebral artery or lateral medullary artery and results in dysphagia, dysphonia, paralysis of the

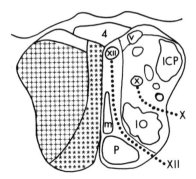

Figure 8.2 Transverse section through the medulla oblongata showing the structures affected in lateral medullary syndrome and medial medullary syndrome. ICP, inferior cerebellar penduncle; IO, inferior olive; M, medial lemniscus; P, pyramid; X, nucleus ambiguus and vagus nerve; XII, hypoglossal nucleus and hypoglossal nerve; ⊞ lateral medullary syndrome; ⊡ medial medullary syndrome.

soft palate, nausea, vomiting and oscillopia (objects visually jump). The brainstem structures affected by lateral medullary syndrome are shown in Figure 8.2.

Lesions to the vagus nerve distal to the branch which supplies the soft palate (the pharyngeal branch) but proximal to the exit of the superior laryngeal nerve, have the same effect on phonation as brainstem lesions. However, such lesions do not produce hypernasality since functioning of the levator veli palatini is not compromised. Lesions limited to the recurrent laryngeal nerves (as may occur as a consequence of damage during thyroidectomy or as a result of compression of the vagus by intra-thoracic masses or aortic arch aneurysms) are also associated with dysphonia. In this latter case, however, the crico-thyroid muscles (the principal tensor muscles of the vocal cords) are not affected and the vocal cords are paralysed closer to the midline (the para-median position). Consequently, the voice is likely to be harsh and reduced in loudness, but with a lesser degree of breathiness, however, than seen in cases with brainstem lesions involving the nucleus ambiguus. Bilateral damage to the recurrent laryngeal nerves is rare. If present, bilateral paralysis of the vocal cords is more likely to have resulted from a brainstem lesion.

The presence or absence of hypernasality in combination with dysphonia, therefore, can provide valuable information about the location of the lesion along the course of the vagus nerve. The major lesion sites which may be associated with disruption of the vagus nerve together with their clinically recognizable effects on speech production are summarized in Figure 8.3.

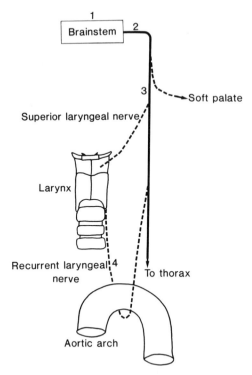

Figure 8.3 Distribution of the vagus nerve to the speech musculature showing the major lesion sites associated with disruption of speech. 1. Lesion in the nucleus ambiguus leading to impaired phonation and resonation. 2. Lesion in the region of the jugular foramen leading to impaired phonation and resonation. 3. Lesion distal to the origin of the pharyngeal nerve associated with impaired phonation and normal soft palate function. 4. Lesion of the left recurrent laryngeal nerve associated with flaccid paralysis of all left laryngeal muscles except crico-thyroid; as a consequence, abduction and adduction of the left vocal cord is impaired.

(c) Trigeminal, facial and hypoglossal nerve lesions

Together the trigeminal, facial and hypoglossal nerves regulate the functioning of the articulators of speech. The trigeminal nerve controls the movements of the mandible. Unilateral trigeminal lesions have only a minor effect on speech, movements such as elevation of the mandible being impaired to only a minor extent. Bilateral trigeminal lesions, however, have a devastating effect on speech production in that the elevators of the mandible (e.g. masseter and temporalis muscles) may be too weak to approximate the mandible and maxilla which may prevent the tongue and lips from making the necessary contacts with oral structures for the production of labial and lingual consonants and vowels.

Unilateral flaccid paralysis of the muscles of facial expression, as occurs following lesions in one or other of the facial nerves, causes distortion of bilabial and labio-dental consonants. As a result of weakness of the lips on the affected side, these patients are unable to seal their lips tightly and air escapes between the lips during the build-up of intra-oral pressure. Consequently, the production of plosives, in particular, is defective. In patients with bilateral facial paralysis or paresis (e.g. in Mobius syndrome), the above situation is exaggerated. Bilateral weakness leads to speech impairments that range from distortion to complete obliteration of bilabial and labio-dental consonants. In severe cases, some vowel distortion may also be evident due to problems with either lip rounding or lip spreading.

Lesions of the hypoglossal nerves cause disturbances in articulation by interfering with normal tongue movements. Both phonation and resonation, however, remain normal. Unilateral hypoglossal lesions as may occur in either brainstem conditions such as medial medullary syndrome (see Figure 8.2) or peripheral nerve lesions such as submaxillary tumours compressing either the left or right hypoglossal nerve cause flaccid paralysis of the tongue on the same side as the lesion. Although this may be associated with mild temporary articulatory imprecision, especially during production of linguo-dental and linguo-palatal consonants, in most cases the patient learns to compensate rapidly for the unilateral tongue weakness or paralysis (usually within a few days post-onset in acute conditions). More serious articulatory impairments, however, are associated with bilateral hypoglossal nerve lesions. Tongue movement in such cases may be severely restricted and speech sounds such as high front vowels and consonants that require elevation of the tongue tip to the upper alveolar ridge or hard palate (e.g. *t, d, n, l,* etc.) may be grossly distorted.

(d) Multiple cranial nerve lesions

The most severe form of flaccid dysarthria results from disruption of several cranial nerves simultaneously. In bulbar palsy the muscles supplied by cranial nerves V, VII, IX, X, XI and XII may simultaneously dysfunction. Consequently, in this condition the lips, tongue, jaw, palate and larynx are affected in varying combinations and with varying degrees of weakness. Disorders evident in the affected person's speech may include: hypernasality with nasal emission, due to disruption of the palato-pharyngeal valve; breathiness, harsh voice, audible inspiration, monopitch and monoloudness associated with laryngeal dysfunction; and distortion of consonant production due to impairment of the articulators.

8.1.4 Speech disorders in myasthenia gravis

Myasthenia gravis has been defined by Penn (1980, p. 382) as 'a disorder of neuromuscular transmission, resulting from an auto-immune attack upon the nicotinic post-synaptic receptor for acetylcholine'. The condition is characterized by muscle weakness that worsens as the muscle is used (fatigability) and rapidly recovers when the muscle is at rest. Females are more frequently affected than males and onset is usually in adult life between the ages of 20 and 50 years.

The abnormal muscular fatigability may, for a long time, be confined to, or predominate in, an isolated group of muscles. Ptosis (drooping) of one or both upper eye-lids caused by weakness of the levator palpebrae is often the first symptom of the condition. Facial, jaw, bulbar and neck muscle weakness ultimately develops in about 50% of cases. Symptoms include diplopia, dysarthria, a tendency for the jaw to hang open and difficulty chewing. There is also dysphagia, drooling and neck muscle weakness. Weakness of all facial muscles is common.

Darley et al. (1975a) regard myasthenia gravis as a special case of flaccid dysarthria because of the progression and increase in severity of speech difficulties with prolongation of speaking activity. As these patients speak, fatigue of the bulbar musculature becomes more and more evident in increased hypernasality, deterioration of articulation, onset and increase of dysphonia and reduction of loudness level (Darley et al., 1975a). Finally, the speech becomes unintelligible. Bannister (1985) suggests that the characteristic fatigability may be readily demonstrated by asking the patient to count up to 50, during which speech becomes progressively less distinct.

8.2 SPASTIC DYSARTHRIA (UPPER MOTOR NEURONE DYSARTHRIA)

Spastic dysarthria is associated with damage to the upper motor neurones. The upper motor neurones convey nerve impulses from the motor areas of the cerebral cortex (primarily the pre-central gyrus and pre-motor cortex) to the lower motor neurones. Lesions of upper motor neurones that can cause dysarthria may be located in the cerebral cortex, the internal capsule, the cerebral peduncles or the brainstem. Clinical signs of upper motor neurone lesions include: spastic

Table 8.4 Clinical signs of upper and lower motor neurone lesions

Upper motor neurone lesions	Lower motor neurone lesions
• Hypertonus (spasticity) • Mild atrophy of disuse • Hyperactive muscle stretch reflexes (e.g. jaw-jerk) • Positive sucking reflex • Positive Babinski sign	• Hypotonus (flaccidity) • Atrophy of individual muscles • Muscle stretch reflexes decreased or absent • Negative sucking reflex • Negative Babinski sign

paralysis or paresis of the involved muscles; little or no muscle atrophy (except for the possibility of some atrophy associated with disuse); hyperactive muscle stretch reflexes (e.g. hyperactive jaw-jerk); and the presence of pathological reflexes (e.g. positive Babinski sign, grasp reflex, sucking reflex, etc.). One of the basic features of upper motor neurone lesions is that reflex arcs remain anatomically intact, whereas in lower motor neurone lesions the reflex arc is disrupted and reflexes become absent or diminished. Table 8.4 compares the major signs associated with upper versus lower motor neurone lesions.

8.2.1 Pyramidal and extra-pyramidal systems

The upper motor neurone system can be divided into two major components, one a direct component and the other an indirect component. In the direct component the axons of the upper motor neurones descend from their cell bodies in the motor cortex to the level of the lower motor neurones without interruption (i.e. without synapsing). The direct component is also known as the pyramidal system. The indirect component, called the extra-pyramidal system, descends to the level of the lower motor neurone by way of a multi-synaptic pathway involving structures such as the basal ganglia, thalamus, reticular formation etc. on the way (see Figure 8.4).

The extra-pyramidal system consists of all those tracts, besides those of the pyramidal system, that transmit motor signals from the cerebral cortex to the lower motor neurones. The pathways of the extra-pyramidal system originate from a number of cortical areas, but especially from Brodmann areas 4 and 6. Many of the extra-pyramidal fibres descend in the internal capsule and cerebral peduncles to the pons and are then relayed to the cerebellum, from which projections then pass to either the brainstem or back to the cerebral cortex via the thalamus. Many other extra-pyramidal fibres descend from the

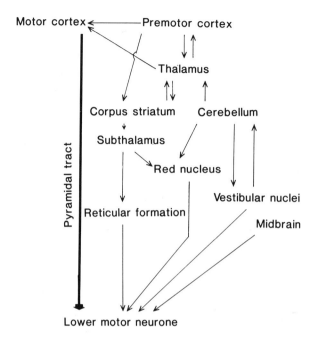

Figure 8.4 Schematic diagram of the pyramidal and extra-pyramidal systems.

cerebral cortex via the internal capsule to the basal ganglia and then pass by a variety of pathways to the excitatory and inhibiting centres of the brainstem. The final pathways for transmission of extra-pyramidal signals to the lower motor neurones include the reticulo-spinal tracts, the rubro-spinal tracts, tecto-spinal tracts and vestibulo-spinal tracts. The extra-pyramidal system appears to be primarily responsible for postural arrangements and the orientation of movement in space, whereas the pyramidal system is chiefly responsible for controlling the far more discrete and skilled voluntary aspects of a movement. Because in most locations (e.g. internal capsule) the two systems lie in close anatomical proximity, lesions that affect one component will usually also involve the other component. The term 'upper motor neurone lesion' is usually not applied to disorders affecting only the extra-pyramidal system (e.g. in basal ganglia lesions). Such disorders are termed 'extra-pyramidal syndromes' and are discussed further in Chapter 9.

The pyramidal system can be sub-divided into those fibres that project to the spinal cord and those that project to the brainstem. In all, three major fibre groups comprise the pyramidal system: the cortico-spinal tracts (pyramidal system proper); the cortico-mesencephalic tracts; and the cortico-bulbar tracts. The cortico-spinal tracts descend from the cerebral cortex to various levels of the spinal cord where they synapse with lower motor neurones. Although the greatest proportion of fibres arise from the motor cortex (primarily the pre-central gyrus), the cortico-spinal tracts originate from both the motor and sensory areas of the cerebral cortex. The cortico-spinal tract in each cerebral hemisphere enters the sub-cortical white matter from the cortex in a fan-shaped distribution of fibres called the corona radiata (radiating crown). The common central mass of white matter in each cerebral hemisphere which contains commissural, association and projection fibres and into which the pyramidal fibres pass, has an oval appearance in horizontal sections of the brain and is, therefore, called the centrum semiovale. From the corona radiata the fibres of the cortico-spinal tracts converge into the posterior limb of the internal capsule (see Chapter 3) and then pass via the cerebral peduncles of the mid-brain to the pons. As the fibres of the cortico-spinal tracts are closely grouped together as they pass through the internal capsule, even small lesions in this area can have a devastating effect on the motor control of the limbs on one half of the body. After traversing the pons, the fibres group together to form the pyramids of the medulla oblongata. It is from the pyramids that the term 'pyramidal tracts' is derived. Near to the junction of the medulla oblongata and the spinal cord, the majority (85–90%) of the fibres in each pyramid cross to the opposite side, interlacing as they do so and forming the decussation of the pyramids. It is this crossing that provides the contralateral motor control of the limbs, the left motor cortex controlling movement of the right limbs and *vice versa*. The fibres that cross then descend in the lateral funiculus of the spinal cord as the lateral cortico-spinal tracts. Of those fibres that remain uncrossed, most descend in the ventral funiculus as the anterior cortico-spinal tracts. Most of these latter fibres decussate to the opposite side further down the spinal cord.

The cortico-mesencephalic tracts are comprised of fibres which descend from the cerebral cortex to the nuclei of cranial nerves III, IV and VI which provide the motor supply to the extrinsic muscles of the eye. These fibres arise from the frontal eye field which is that part of the cerebral cortex of the frontal lobe that lies immediately anterior to the pre-motor cortex.

The fibres of the cortico-bulbar tracts start out in company with those of the cortico-spinal tracts but take a divergent route at the level of the mid-brain. They terminate by synapsing with lower motor neurones in the nuclei of cranial nerves V, VII, IX, X, XI and XII. For this reason, they form the most important component of the pyramidal

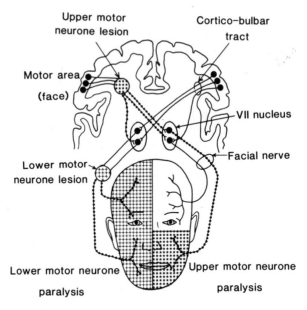

Figure 8.5 The effects of unilateral disruption to the upper and lower motor neurone supply to the muscles of facial expression.

system in relation to the occurrence of spastic dysarthria. Although the majority of cortico-bulbar fibres cross to the contralateral side, uncrossed (ipsilateral) connections also exist. In fact, most of the motor nuclei of the cranial nerves in the brainstem receive bilateral upper motor neurone connections. Consequently, although to a varying degree the predominance of upper motor neurone innervation to the cranial nerve nuclei comes from the contralateral hemisphere, in most instances there is also considerable ipsilateral upper motor neurone innervation.

One important exception to the above upper motor neurone innervation of the cranial nerve nuclei is that part of the facial nucleus that gives rise to the lower motor neurones that supply the lower half of the face. It appears to receive only a contralateral upper motor neurone connection (Sears and Franklin, 1980; Snell, 1980) (see Figure 8.5).

Clinically, the presence of a bilateral innervation to most cranial nerve nuclei has important implications for the type of speech disorder that follows unilateral upper motor neurone lesions. Although a mild and usually transient impairment in articulation may occur subsequent to unilateral cortico-bulbar lesions, in general bilateral cortico-bulbar lesions are required to produce a permanent dysarthria.

Unilateral upper motor neurone lesions located in either the motor cortex or internal capsule, etc. cause a spastic paralysis or weakness in the contralateral lower half of the face but not the upper part of the face which may be associated with a mild, transient dysarthria due to weakness of orbicularis oris. There is no weakness of the forehead, muscles of mastication, soft palate (i.e. no hypernasality), pharynx (i.e. no swallowing problems) or larynx (i.e. no dysphonia). A unilateral upper motor neurone lesion may, however, produce a mild unilateral weakness of the tongue on the side opposite the lesion. In the case of such a unilateral lesion it appears, therefore, that the ipsilateral upper motor neurone is adequate to maintain near normal function of most bulbar muscles, except those in the tongue. Although most authors agree that the hypoglossal nucleus receives bilateral upper motor neurone innervation, for some reason, the ipsilateral connection appears to be less effective than in the case of other cranial nerve nuclei. Snell (1980) suggested that the part of the hypoglossal nucleus that supplies the genioglossus muscle (the only muscle that can protrude the tongue) receives upper motor neurone innervation from only the contralateral cerebral hemisphere.

8.2.2 Neurological disorders associated with upper motor neurone lesions

Persistent spastic dysarthria is caused by bilateral disruption of the upper motor neurone supply to the bulbar cranial nerve nuclei. The general name given to spastic paralysis affecting the bulbar musculature as a result of bilateral upper motor neurone lesions is pseudobulbar palsy (supra-nuclear bulbar palsy). This syndrome is the neurological disorder with which spastic dysarthria is most commonly associated and takes its name from its clinical resemblance to bulbar palsy (pseudo = 'false').

Pseudo-bulbar palsy is characterized by features such as bilateral facial paralysis, dysphagia, hypophonia, bilateral hemiparesis, incontinence and bradykinesia. Drooling from the corners of the mouth is common and many of these patients exhibit excessive emotional responses (e.g. uncontrolled outbursts of laughing or crying) to otherwise normal emotional or environmental stimuli. A hyperactive jaw reflex and a positive sucking reflex are also evident.

All aspects of speech production including phonation, resonation, articulation and respiration are affected in pseudo-bulbar palsy but to varying degrees. Bilateral spastic paralysis of the laryngeal muscles causes narrowing of the glottis thereby increasing the resistance to air-flow at this point. It should be noted that hypertonic changes in the vocal cords cannot easily be visualized so that laryngoscopy of patients with bilateral upper motor neurone lesions often does not reveal any obvious abnormality in their structure or function. Evidence of vocal cord dysfunction, however, will be present in the patient's voice as a harsh voice quality and a strained-strangled sound, associated with the exhaled breath during speech being squeezed with difficulty through the narrow glottis. This is the reverse of what is found in flaccid dysarthria. The range of movement of the vocal cords may also be reduced by hypertonus, thereby causing changes in prosody. The vocal pitch of patients with pseudo-bulbar palsy is lower than in normals (Aronson, 1981).

The major articulatory disorder evidenced by patients with pseudo-bulbar palsy is the production of imprecise consonants, although in severe cases vowel distortions may also be present. A slow rate of articulation is frequently observed in these cases. An oro-motor examination usually reveals weakness of the tongue and lips. Although the tongue is of normal size and not atrophied, movement of the tongue in and out of the mouth is performed slowly and the extent of tongue protrusion limited. In some cases, the patient may be unable to protrude the tongue beyond the lower teeth and lateral movements are also restricted. Voluntary lip movements are also slow and restricted in range. According to Darley *et al.* (1969a,b), the production of syllable repetitions is usually slow but rhythmic.

Hypernasality is a usual finding in pseudo-bulbar palsy. During phonation the soft palate can be seen to elevate symmetrically; however, the elevation is slow and may be incomplete. Swallowing problems are also a common feature of the disorder and there is a definite danger of choking in the more severe cases.

Pseudo-bulbar palsy may be associated with a variety of neurological disorders which bilaterally affect the upper motor neurones anywhere from their cell bodies, located in the motor cortex, through to their synapses with the appropriate lower motor neurones. Bilateral cerebrovascular accidents, multiple sclerosis, motor neurone disease, extensive neoplasms, congenital disorders, encephalitis and severe brain trauma are possible causes of this syndrome.

8.2.3 Clinical characteristics of spastic dysarthria

Darley *et al.* (1975a) identified four major symptoms of muscular dysfunction subsequent to disruption of the upper motor neurone supply to the speech musculature that reflect in the speech output: spasticity; weakness; limited range of movement; and slowness of movement. Consequently spastic dysarthria is characterized by slow, dragging, laboured speech which is produced with some effort.

The most prominent perceptible speech deviations reported by Darley et al. (1969a,b) to be associated with spastic dysarthria include: imprecise consonants, monopitch, reduced stress, harsh voice quality, monoloudness, low pitch, slow rate, hypernasality, strained-strangled voice quality, short phrases, distorted vowels, pitch breaks, continuous breathy voice and excess and equal stress. The deviant speech characteristics cluster primarily in the areas of articulatory–resonatory incompetence and prosodic insufficiency. Using the Frenchay Dysarthria Assessment (Enderby, 1983), Enderby (1986) identified the major aspects of spastic dysarthria (in decreasing order of frequency of occurrence) as including: poor movement of the tongue in speech; slow rate of speech; poor phonation and intonation; poor intelligibility in conversation; reduced alternating movements of the tongue; poor lip movements in speech; reduced maintenance of palatal elevation; poor intelligibility of description; hypernasality and lack of control of volume.

In recent years, research into the nature of spastic dysarthria has focused on quantifying the perceptual features observed by Darley et al. (1969a,b). Using acoustic and physiological measurements, spastic dysarthria speakers have been found to have a slow rate of speech when assessed by reading a standard passage or by syllable repetition (Portnoy and Aronson, 1982; Linebaugh and Wolfe, 1984; Dworkin and Aronson, 1986; Zeigler and von Cramon, 1986). Zeigler and von Cramon (1986) aimed to make Darley's concepts of 'slow rate', 'imprecise consonants' and 'distorted vowels' more precise and quantifiable using computerized signal processing. They found an increase in word and syllable duration (indicative of a slow rate), a reduction of sound pressure level contrast in consonant articulation (indicative of imprecise consonants) and centralization of vowel formants (indicative of distorted vowels).

Portnoy and Aronson (1982) investigated the feature of slow diadochokinetic rate in a group of patients with spastic dysarthria using computerized instrumentation. They found that patients with spastic dysarthria not only had significantly slower syllable repetition rates than normal subjects, but also significantly more variable than normal syllable repetition rates. This later quantitative result is inconsistent with the perceptual analysis of Darley et al. (1969a,b) who described spastic dysarthric subjects as having regular rhythm of syllable repetition. Dworkin and Aronson (1986) also found significantly slower rates of syllable repetition between spastic and normal speakers.

The articulation rate of spastic dysarthric speakers was also investigated by Linebaugh and Wolfe (1984). As a measure of articulation rate, they used the mean syllable duration which was obtained by dividing the audible speech emission time by the number of syllables produced during a standard reading passage. Linebaugh and Wolfe also found that spastic dysarthric speakers had statistically significantly longer mean syllable durations than normals and that the mean syllable duration significantly correlated with both intelligibility and naturalness for the spastic dysarthric speakers. This latter finding has clinical implications since identification of the features that most affect the outcomes of intelligibility and naturalness should improve the efficiency of intervention.

Hirose (1986) used cine-radiography and X-ray microbeam systems, fibreoptic and photo-glottographic recording, ultrasonic techniques, position-sensitive detector and electromyographic assessment to investigate a variety of dysarthric patients. Pseudo-bulbar palsy patients showed a reduced range of articulatory movement and a slow rate of speech. The dynamic pattern of articulatory movements and maximum velocity values were consistent and did not show the variability found by Portnoy and Aronson (1982).

The trend of quantifying the perceptual features of dysarthric speech was continued by Murry (1983) who measured five aspects of stress: peak intra-oral pressure and integrated pressure time (both related to consonant production), and fundamental frequency, vowel duration and vowel intensity (related to production of vowels). Their results suggested that spastic dysarthric speakers mainly use articulatory effort to stress the initial word of a phrase. Strategies used by normal speakers, such as increases in fundamental frequency and intensity of the stressed words and increased vowel duration, were rarely used to stress the initial word of a phrase. For the final word of a phrase, spastic dysarthric patients increased fundamental frequency and intensity of the stressed word, but articulatory effort was reduced. The shift in strategies from the initial word of a phrase to the final word position is consistent with the perceptual finding of prosodic insufficiency. Caligiuri and Murry (1983) used similar instrumentation as Murry to feed back the acoustic signals of prosody to three dysarthric patients including a spastic dysarthia speaker. Nine weeks of visual feedback improved the speaking rate, prosodic control and overall severity of all subjects.

Platt, Andrews and Howie (1980) and Platt et al. (1980) carried out a study of phonological production in 50 cerebral palsy adults, 32 of whom had spastic dysarthria. Sounds which proved most difficult were post-alveolar fricatives s, z, affricates ts, dz, and the labio-dental consonant v. There were more word final consonant errors than word initial consonant errors and the three vowels which represent the extremes of the vowel quadrilateral i, a, u were more difficult. Diadochokinetic rates were about half of that expected of a normal population. It is not known whether the speech patterns of cerebral palsy adults who have had spastic dysarthria from birth are similar to patients who have acquired a pseudo-bulbar palsy much later in life. Therefore, these phonological findings may not be directly applicable to the type of acquired disorders discussed so far.

In general, acoustic and physiological measures support the perceptual analysis of spastic dysarthria by Darley *et al.* (1969a,b). Instrumental studies have shown that spastic dysarthric speakers have a slow rate of speech which appears to be caused by longer durations of syllables and, perhaps, longer pauses within and between words. Reduced tongue strength and range and velocity of articulatory movement as well as prosodic insufficiency are also present in this group.

CHAPTER NINE

Dysarthrias associated with extra-pyramidal syndromes

The term 'extra-pyramidal system' was first used by Wilson (1912) to refer to those parts of the central nervous system concerned with motor function but which are not a part of the pyramidal system. The extra-pyramidal system, as described in Chapter 8, consists of a complex series of multi-synaptic pathways which indirectly connect the motor areas of the cerebral cortex to the level of the lower motor neurones. The major components of the extra-pyramidal system include the basal ganglia (see Chapter 3) within the cerebral hemispheres plus the various brainstem nuclei that contribute to motor functioning. These latter nuclei include the paired substantia nigra, the red nuclei and the sub-thalamic nuclei.

Diseases which selectively affect the extra-pyramidal system without involving the pyramidal pathways are referred to as 'extra-pyramidal syndromes' and include a number of clinically defined disease states of diverse aetiology and often obscure pathogenesis. Extra-pyramidal syndromes share a number of related symptoms and the major pathological changes noted in these disorders are located within the various extra-pyramidal nuclei. Movement disorders are the primary features of the extra-pyramidal syndromes and, where the muscles of the speech mechanism are involved, disorders of speech may occur. The clinical signs and symptoms that characterize extra-pyramidal syndromes and help tie these various disorders together fall into the following four groups: (1) hypokinesia (akinesia) – slowness and poverty of spontaneous movement; (2) hyperkinesia – abnormal involuntary movements; (3) rigidity of the muscles; and (4) loss of normal postural reactions.

Overall, the extra-pyramidal system appears to control muscle tone for the maintenance of posture and for supporting movements (i.e. those muscle actions which provide a firm base of support against which skilled voluntary acts can take place). Depending upon the specific part of the extra-pyramidal system affected, extra-pyramidal disorders may be associated with two types of dysarthria – hypokinetic dysarthria and hyperkinetic dysarthria.

9.1 HYPOKINETIC DYSARTHRIA

9.1.1 Neurological disorders associated with hypokinetic dysarthria

Hypokinetic dysarthria is most commonly associated with Parkinson's disease. In fact, the term 'hypokinetic dysarthria' was first used by Darley, Aronson and Brown (1969a,b) to describe the resultant complex pattern of perceptual speech characteristics associated with parkinsonism. More recently, hypokinetic dysarthria has also been noted in progressive supra-nuclear palsy (Hanson and Metter, 1980). Although the pathological signs in progressive supra-nuclear palsy and Parkinson's disease are more different, Metter and Hanson (1986) found that acoustically, the dysarthria in each of these two types of disease are similar.

(a) Parkinson's disease

Hypokinetic dysarthria has been reported to occur in between 49% (Oxtoby, 1982) and 73% (Selby, 1968) of patients suffering from Parkinson's disease although it is generally accepted that speech disturbances occur in half of all cases and become more prevalent as the disease progresses (Uziel et al., 1975). Parkinsonism occurs most often in persons in their 50s or 60s. The disease most often begins insidiously and is slowly progressive in virtually all patients. The degree and rate of progression, however, does vary from patient to patient. Affected persons may complain of rigidity and tremor, immobility of facial expression, slowness of movements and diminished swinging of the arms and heaviness of the limbs when walking. Posture is commonly stooped forward, with the arms at the sides, elbows slightly flexed and fingers adducted. Parkinson's disease patients often exhibit a characteristic gait which involves short, slow, shuffling steps. In addition, as they walk, affected persons tend to stoop forward and develop increasing speed (festinating gait) which at times may lead to a loss of balance with subsequent falls and injuries. Dementia occurs in approximately 30–39% of parkinsonian cases (Bayles, 1984).

The most common form of parkinsonism is idiopathic Parkinson's disease, a form of the disorder in which no immediate cause is obvious. In these cases the disease has been ascribed by various authors to degeneration of nerve cells and tracts in either the corpus striatum or substantia nigra (including the nigra-striatal pathways), although it is now agreed that the most important lesions are those located in the substantia nigra. The degenerative changes in these structures appear to be associated with a deficiency of a neuro-transmitter substance called dopamine. Abnormally low concentrations of dopamine in the basal ganglia and substantia nigra have been reported upon post-mortem examination of brains of patients with idiopathic parkinsonism. In addition to the idiopathic variety, Parkinson's disease may also be precipitated by an attack of epidemic encephalitis or may be the outcome of cerebral arteriosclerosis, carbon monoxide or manganese poisoning, traumatic head injury, neurosyphilis, cerebrovascular accidents or drug toxicity.

Four separate major groups of symptoms are usually described as part of the symptom complex of parkinsonism. These groups include tremor, rigidity of the muscles, akinesia and loss of normal postural fixing reflexes. The degree to which each of these four signs occur in parkinsonism varies considerably from patient to patient. The tremor consists of the rhythmically alternating contraction of a given muscle group and of its antagonists. The rate of tremor averages about two to six oscillations per second and is usually more obvious in the distal parts of the limbs than in proximal parts. The tremor, however, does not only affect the limbs but also in some cases the jaw, tongue, face and pectoral structures. Although the limb tremor is most commonly of the 'pill-rolling' type (involving alternate movement of the thumb against the opposing index finger), the rotary component may not be present so that the term 'to and fro' may be more applicable in some cases. One of the classic features of the muscle tremor seen in Parkinson's disease is that it is most obvious during rest. In contrast the tremor disappears during movement but may re-emerge when a posture is maintained. Characteristically the tremor is absent during sleep.

Rigidity of the muscles is another major component of Parkinson's disease. Although the rigidity is the product of an increase in muscle tone, clinically it is manifestly different from the muscle spasticity seen in association with upper neurone lesions (see Chapter 8). In spasticity, although when a patients limb is move passively by the examiner there is an initial resistance to that movement, that resistance suddenly gives out ('clasp-knife' phenomenon). In contrast, as a result of the associated rigidity in Parkinson's cases there is usually a resistance to passive movement present throughout the entire range of motion, thereby serving to differentiate rigidity from spasticity.

When the intensity of the rigidity is similar throughout the range of motion, the rigidity is called 'lead-pipe' rigidity. In some Parkinson's cases, however, the hypertonic muscles when passively stretched exhibit an irregular jerkiness in resisting the movement as if they were being pulled over a ratchet, a condition referred to as 'cog-wheel' rigidity. Rigidity may be the initial symptom of Parkinson's disease in some patients.

Although tremor at rest is the most dramatic symptom of Parkinson's disease, the most disabling symptom of the condition is akinesia. The term 'akinesia', when taken generally, refers to three related symptoms: (1) marked poverty of spontaneous movements; (2) loss of normal associated movements (e.g. loss of arm swing during walking); and (3) slowness in the initiation and execution of all voluntary movements. Overall akinesia is manifest clinically in signs, such as a 'mask-like' facial expression and a reduction in a wide variety of spontaneous movements seen in normal individuals. The muscles of the face show a marked poverty of movement in both volitional and emotional activities. Where emotional responses do occur they tend to be slow in developing and may become prolonged (e.g. frozen smile). Further, Parkinson's patients often sit immobile, seldom carrying out activities such as crossing their legs or folding their arms. In particular patients with Parkinson's disease have great difficulty initiating gross motor actions such as standing up from a chair.

Postural fixation reflexes are those reflexes that cause sufficient muscular contractions to support part of the body to maintain a particular posture. The postural fixation of the head of Parkinson's patients is often abnormal such that the affected individual's head may fall forward while the patient is in an upright position. Persons with Parkinson's disease may also have problems with postural fixation of the trunk so that they may be unable to maintain an upright position while seated, standing or walking.

It has been suggested that muscle rigidity and tremor seen in Parkinson's disease are the result of areas of the extra-pyramidal system being released from the control normally exerted over them by other areas of the central nervous system. This release from control causes the affected part of the extra-pyramidal system to become over-active. On the other hand, disorders of postural fixation are believed to result from destruction within the globus pallidus.

(b) Progressive supra-nuclear palsy

Progressive supra-nuclear palsy is a progressive neurological disorder which occurs in middle to later life. The condition was first described by Steele, Richardson and Olszewski (1964). The initial symptoms of progressive supra-nuclear palsy have been described as feelings of unsteadiness, vague visual difficulties, unclear speech and minor changes in personality. As the disease progresses, symptoms include supra-nuclear opthalmoplegia affecting chiefly vertical gaze, pseudo-bulbar palsy, dysarthria, dystonic rigidity of the neck and upper trunk and mild dementia as well as other cerebellar and pyramidal symptoms. The disease is rapidly progressive resulting in marked incapacity of the patient in two to three years.

Patients with progressive supra-nuclear palsy tend to have mask-like faces and akinesia as seen in Parkinson's disease. They do not, however, exhibit tremor and have relatively good associated move-ments (e.g. arm swinging when walking). Affected individuals have a peculiar erect posture with backward retraction of the neck. Although these patients have been reported by Steele, Richardson and Olszewski (1964) to have only minimal rigidity in the extremities, they do have severe rigidity of the axial musculature, especially in the later stages of the disorder. Pathologically progressive supra-nuclear palsy patients exhibit degenerative changes in several specific regions of the brain. In particular these sites include the globus pallidus, sub-thalamic nucleus and pre-tectal region of the brainstem including: the superior corpora quadrigemina; red nuclei; tegmentum of the mid-brain and pons; substantia nigra; oculomotor, trochlear and vestibular nuclei; and the dentate nucleus of the cerebellum (see Chapter 10). The cerebral and cerebellar cortices, however, are remarkably spared. The aetiology of progressive supra-nuclear palsy is unknown. Steele, Richardson and Olszewski (1964), however, speculated that the cause may be primarily degenerative or a degenerative process initiated by an inflammatory event such as encephalitis lethargica.

9.1.2 Clinical characteristics of hypokinetic dysarthria

The speech characteristics associated with hypokinetic dysarthria follow largely from the generalized pattern of hypokinetic motor dis-order which includes marked reduction in the amplitude of voluntary movements, slowness of movement, initiation difficulties, muscular rigidity, loss of the automatic (associated) aspects of movement and tremor at rest (Darley, Aronson and Brown, 1975a). According to these authors marked limitation of the range of movement of the muscles of the speech mechanism is the outstanding characteristic of hypokinesia as it affects speech. These workers stated that reduced mobility, restricted range of movement and supra-normal rate of the repetitive movements of the muscles involved in speech production lead to the various manifestations of hypokinetic dysarthria. Although a confirmatory sign of hypokinesia, tremor does not appear to have an effect on speech as do the other features listed above.

The speech loudness level is reduced in most cases of Parkinson's disease. In addition, although some Parkinson's patients speak slowly, others speak slightly more rapidly than normal (Canter, 1963; Critchley, 1981; Darley et al., 1975a). In the Mayo Clinic study (Darley et al., 1969a,b) parkinsonian dysarthrics were reported to be unique among dysarthric patients in being perceived as having a slightly faster than average speaking rate. All other dysarthric groups showed a slower than normal speaking rate. A few individuals with parkinsonism demonstrate a progressive acceleration of rate towards the end of a sentence similar to the festination in gait, leading some writers to label the speech of Parkinson's subjects as festinant. Some Parkinson's patients also exhibit fluctuations in their level of intelligibility and others have a special difficulty in initiating speech.

A number of authors have documented the perceptual speech characteristics of hypokinetic dysarthria (Enderby, 1986; Darley et al., 1969a,b; Zyski and Weisiger, 1987; Chenery, Murdoch and Ingram, 1988). The most prominent deviant speech characteristics of hypokinetic dysarthria reported by Darley et al. (1969a,b) included monopitch, reduced stress and monoloudness, all of which represent alterations in the prosodic aspects of speech. Other deviant speech characteristics noted by Darley and co-workers included imprecise articulation, inappropriate silences, short rushes, harsh and breathy voice quality and variable rates of speech.

Chenery, Murdoch and Ingram (1988) reported that deficits in the phonatory system, including the presence of hoarseness and a strained-strangled phonation with intermittent breathiness, to be the most frequently occurring speech deviations exhibited by their group of Parkinson's subjects. Disturbed prosodic features were also noted by these authors including a disturbed general stress pattern and a lack of variation in both pitch and loudness levels. Based on their performance on the Frenchay Dysarthria Assessment (Enderby, 1983), Enderby (1986) identified the most frequently occurring characteristics of extra-pyramidal dysarthria to be reduced phonation and intonation, increased rate of speech, reduced intelligibility of conversation, reduced control over volume, reduced phonation time, reduced ability to elevate the tongue, reduced intelligibility of description, inadequate tongue movements in speech, reduced alternating movements of the tongue and dribbling.

In combination, the above-listed deviant speech characteristics make for a distinctive type of dysarthria that can be perceptually distinguished from other dysarthria types. Even though a heterogeneity in the characteristics of the speech of Parkinson's patients has been reported (Enderby, 1986), in a study of the accuracy of perceptual analysis for identifying types of dysarthria, Zyski and Weisiger (1987) found that hypokinetic dysarthria could be perceptually identified with greater accuracy than any other dysarthria type.

Although according the Darley *et al.* (1969a,b) prosodic changes tend to dominate the speech disorder of Parkinson's patients, it is evident, not only from their findings, but also from the results of other perceptual studies, that disturbances also occur in some Parkinson's patients in the other aspects of speech production, including articulation, phonation and resonation. Logemann *et al.* (1978) in a study of the vocal tract characteristics of 200 Parkinson's patients identified disorders of phonation, rate and articulation as well as occasional instances of hypernasality. Overall, laryngeal dysfunction was present in 89% of their 200 subjects, articulation disorders in 45%, rate disorders in 20% and hypernasality in only 10%. Of the laryngeal dysfunctions, hoarseness was the major perceived characteristic reported by Logemann *et al.* (1978). A more detailed analysis of the articulatory errors of the same patients with Parkinson's disease examined by Logemann *et al.* (1978) was conducted by Logemann and Fisher (1981). They found that changes in the manner of articulation predominated over changes in place of articulation. Stop-plosives, affricatives and fricatives were most affected, as were features of continuancy and stridency. The incomplete contact for stops (spirantization) and partial constriction for fricatives were interpreted as resulting from inadequate narrowing of the vocal tract at the point of articulation as a result of inadequate tongue elevation. Overall, the work of Logemann and co-workers indicated that typical Parkinson's patients can be expected to exhibit a voice-quality disorder and may also have a problem with articulation. Alterations in speech rate and hypernasality, however, are less likely to be symptoms of the associated dysarthria.

The major limitation of perceptual assessments of dysarthric speech is that in most circumstances they are unable to accurately and reliably identify which part of the motor speech mechanism is responsible for the perceived speech deficit. For example, it is possible that a perceived phonatory disturbance could be the result of a deficit in either the respiratory or laryngeal systems, or both. By means of acoustic and/or physiological techniques, a number of researchers have attempted to confirm the features of hypokinetic dysarthria identified by perceptually based studies. Indeed, many authors and clinicians regard the instrumental approach as important if the speech pathologist is to be in a position to formulate an approach to therapy based on an understanding of the nature and cause of the perceived speech disorders exhibited by their clients.

Canter (1963, 1965a,b) made a number of objective measures of the speech of patients with Parkinson's disease including: vocal intensity; vocal duration; diadochokinetic rate; pitch; and syllable duration. He found that Parkinson's patients exhibited significantly higher pitch levels than control subjects but had similar vocal intensity levels and a similar rate of speaking. The Parkinson's cases also demonstrated an impaired ability to perform rapid movements of the tongue tip, back of the tongue, lips and vocal cords. Imprecise production of plosive consonants and incoordination of phonatory and articulatory activity was also documented. Canter (1965a, p. 49) concluded that 'many of the deficiencies in the conversational speech of persons with Parkinson's disease derive, at least in part, from reduced physiological support for speech'.

The sustained-vowel and syllable repetition abilities of ten Parkinson's subjects were determined by Mueller (1971) and compared to those of a group of control subjects. All of the Parkinson's patients had a reduced phonation time and expended a lower volume of air than the controls, during the sustained production of the vowel *a*. In addition, the Parkinson's subjects also showed a reduced phonation time, a reduction in the total number of syllables produced and a reduced intra-oral pressure during repeated utterance of the syllable *sʌ*. The rate of syllable production, however, was the same as for the control group. Mueller (1971) concluded that in Parkinson's disease, the neuromuscular impairment prevents the individual from generating sufficient amounts of aerodynamic energy for normal speech production.

The sustained phonation, syllable diadochokinesis and reading rates of Parkinson's subjects were compared with those of normal young adults and normal elderly adults by Kreul (1972). Consistent with the findings of Meuller (1971), he found no significant difference between the controls and Parkinson's patients in the syllable diadochokinesis task. He did, however, find a statistically significant difference between the Parkinson's and control subjects in the diadochokinetic rates for repetition of interrupted vowels (e.g. *i*) and vowel glides (e.g. *u-i*). Kreul (1972) suggested that the difference in diadochokinetic rate for syllables versus interrupted vowels and vowel glides is that the Parkinson's subjects are able to utilize their muscle tremor in the production of reciprocal movements as occur in the repetition of *pʌ* but not in the production of interrupted vowels or vowel glides. The repetition of interrupted vowels requires discrete control of the larynx while the production of vowel glides requires controlled oral movements between vowels, movements which cannot make use of the muscle tremor.

Ludlow and Bassich (1984) developed a set of objective acoustic measures which reflected the perceptual attributes reported to be associated with hypokinetic dysarthria by Darley *et al.* (1969a,b). By use of this acoustic measurement system, these authors determined that two dysfunctions are the principal contributors to the speech disturbance seen in Parkinson's disease: (1) impaired laryngeal control; and (2) impaired rate and stress control. They suggested that both of these factors are indicative of the rigidity exhibited by Parkinson's patients, particularly as it affects these patients' abilities to manoeuvre their larynx and articulators to provide the prosodic aspects of speech. Ludlow and Bassich (1984) also carried out perceptual ratings of their subjects' speech abilities similar to those used by Darley *et al.* (1969a,b). The results of their perceptual analysis showed that reduced stress and short rushes were less severe in their Parkinson's subjects than in the patients of Darley *et al.* (1969a,b). Conversely, the deviant speech dimensions of variable rate and harsh voice were more prominent in the subjects of Ludlow and Bassich (1984). These differences suggest that the patients of Ludlow and Bassich were more similar to patients with hyperkinetic dysarthria (see below), such as occurs in chorea and dyskinesia, than those with hypokinetic dysarthria. Ludlow and Bassich (1984) attributed this finding to the drug therapy that their patients were receiving. They suggested that any rapid rushes of speech, increase in speech rate and reductions in stress exhibited by their patients may have been relieved by dopamine enhancement therapy. As most Parkinson's patients attending speech pathology clinics are also receiving drug therapy, then the appropriateness of the classical speech symptoms described by Darley *et al.* (1975a) could be questioned.

To investigate the factors associated with speech timing, Ludlow, Connor and Bassich (1987) compared patients with Parkinson's disease, Huntington's disease and control subjects on three aspects of speech timing: planning; initiation; and production. They found that the most affected aspect of speech timing in both the extra-pyramidal disorders was the rate of speech movements and their controlled alteration during speech production. Neither the initiation nor planning of speech were observed to be impaired in either disease. Further, it is also of interest to note that these authors found no evidence of an abnormally fast syllable repetition or sentence production rate to support the clinical impression of accelerated rate (festinant speech) in their Parkinson's cases. Although there were differences between the Parkinson's and Huntington's patients, they suggested that the different patterns of impairment exhibited by these two groups may only be a matter of degree, with the Huntington's patients being more severely impaired. Enderby (1983) classified these two groups together under the heading of 'extra-pyramidal dysarthria' rather than dividing them into the hypokinetic and hyperkinetic types used by Darley *et al.* (1975a).

A number of studies based on physiological measurements have indicated that persons with Parkinson's disease have reduced articulatory displacements (Hirose *et al.*, 1981; Hirose, Kiritani and Sawashima, 1982b; Hunker, Abbs and Barlow, 1982). Using an X-ray micro-beam system and electromyography, Hirose *et al.* (1981) found that the range of movements of the articulators was limited in Parkinson's patients and that the frequency of repetitive production of a monosyllable tended to increase gradually. This finding contradicted those of several other authors (Mueller, 1971; Kreul, 1972; Kent and Rosenbek, 1982; Ludlow *et al.*, 1987) who found no acoustic or physiological evidence of acceleration of speaking rate in patients with Parkinsonism. Hirose *et al.* (1981) suggested that the festinant speech is related to a disturbance of the inhibitory function of the extra-pyramidal system and that the reduction in the range of movement of the articulators can be attributed to a deterioration in the reciprocal adjustment of the antagonistic muscles.

Kent and Rosenbek (1982) provided spectrographic evidence to support the perceptual identification of spirantization in the speech of Parkinson's cases reported by Logemann and Fisher (1981). These authors also found spectrographic evidence of the presence of continuous voicing, hypernasality and poor consonant articulation which they suggested could possibly be misperceived in a perceptual analysis as a faster than normal speaking rate. Weismer (1984) using acoustic analysis techniques also found evidence of spirantization and continuation of vocal fold vibration into voice-less stop closure in the speech of persons with Parkinson's disease. He, however, attributed the continuous voicing characteristic to the influence of the age of his subjects rather than to the effects of the disease itself. Using measurements of speech waveform durations derived from storage oscilloscope displays, Weismer found that the segment durations and phrase-level durations were reduced in Parkinson's subjects relative to age-matched controls, which he suggested might contribute to the often-cited perception that speech rate is increased in Parkinson's patients. Since Parkinson's subjects in general fall into the gerontological age group, Weismer suggested that, during perceptual analysis, listeners may compare the voice characteristics of Parkinson's cases with the expected speech characteristics of aged persons. Consequently, even though individuals with Parkinson's disease may speak at a rate similar to young adults, they may be perceived as having a faster than normal speech rate when compared to the slower speech rate of geriatrics.

Abbs, Hunker and Barlow (1983) argued that perceptual and acoustic measures have only a limited potential for determining the pathophysiological basis of the various dysarthrias. They stressed the importance of physiological measures in determining the location and nature of the motor impairment in the speech mechanism. These authors reported evidence of differential sub-system impairment in their Parkinson's subjects. The degree of muscle rigidity and impairments in the range of movement were found to be differentially involved in the upper and lower lips. The patient's ability to sustain a steady force was also found to be differentially impaired in the lips, tongue and jaw. Hunker and Abbs (1984) observed different forms of tremor in the muscles of the oro-facial system of Parkinson's patients which were dependent upon whether the structures were being rested, postured or actively or passively moved. Further, they reported that the different forms of tremor could be present simultaneously in different muscles of the oro-facial system. On the basis that each of the different types of tremor might influence the quality of motor performance in a different manner, Hunker and Abbs (1984) suggested that tremor analysis might have a role in clinical dysarthria assessment procedures.

In summary, acoustic and physiological measures have questioned some of the perceptually recognized characteristics of Parkinson's disease. These include the perception of accelerating speech rate and the assumption that all sub-systems of the speech mechanism are similarly affected in Parkinson's disease.

9.2 HYPERKINETIC DYSARTHRIA

Hyperkinetic dysarthria is seen in association with a variety of extra-pyramidal disorders in which abnormal involuntary movements of the limbs, trunk, neck, face, etc. disturb the rhythm and rate of motor activities, including those involved in speech production. Darley *et al.* (1975a) distinguished between two categories of hyperkinetic disorders – quick hyperkinesias and slow hyperkinesias. They noted, however, that such a dichotomy is somewhat artificial in that any one patient with a hyperkinetic extra-pyramidal disorder may exhibit some elements of both quick and slow hyperkinetic movements. In decreasing order of quickness, quick hyperkinesias include myoclonic jerks, tics, chorea and ballismus while slow hyperkinesias include athetosis, dyskinesia and dystonia.

9.2.1 Quick hyperkinesias

The rapid abnormal involuntary movements that fall into this category are either unsustained movements or sustained only very briefly and are random in occurrence in terms of the body part affected.

(a) Myoclonic jerks

Myoclonic jerks are abrupt, sudden, unsustained muscle contractions which occur irregularly. Muscles of the limbs, face, oral cavity, soft palate, larynx and diaphragm may by affected among others. Although isolated muscles may be involved, myoclonic jerks may also occur simultaneously in larger groups of muscles. In the latter case these jerks may produce joint movements of sufficient violence to throw the patient to the ground. On the other hand, if individual muscles are involved, the sudden involuntary contraction may be only sufficient to cause a small movement such as the twitch of a finger.

Myoclonus may be associated with convulsive disorders (epilepsy) or may occur in a variety of extra-pyramidal diseases involving the basal ganglia or ventro-lateral nuclei of the thalamus. Myoclonic jerks may also be seen in association with diffuse metabolic, infectious or toxic disturbances of the nervous system such as diffuse encephalitis and toxic encephalopathies. In one form of myoclonus, called 'synchronous myoclonus', a number of different muscle groups may jerk simultaneously. For instance, both arms or both legs may jerk at the same time. In these cases there may be a single simultaneous myoclonic jerk or repetitive rhythmic jerking. Another form of myoclonus, 'asynchronous myoclonus', involves muscle groups contracting one after the other in a random fashion.

Myoclonic jerks may affect the muscles of the speech mechanism in the same way as they affect the limbs. The muscles of the face, soft palate, larynx and diaphragm may be either involved individually or in combination (e.g. palato-pharyngo-laryngeal myoclonus). Palatal myoclonus usually occurs rhythmically from one to four times per second and may result in hypernasality and a lack of precision in phoneme production coincident with the occurrence of the myoclonic jerks. Laryngeal myoclonus is evidenced in the patient's speech as momentary interruptions of phonation while jerks in the muscles of the diaphragm if present, may produce momentary irregularities in the respiratory airflow during the sustained phonation of vowels. The hyperkinetic dysarthria resulting from myoclonus is particularly deceptive in that patients may not demonstrate a speech deficit in their contextual speech. When the patient is asked to sustain the production of a vowel, however, deficits such as brief interruptions of voice resulting from laryngeal myoclonus become evident.

(b) Tics

Tics are brief, unsustained, recurrent, compulsive movements that involve a relatively small part of the body. Although tics may be briefly controlled voluntarily, such periods of suppression are often followed by a period of more intense involuntary contraction.

Multiple tics occur in the Gilles de la Tourette's syndrome, a distinctive childhood disease characterized by the development of tics progressively involving the face, neck, upper limbs and eventually the entire body. Uncontrolled vocalizations often occur as a result of the involuntary contractions of the muscles of the speech mechanism. These include grunting, coughing, barking, hissing and snorting. In addition, stuttering-like repetitions, unintelligible sounds and echolalia have also been reported (Field et al., 1966). Coprolalia (involuntary swearing) although not universal, is also an important distinguishing feature of Gilles de la Tourette's syndrome. Occasionally soft neurological signs, such as mild incoordination in motor skills and slight asymmetry of motor function, including deep tendon reflexes, are also evident on examination of the child. Intelligence is normal and no deterioration in either mental or psychologic function occurs.

Gilles de la Tourette's syndrome runs a prolonged and non-fatal course and affects primarily boys. The age at onset is usually between 2 and 15 years. Patients are often able to voluntarily control the symptoms for short periods of time (up to an hour or so) such as that required for neurological or psychiatric evaluation. The cause of the condition is unknown and no specific lesion has been identified. It has been suggested, however, that the pathophysiological basis of the disease may be increased dopamine activity (Sweet et al., 1973).

(c) Chorea ✓

The term 'chorea' is derived from the Greek word for dance and was originally applied to the dance-like gait and continual limb movements seen in acute infectious chorea. A choreic (or choreiform) movement consists of a single, unsustained, isolated muscle action producing a short, rapid, uncoordinated jerk of the trunk, limb, face, tongue, diaphragm, etc. They are random in their distribution and their timing is irregular and unpredictable. Choreic contractions are slower than myoclonic jerks, each lasting from 1/10th to 1 second.

The simultaneous or successive occurrence of two or more choreic movements can result in complex movement patterns, while the super-imposition of the abnormal involuntary movements on normal movements can cause characteristic symptoms such as a dance-like gait. When super-imposed on the normal movements of the speech mechanism during speech production, choreiform movements can cause momentary disturbances to the course of contextual speech.

Huntington's chorea (Huntington's disease) and Sydenham's chorea are the two major diseases in which choreic movements are manifested. Chorea, however, may also occur in a variety of other conditions such as hyperthyroidism, systemic lupus erythematosus, Wilson's disease, pregnancy and hysterical reaction and also as a complication of taking oral contraceptives. Huntington's chorea is a chronic degenerative neurological disorder that is manifested by progressive chorea, or at times, other extra-pyramidal symptoms as well as progressive intellectual deterioration. Either the mental changes or the choreiform movements may appear first. In most cases the chorea appears first, although the clinical picture is variable.

Huntington's chorea commonly begins in adult life and is usually fatal within 10 to 15 years post-onset. The condition is inherited as an autosomal dominant trait. Pathologically, Huntington's chorea is marked by a loss of neurones in the caudate nucleus and putamen and these structures are grossly shrunken and atrophic. The globus pallidus is usually well preserved as is the substantia nigra. In addition to the caudate nucleus and putamen, atrophy also occurs in the cerebral cortex, particularly in the frontal lobes. Although major pathological changes are seen in the striatum and cerebral cortex, some changes have also been noted in the cerebral white matter (including the inter-hemispheric fibre systems and association fibre systems), the thalamus and the hypothalamus.

As Huntington's chorea is a progressive disorder, the degree of choreiform movements gradually beomes greater and greater as the condition advances. The clinical picture in the advanced stages of the condition include facial grimacing (involving the lips, tongue and cheeks), jerks of the head, weaving movements of the arms and shoulders, twists and jerks of the body as well as super-imposed voluntary movements (e.g. an involuntary upward jerk of the arm may be fused into a voluntary scratching of the head). The patient's gait at this stage is often markedly involved, being comprised of jerky lurching steps that represent a combination of voluntary and involuntary movements. Muscular strength, however, is unimpaired. Unlike Parkinson's disease, the ability to initiate voluntary movements is intact. However, the conduct of a continuous movement (e.g. walking or speech production) is frequently impeded by super-imposed muscle jerks. Increased muscle tone is often noted in Huntington's chorea and can in some cases dominate the extra-pyramidal features.

Currently there is no way to prevent the progression of Huntington's chorea. The pathogenesis of the neuronal degeneration seen in the caudate nucleus, putamen and cerebral cortex is unknown. The pathology of Huntington's chorea appears almost opposite to that in Parkinson's disease. Pharmacologically those drugs that are able to produce Parkinsonism are helpful in the control of choreiform movements seen in Huntington's chorea. These drugs decrease the effectiveness of dopamine.

Sydenham's chorea usually occurs in childhood or adolescence with onset usually occurring between the ages of 5 and 10 years of age. Females are affected more than males and the condition is characterized by either acute or gradual onset of choreic involuntary movements. Although the prognosis for Sydenham's chorea is good and recovery is the general rule, the course is extremely variable. In some cases there is recovery within a matter of weeks while in others the abnormal movements can persist for years. Some patients have frequent relapses.

In many instances, Sydenham's chorea appears to be associated with either streptococcal infection (strep-throat) or rheumatic heart disease. Pathological changes have been variously described in the cerebral cortex, cerebellum, thalamus, caudate nucleus, putamen and mid-brain. These changes consist of widespread neuronal degeneration, vascular changes and, rarely, focal brain lesions from embolization resulting from endocarditis. No consistent neuro-pathologic lesion as found in Huntington's chorea, however, has been identified.

All aspects of speech production can be disrupted in patients with chorea. During contextual speech, the choreiform movements are super-imposed on the normal movements of the speech mechanism. Consequently, the hyperkinetic dysarthria of chorea is characterized by a highly variable pattern of interference with articulation, phonation, resonation and respiration. Choreiform movements of the respiratory muscles are manifested in the patient's speech as abrupt inhalation and expiratory sighs and sniffing noises. Phonatory disturbances include momentary voice arrests, strained-strangled voice, grunting and transient breathiness. Hypernasality occurs only in some choreic patients as evidence of abnormal involuntary movements of the soft palate. When present, the hypernasality is mild and fluctuates in accordance with the occurrence of the choreiform movements of the soft palate. Articulation may be normal for some periods during contextual speech but becomes distorted in an erratic manner by the sudden onset of choreiform movements of the lips, tongue and mandible. The flow of speech is, therefore, often jerky, being generated in fits and starts. Consequently a prosodic disturbance constitutes a significant part of the perceived speech deficit of the choreic patient. In addition these patients are usually unable to sustain syllable repetitions or prolong vowels because of an involuntary impulse to abort such efforts.

The ten most deviant speech dimensions observed in patients with chorea by Darley *et. al.* (1969a) were, in rank order: (1) imprecise consonants; (2) prolonged intervals; (3) variable rate; (4) monopitch; (5) harsh voice quality; (6) inappropriate silences; (7) distorted vowels; (8) excess loudness variation; (9) prolonged phonemes; (10) monoloudness. In particular, they reported that choreic patients were most distinctive from other dysarthrics in the area of prosodic alterations, being more deviant than any other neurological group on the speech dimensions of excess loudness variation, variable rate and prolongation of intervals. According to Darley *et al.* (1975a) the prosodic changes appear to represent an attempt by the choreic speaker to avoid articulatory and phonatory interruptions by variably altering their rate of speech, prolonging their phonemes and prolonging the intervals between words, equalizing the stress on syllables and introducing inappropriate silences.

(d) Ballism (hemiballismus)

Ballism is a rare involuntary movement disorder characterized by wild flailing movements on one side of the body, most marked in the arm. Facial muscles may also be affected. The term 'hemiballismus' derives from the fact that the condition usually affects only one side. The aetiology in most cases is vascular with the pathology involving either a haemorrhage or infarct in the contralateral sub-thalamic nucleus. Of all the quick hyperkinesias, ballism is the least important in relation to the occurrence of hyperkinetic dysarthria.

9.2.2 Slow hyperkinesias

The three main varieties of slow hyperkinesias are athetosis, dyskinesia and dystonia. Unlike the unsustained, abnormal involuntary movements seen in association with quick hyperkinesias, the abnormal involuntary movements seen in slow hyperkinesias build up to a peak slowly and are sustained for at least one second or longer. On occasions the abnormal muscle contractions are sustained for many seconds or even minutes. Muscle tone waxes and wanes producing a variety of distorted postures.

(a) Athetosis

Athetoid movements are continuous, arrhythmic, slow, writhing-type movements that are always the same in the same patient and cease only during sleep. The affected muscles are always hypertonic and may show transient stages of spasms. Although these movements especially involve the distal musculature of the limbs, the muscles of the face, neck and tongue may also be affected causing facial grimacing, protrusion and writhing of the tongue and difficulty in speaking and swallowing. Athetoid movements disrupt these functions by interfering with the normal contraction of the muscles involved. Athetosis is often part of a congenital complex of neurological signs, including those of cerebral palsy, that result from disordered development of the brain, birth injury, or other aetiological factors (Bannister, 1985). Although a critical lesion site has not been located, athetosis has been reported in association with lesions in the mid-brain tegmentum, subthalamic nucleus, ventro-lateral nucleus of the thalamus, globus pallidus, corpus striatum and cerebral cortex. Most commonly the condition is associated with pathological changes in the corpus striatum and cerebral cortex.

The muscles of the speech mechanism can be involved if the athetosis is bilateral, in which case the condition may also be associated with hyperkinetic dysarthria. Respiratory anomalies with the potential to cause speech problems, including irregular respiratory cycling, shallow rapid breathing, reduced vital capacity, reduced expiratory reserve volume and difficulty controlling prolonged exhalation have been reported in athetoid cerebral palsy children (Palmer, 1952; Hardy, 1964). In addition, problems associated with athetoid movements at other levels of the speech mechanism have also been noted in child athetoid cerebral palsy cases, including phonatory disorders resulting from spasm of the laryngeal abductor and adductor muscles (Palmer, 1952), phonation out of phase with articulation, monotony of pitch and breathiness (Rutherford, 1944) and articulatory disorders caused by aberrant control of the tongue, lips and mandible (Palmer, 1952).

Few studies have examined the speech characteristics of adult athetoid clients. The articulation of adult athetoid patients was investigated by Kent and Netsell (1978) and Platt, Andrews and Howie

(1980). The findings of these studies showed that articulatory disturbances decrease the intelligibility of athetoid patients. Further, the athetoid adults were found to exhibit disordered movements of the articulators during speech including features such as inaccuracy of tongue placements, reduced precision of fricative and affricative production, problems with vowel formation and a large range of jaw movements. Most of the adult athetoid patients examined by Kent and Netsell (1978) also had trouble achieving normal velo-pharyngeal closure.

(b) Dyskinesia (lingual-facial-buccal dyskinesias)

Dyskinesia is defined as 'impairment of the power of voluntary movements' (Miller and Keane, 1978). Although by definition all involuntary movements could be described as dyskinetic, it is usual that only two dyskinetic disorders are considered under this heading: tardive dyskinesia and levodopa-induced dyskinesia. The basic pattern of abnormal involuntary movement in both of these conditions is one of repetitive, slow, writhing, twisting, flexing and extending movements, often with a mixture of tremor. As both conditions may be limited to the bulbar musculature, they are sometimes referred to as 'focal dyskinesias'. In particular because the muscles of the tongue, face and oral cavity are most often affected, these two disorders are also termed 'lingual-facial-buccal' dyskinesias.

Tardive dyskinesia is a well-recognized side-effect of long-term neuroleptic treatment (treatment with a pharmacological agent having anti-psychotic action) (Crane, 1968; Maxwell, Massengill and Nashbold, 1970). The most prominent manifestation is lingual-facial-buccal dyskinesia, although occasionally limb, trunkal and respiratory chorea may accompany the facial movements. Maxwell, Massengill and Nashold (1970) described the symptoms of two cases of tardive dyskinesia and reported the nature of the associated speech disturbance in each case. The connected speech of one case was described as sounding muffled and was unintelligible even though individual phonemes were correctly produced. Hypernasality accompanied minimal movement of the soft palate and lateral pharyngeal walls and the patient also demonstrated tongue thrust behaviour. The connected speech of the second patient was also described as muffled and unintelligible. As in the first case, she also produced some individual phonemes correctly and some single words could be understood. Tongue thrust behaviour was also exhibited but no hypernasality, however, was perceived in her speech.

Tardive dyskinesia occurs late in the course of neuroleptic therapy, often after a decrease in the drug dosage or discontinuation of the drug therapy. The involuntary lingual–facial–buccal movements often persist for months to years after the neuroleptic therapy has been discontinued. If recognized early, however, the symptoms of tardive dyskinesia (including the associated hyperkinetic dysarthria) may recede with the discontinuation of neuroleptic treatment (Portnoy, 1979).

Combined lingual–facial–buccal dyskinesia is not unique to tardive dyskinesia but may also be seen as part of the hyperkinesia of Huntington's disease and following high-dose levodopa therapy in Parkinson's disease. Again the abnormal movements seen in this latter condition are typically choreic and also characteristically involve the muscles of the tongue, face and mouth. The lips may pucker and retract while the tongue may demonstrate 'fly-catcher' movements in which it involuntarily moves in and out of the mouth. In addition the jaws may open and close spontaneously or move from side to side and the palate also may elevate and lower involuntarily. If the limbs are affected it is interesting to note that the limbs least affected by the Parkinsonism show the most prominent choreiform movements following levodopa therapy, suggesting that Parkinsonism and chorea represent opposite dysfunctions of the neostriatum. The induction of lingual–facial–buccal dyskinesia by levodopa therapy is consistent with the proposal of Ludlow and Bassich (1984) that Parkinson's patients treated with levodopa have a speech disorder resembling a hyperkinetic rather than a hypokinetic dysarthria.

(c) Dystonia

Dystonic movements are abnormal involuntary movements which are slow and sustained for prolonged periods of time. Affected muscles are hypertonic and the dystonia tends to involve large parts of the body, particularly the muscles of the trunk, neck and proximal parts of the limbs. The involuntary movements tend to have an undulant, sinuous character which may produce grotesque posturing and bizarre writhing movements. The muscles of the speech mechanism may also be involved in which case the patient may exhibit: spasms of the face producing facial grimacing, eye closing (blepharospasm) and pursing of the lips, jaw spasm, causing the jaws to close or open widely; involuntary twisting and protrusion of the tongue; and respiratory irregularities.

The maintenance of an abnormal or altered posture, whether it involves a single focal part of the body (e.g. focal cranial dystonia) or a diffuse region of the body, is the most important distinguishing feature of dystonia. The abnormal involuntary contractions build up slowly, produce a prolonged distorted posture such as twisting of the trunk about the long axis (torsion spasm) and then gradually recede. Occasionally dystonic movements begin with a jerk and then build up to a peak before subsiding.

A variety of conditions may lead to dystonia including encephalitis, head trauma, vascular diseases and drug toxicity (especially the more potent tranquillizers). In addition, various progressive degenerative diseases of the central nervous system such as Wilson's disease and Huntington's disease often manifest dystonic features at some time in their course. One type of dystonia, dystonia musculorum deformans, is an inherited disease. In many cases, however, the cause of dystonia is unknown. Although it is often reported that dystonia is a disorder of the basal ganglia, lesions in the corpus striatum and globus pallidus having been described, no consistent and specific pathophysiology or pathomorphologic alteration in the brain has been identified.

Dystonia can interfere with speech production in that the prolonged, involuntary contractions (dystonic movements) commonly delay the start of voluntary movements, including those movements required for speech. In addition, the voluntary movements of dystonic patients tend to be slow in attaining their desired excursion and repetitive movements are restricted in range. Articulation, phonation and prosody may all be significantly disturbed in patients with dystonia. In a study of 30 dystonia cases, Darley *et al.* (1969a) found that the deviant speech dimensions of imprecise consonants, disturbed vowels, harsh voice quality, irregular articulatory breakdown, strained–strangled voice quality, monopitch, monoloudness, inappropriate silences, short phrases and prolonged intervals to be the ten major features of the hyperkinetic dysarthria seen in dystonia. Three of the four most prominent deviant characteristics were related to articulatory disturbance. Darley *et al.* (1969a) also reported that most of their dystonic patients also had some impairment in intelligibility while hypernasality was only observed in about one-third of their subjects with dystonia. Probably as a compensatory device to avoid articulatory and phonatory breakdowns, dystonic patients tend to slow their speech down by prolonging their production of phonemes and the intervals between syllables and inserting inappropriate silences (Darley *et al.* 1975a).

Golper *et al.* (1983) documented the clinical features of 10 cases with focal cranial dystonia. They reported that those subjects with oro-mandibular dystonia who also had a clinically significant dysarthria exhibited the speech characteristics as predicted to be found among slow hyperkinetic dysarthrias by Darley *et al.* (1969a, 1975a).

9.2.3 Essential tremor (organic voice tremor)

Tremors are involuntary movements resulting from the contraction of opposing muscle groups, which produces rhythmic or alternating movements of a joint or group of joints. A number of different types of tremor are recognized which include: physiological tremor (e.g. tremor associated with cold or nervousness); essential tremor (e.g. familial and senile); toxic tremor (e.g. tremor in thyrotoxicosis and alcoholism); and pathological tremor (e.g. intention tremor in cerebellar disorders and rest tremor in Parkinson's disease).

Essential tremor, also called 'hereditary', 'familial', 'action' or 'senile' tremor, is usually regarded as an exaggeration of normal physiological tremor. This type of tremor is absent at rest and appears when muscles act to move or support a body part (hence the name 'action tremor'). Essential tremor is not intensified towards the end of a movement as is intention tremor and is not associated with other evidence of neurologic disease. The tremor may occur sporadically or have a familial incidence and onset may occur at any age. The condition tends to be slowly progressive and the hands and head are the regions of the body most commonly affected. Although the location of the lesion causing essential tremor is unknown, a variety of sites involving components of the extrapyramidal system such as the caudate nucleus and putamen have been suggested (Critchley, 1949).

Essential tremor of the laryngeal muscles produces a condition called 'organic voice tremor'. In this condition the intrinsic and extrinsic laryngeal muscles may show tremor either independently or in combination with other parts of the body, such as the hands, jaw or face. Severe cases of organic voice tremor have been described as having acoustic features similar to cases of spastic dysphonia, including regular voice arrests (Brown and Simonson, 1963; Aronson *et al.*, 1968). Other features of laryngeal dysfunction observed by these authors in cases of organic voice tremor included excessively low pitch, monopitch, strained–strangled harshness and pitch breaks.

CHAPTER TEN

Dysarthrias associated with lesions in other motor systems

10.1 ATAXIC DYSARTHRIA

The cerebellum is responsible for the co-ordination of muscular activity throughout the body. Although it does not itself initiate any muscle contractions, the cerebellum monitors those areas of the brain that do in order to co-ordinate the action of muscle groups and time their contractions so that movements involving the skeletal muscles are performed smoothly and accurately. Damage to the cerebellum or its connections leads to a condition called 'ataxia' in which movements become uncoordinated. If the ataxia affects the muscles of the speech mechanism, the production of speech may become abnormal leading to a cluster of deviant speech dimensions collectively referred to as 'ataxic dysarthria'.

To understand why the disturbed speech features characteristic of ataxic dysarthria following damage to the cerebellum occur, a knowledge of the basic neuroanatomy and functional neurology of the cerebellum is essential.

10.1.1 Neuroanatomy of the cerebellum

Located posterior to the brainstem, the cerebellum occupies most of the posterior cranial fossa and is separated from the occipital and temporal lobes of the cerebrum by the tentorium cerebelli.

The cerebellum is comprised of two large cerebellar hemispheres which are connected by a mid-portion called the vermis (worm-like) (see Figure 10.1). The cerebellar surface or cerebellar cortex consists of complexly folded ridges of grey matter while the central core of the cerebellum consists of white matter in which are located several nuclear grey masses called the deep (cerebellar) nuclei. A series of

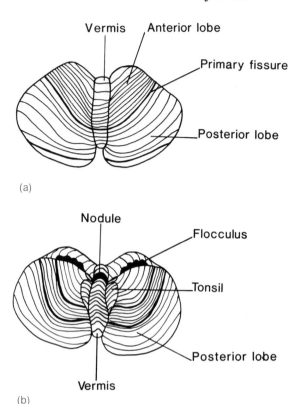

Figure 10.1 (a) Superior view of cerebellum and (b) inferior view of cerebellum.

deep and distinct fissures divide the cerebellum into a number of lobes. Although different authors have classified the cerebellar lobes in different ways, most neurologists recognize three different cerebellar lobes. These include: the anterior lobe; the posterior lobe; and the flocculonodular lobe (includes the paired flocculi and the nodulus) (see Figure 10.1). The anterior lobe, which can be seen from a superior view of the cerebellum, is that portion of the cerebellum that lies anterior to the primary fissure. It roughly corresponds to that part of the cerebellum called the paleocerebellum. The anterior lobe has a significant role in the regulation of muscle tone and receives its primary input from proprioceptors and exteroceptors in the head and body, including some from the vestibular system.

The largest portion of the cerebellum is the posterior lobe, also referred to as the neocerebellum. It is located between the other two lobes and is phylogenetically the newest portion of the cerebellum, being best developed in those animals such as primates that have a well-developed cerebral cortex. It functions in close association with the cerebral cortex and is most concerned with the regulation of voluntary movements. In particular it plays an essential role in the co-ordination of phasic movements and is the most important part of the cerebellum for co-ordination of speech movements. The flocculo-nodular lobe is comprised of the nodulus and the paired flocculi and occupies the inferior-rostal region of the cerebellum. The nodulus represents the rostral portion of the inferior vermis while the flocculi are two small irregular-shaped appendages attached to the inferior region of the cerebellum. Phylogenetically, the flocculonodular lobe represents the oldest portion of the cerebellum and is also known as the archicerebellum. It functions in close association with the vestibular system and is therefore important in maintaining equilibrium and keeping the individual oriented in space.

The cerebellum is made of both grey and white matter. As in the cerebral hemisphere, most of the grey matter is found covering the surface of the cerebellum as cortex. The cerebellar cortex is highly folded into thin transverse folds or folia. As a consequence of this extensive folding, in the region of 85% of the cerebellar cortex is concealed and its surface area is much larger than might be expected (about three-quarters of that of the cerebrum). Unlike the cerebral cortex, the cerebellar cortex is uniform throughout its structure.

The central core of the cerebellum is made up of white matter in which are embedded four grey masses on either side of the midline. These grey masses are referred to as the cerebellar or deep nuclei. The largest and most medially placed of these nuclei is the dentate nucleus. Medial to the dentate nucleus are two smaller nuclei, the globose and emboliform nuclei (taken together called the inter-positus), and most medial of all the deep nuclei is the fastigial nucleus. Most of the Purkinje cell axons which carry impulses away from the cerebellar cortex terminate in these nuclei. Some fibres from the cortex of the flocculonodular lobe, however, by-pass the deep nuclei and proceed to the brainstem.

The cerebellum is attached to the brainstem on either side by three structures called the cerebellar peduncles (see Figure 10.2). These peduncles are comprised of bundles of nerve fibres which convey impulses either to or from the cerebellum. The largest of the cerebellar peduncles is the middle peduncle or brachium pontis, which passes between the cerebellum and the pons. It is formed by the transverse fibres of the pons which arise from the pontine nuclei. These fibres cross the mid-line to pass through the middle peduncle on the side opposite to their nucleus of origin to reach the cortex of

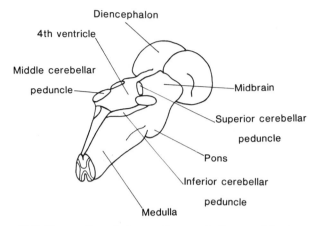

Figure 10.2 Connections between the cerebellum and brainstem.

the neocerebellum of the contralateral cerebellar hemisphere. The middle peduncle is comprised of afferent fibres belonging to the cortical–pontine–cerebellar pathway which conveys information from the frontal and temporal lobes of the cerebrum to the cerebellum via the pons.

Fibres running between the medulla oblongata and the cerebellum comprise the inferior cerebellar peduncle or restiform body. This peduncle is made up largely of afferent fibres which pass to the cerebellum and include the posterior spino-cerebellar tracts, the cuneo-cerebellar tracts, the olivo-cerebellar tract, the reticulo-cerebellar tract and the vestibulo-cerebellar tract.

The superior cerebellar peduncle, or brachium conjunctivum, is connected to the mid-brain at its junction with the pons. It carries most of the fibres that leave the cerebellum. These fibres originate in the dentate, globose and emboliform nuclei and pass to structures such as the contralateral red nucleus and the ventro-lateral nucleus of the thalamus. Although the majority of the superior peduncle is comprised of efferent fibres, a small portion of it carries afferent fibres which include the anterior spino-cerebellar tract, the rubro-cerebellar tract and the tecto-cerebellar tract.

In order to co-ordinate muscular contractions, the cerebellum is linked to a large number of other parts of the nervous system by an extensive system of afferent and efferent fibres. Damage to these pathways can cause cerebellar dysfunction and possible ataxic dysarthria the same as damage to the cerebellum itself. All the afferent fibres to the cerebellum (i.e. those fibres conveying impulses to the cerebellum) terminate in the cerebellar cortex, giving off collateral fibres to the cerebellar nuclei on their way. The principal afferent projections to the cerebellum are shown in Table 10.1.

The afferent fibres originate from three major sources: the cerebrum and brainstem; the vestibular nuclei; and the spinal cord. All of the afferent pathways arising from the cerebrum and many of those originating in the spinal cord undergo synaptic interruption in the brainstem, so that secondary fibres are relayed to the cerebellum from brainstem centres which receive primary descending fibres from the cerebral cortex and ascending fibres from the spinal cord. The major afferent pathway originating from the cerebral cortex is the cortical–pontine–cerebellar pathway. Although virtually all areas of the cerebral cortex contribute fibres to this pathway, it originates primarily in the motor cortex and projects to the ipsilateral pontine nuclei. From here, secondary fibres project primarily to the cortex of the neocerebellum in the contralateral cerebellar hemisphere. It is via the cortical–pontine–cerebellar tract that the cerebellum receives information regarding volitional movements that are in progress or are about to take place.

Afferent pathways to the cerebellum which also project from the brainstem arise in structures such as the olive (olivo-cerebelllar tract), the red nucleus (rubro-cerebellar tract), the reticular formation (reticulo-cerebellar tract), the mid-brain (tecto-cerebellar tract) and the cuneate nucleus (cuneo-cerebellar tract). The inferior olivary nucleus in the medulla oblongata receives major input from the cerebral cortex, caudate nucleus, globus pallidus, red nucleus, brainstem reticular formation and spinal cord and projects mainly crossed fibres to the neocerebellum. The cuneate nuclei receive input from the spinal cord and project to the ipsilateral paleo- and neo-cerebellum. Fibres from the vestibular nuclei project ipsilaterally to the flocculonodular lobe.

Cutaneous and proprioceptive information from the entire body also reaches the cerebellum. That from the face reaches the cerebellum either directly or via the trigeminal nucleus (trigeminal–cerebellar tract) while that from the remainder of the body is carried in the spinal–cerebellar tracts to the anterior lobe of the cerebellum. The signals transmitted in these tracts originate in various receptors such as muscle spindles, Golgi tendon organs, joint receptors and skin receptors. They inform the cerebellum of the momentary status of muscle contraction, degree of tension on the muscle tendons, positions of parts of the body and forces acting on the surfaces of the body.

All of the efferent tracts from the cerebellum arise in the deep nuclei. None arises directly from the cerebellar cortex which transmits

Table 10.1 Principal afferent projections to the cerebellum

Origin in Central nervous system	Pathway	Specific site of origin	Peduncle	Termination in cerebellum
Cerebral cortex	Cerebro-pontine-cerebellar	Primarily the motor cortex of the frontal lobe. Secondary fibres project from the pontine nuclei	Middle	Posterior lobe
Brainstem	Olivo-cerebellar	Inferior olivary nuclear complex	Inferior	All lobes and deep nuclei
	Rubro-cerebellar	Red nucleus in mid-brain	Superior	Posterior lobe and deep nuclei
	Reticulo-cerebellar	Reticular formation in brainstem	Inferior and middle	All lobes and deep nuclei
	Tecto-cerebellar	Mid-brain quadrigeminal bodies	Superior	Anterior and posterior lobe
	Cuneo-cerebellar	Cuneate nuclei	Inferior	Primarily anterior lobe
	Vestibulo-cerebellar	Vestibular nuclei	Inferior	Flocculonodular lobe
	Trigemino-cerebellar	Trigeminal nucleus	Superior	Anterior lobe
Spinal cord	Spinal-cerebellar	Various peripheral receptors such as muscle spindles and Golgi tendon organs	Inferior and superior	Anterior lobe

Table 10.2 Principal efferent projections from the cerebellum

Pathway	Deep nuclei of origin	Destination	Peduncle
Cerebello-thalamic	Dentate nucleus and interpositus nucleus	Contralateral thalamus (ventro-lateral and ventro-anterior nuclei). Secondary fibres project from here to the motor cortex of the frontal lobe	Superior and inferior
Cerebello-rubral	Interpositus nucleus and dentate nucleus	Contralateral red nucleus	Superior
Cerebello-reticular	Fastigial nucleus, dentate nucleus and interpositus nucleus	Ipsilateral and contralateral reticular nuclei	Superior and inferior
Cerebello-vestibular	Fastigial nucleus	Ipsilateral and contralateral vestibular nuclei	Inferior

its output signals only through the deep nuclei. Efferent signals are transmitted to many portions of the central nervous system, including: the motor cortex via the thalamus; the basal ganglia; the red nucleus; the reticular formation of the brainstem; and the vestibular nuclei. The principal cerebellar efferent fibres are summarized in Table 10.2.

Most fibres from the dentate nucleus project to the thalamus with some fibres also passing to the red nucleus, the olive and the reticular formation. The globose and emboliform nuclei project mainly to the red nucleus, the olive and brainstem reticular formation. The major projection of the fastigial nucleus is to the vestibular nuclei and to a lesser extent to the reticular formation.

Efferent fibres from the dentate nuclei and interpositus nuclei project to the contralateral ventro-lateral (and ventro-anterior nucleus) of the thalamus. These thalamic nuclei, in turn, project fibres to the motor areas of the frontal lobe. By way of this pathway, the cerebellum can influence the functioning of the motor cortex. Important fibres passing from the interpositus nuclei and dentate nucleus also project to the contralatral red nucleus where they influence the activity of the rubro-spinal tract. In the mid-brain the rubro-spinal tract decussates to the opposite side and passes through the brainstem to the lateral funiculus of the spinal cord. Each cerebellar hemisphere is thus linked principally with the same side of the body by means of a double decussation in the mid-brain. Other fibres from the red nucleus project to the thalamus and thus may connect the cerebellum to the basal ganglia and cerebral cortex. The fastigial nucleus sends fibres to the vestibular nuclei in the brainstem to modify the activity of the vestibulo-spinal tracts and thereby influence muscle tone, important for maintaining body posture and equilibrium. Projections from the cerebellar nuclei also travel to the reticular formation where they enable cerebellar modification of the activity of the reticulo-spinal tracts.

Overall the extensive afferent and efferent connections of the cerebellum form a feedback loop by which the cerebellum can both monitor and modify motor activities taking place in various parts of the body to produce a smooth, co-ordinated motor action. Through its afferent supply, the cerebellum receives input from a number of different regions of the nervous system regarding motor activities either already in progress or about to occur. Further, by means of its efferent connections, the cerebellum is able to modify motor actions initiated elsewhere in the central nervous system.

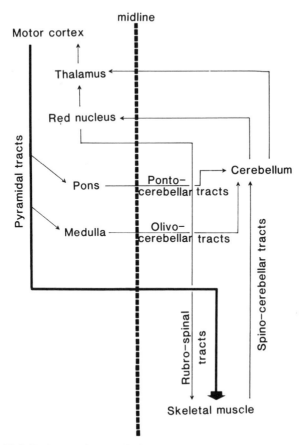

Figure 10.3 Pathways for cerebellar control of voluntary movements.

10.1.2 Function of the cerebellum in voluntary motor activities

The cerebellum does not, itself, initiate any muscular activities. Rather it functions only to co-ordinate those motor activities initiated elsewhere in the central nervous system. Voluntary movements can take place in its absence, but such movements are clumsy and disorganized. Although the role of the cerebellum in motor activities is not known with certainty, a variety of functions have been proposed by different authors. Some researchers have suggested that the function of the cerebellum is to exercise revisory control over the motor commands issued by the motor cortex (Ito, 1970; Eccles, 1973). Figure 10.3 illustrates the basic cerebellar pathways that may be utilized by the cerebellum to co-ordinate voluntary movements. Briefly, it is proposed that, when motor impulses are transmitted from the motor cortex to the lower motor neurones via the pyramidal and extra-pyramidal pathways (see Chapter 8), collateral impulses are also transmitted simultaneously to the cerebellum primarily via the ponto-cerebellar tract and, to a lesser extent, by way of the olivo-cerebellar tract. Consequently, for every voluntary motor movement performed, not only do the muscles receive activating signals, but the cerebellum receives the same signals at the same time.

Once the muscles respond to the motor impulses by contracting, various proprioceptive receptors such as muscle spindles, Golgi tendon organs and joint receptors transmit information back to the cerebellum by way of the spino-cerebellar and trigemino-cerebellar tracts. It has been suggested that one function of the cerebellum may be to regulate the sensitivity of the muscle spindles so that optimum feedback regarding the state of muscle contraction can be supplied to the higher centres as a movement is performed (Gilman, 1969). The cerebellum interprets the afferent proprioceptive information and inte-grates it with the information received from the motor cortex (Konorski, 1967). Subsequently, impulses are transmitted from the cerebellar cortex, where the integration takes place, to the deep nuclei, particularly the dentate nucleus, and from here via the superior peduncle to the ventro-lateral nucleus of the thalamus. The thalamus in turn relays the cerebellar output back to the motor cortex where the stimulus first originated to modify the activity of the cortico-spinal and cortico-bulbar tracts. Thus it can be readily recognized that the circuit described represents a complicated feedback loop beginning in the motor cortex and returning to the motor cortex. Eccles (1973) suggested that the commands from the motor cortex are imprecise and provisional, requiring for their refinement the continuous action of the cerebellar feedback loop. In this way, the motor cortex may depend on the cerebellum and its connections to modify its grossly formed neural instructions to the muscles.

The cerebellum acts as a servo-mechanism to damp muscle move-ments and stop them at their point of intention. It appears to compare the intentions of the motor cortex with the actual performances by the various muscles and if the body parts are not moving in accordance with the intentions of the motor cortex, the 'error' between these two is calculated by the cerebellum so that appropriate and immediate corrections can be made. Ordinarily, the motor cortex transmits a greater volley of impulses than needed to perform each intended movement. Consequently, the cerebellum must act to inhibit the motor cortex at the appropriate time after the muscle has begun to move. The rate of movement is automatically assessed by the cere-bellum and the length of time required for the body part to reach the point of intention calculated. Appropriate inhibitory impulses are then transmitted by the cerebellum to the motor cortex to inhibit the agonist muscle (the muscle responsible for producing the movement) and to excite the antagonist muscle (the muscle that opposes or has the opposite action to the agonist muscle). In this way the cerebellum is able to apply a 'braking' or damping mechanism to stop the movement at the precise point of intention. If the cerebellum is damaged, then the damping action is impaired and over-shooting of the point of intention occurs. The conscious centres at the cerebrum recognize that this over-shooting has occurred and initiate movement in the opposite direction. However, over-shooting may then occur in the opposite direction and thus the body part involved in the movement may oscillate back and forth past the intended point for several cycles. This effect is called 'intention tremor'.

In addition to being able to initiate damping actions, the cerebellum is also able to predict the future positions of moving parts of the body. Cerebellar damage leads to a disturbance in this ability and consequently the patient may be unable to gauge the distance and speed of body movements, an inability referred to as 'dysmetria'. Dysmetria, in combination with the inability to apply appropriate damping mechanisms, leads to the decomposition of body movements in patients with cerebellar disorders, especially those movements that require the co-ordinated action of a number of individual muscles. In complex motor acts, contraction of the individual muscles must be timed in such a way that one movement leads to the next in a smooth and orderly fashion. When the cerebellum becomes dysfunctional, the patient is unable to correctly time the beginning of the individual sequential movements that comprise the overall motor act. As a result, sequential movements may begin too early or too late and the overall movement pattern breaks down. Speech is an extremely complex motor act, requiring the co-ordinated contraction of a large number of individual muscles. It is not surprising, therefore, that speech production may be seriously affected in patients with disorders affecting the cerebellum or its connections.

Involuntary or postural movements controlled by the extra-pyramidal system are also modified by the cerebellum in a manner similar to voluntary movements. Even reflex actions mediated through the spinal cord or brainstem are affected by the cerebellum. The major difference in the cerebellum's role in co-ordinating involuntary as opposed to voluntary motor activities is that different pathways are used. Motor impulses that originate in the basal ganglia and reticular formation and pass to the lower motor neurones via the extra-pyramidal system are conveyed to the cerebellum via the olivo-cerebellar trace. Corrective signals from the cerebellum are passed to the muscles via the major descending pathways of the extra-pyramidal system, namely the reticulo-spinal, rubro-spinal and vestibulo-spinal tracts (see Chapter 8) rather than via the cortico-spinal and cortico-bulbar pathway as in the case of voluntary movements.

10.1.3 Clinical signs of damage to the cerebellum

Disorders of the cerebellum and/or the fibres leading to and from the cerebellum are accompanied by a number of characteristic signs which include disorders of movement, posture and muscle tone. Clinical signs usually appear on the same side of the body as the cerebellar lesion. Different symptoms result from cerebellar damage depending on the part of the cerebellum involved. Broadly, cerebellar disorders can be divided into those that affect the vestibulo-cerebellum (flocculonodular lobe) and those affecting the main mass of the cerebellum (the corpus cerebelli, which includes both the anterior and posterior lobes). Obviously, in many disorders both the corpus cerebelli and the flocculonodular lobe are involved.

Isolated damage to the flocculonodular lobe is most commonly associated with the presence of a tumour, usually a medulloblastoma, and is associated with a disturbance in equilibrium called 'archi-cerebellar syndrome'. Patients are unsteady and have a tendency to fall either backwards, forwards or to one side when standing on a narrow base with their eyes open. Some patients, in fact, are unable to maintain an upright posture. Their gait is staggering and they tend to walk on a wide base. Other signs may also be present if the tumour later invades other parts of the cerebellum.

Damage to the corpus cerebelli or its connections is associated with a group of symptoms commonly collectively called 'neocerebellar syndrome', although the paleocerebellum is also usually involved. Destruction of small portions of the corpus cerebelli causes no detectable abnormality in motor function. It appears that the remaining areas of the cerebellum can compensate for the damaged part. More severe and enduring dysfunction of the cerebellum, however, occurs if either the deep nuclei or superior cerebellar peduncle are involved. When the lesion involves the cerebellum unilaterally, as is usually the case, the motor disturbance occurs on the same side as the lesion. The characteristic signs of neocerebellar syndrome are listed in Table 10.3.

As indicated above, complex movements, such as required for speech production or the movement of an entire limb to a new position, depend on the proper sequencing of composite simple move-

Table 10.3 Clinical signs of neocerebellar syndrome

- Ataxia
- Dysmetria
- Decomposition of movement
- Dysdiadochokinesia
- Hypotonia
- Asthenia
- Tremor
- Rebound phenomenon
- Disturbance of posture and gait
- Nystagmus
- Dysarthria

ments (composition). Further, they are also dependent on the contraction of synergistic muscles to provide postural fixation of certain joints to allow for the precise movement of other joints (synergia meaning 'co-operative action of muscles'). Cerebellar dysfunction causes errors in both of these parameters leading to slowing, dysmetria, dyssynergia and decomposition of movement. In the case of alternating movements, dysdiadochokinesia also results. The resulting uncoordinated, clumsy and disorganized muscular activity is termed 'ataxia'. Ataxia is the principal sign of cerebellar dysfunction.

The presence of dysmetria is evidenced by the patient's inability to stop a movement at the desired point. For example, when reaching for an object, the patient's hand may either over-shoot the intended point or stop short of the intended point. Dyssynergia is reflected in the separation of a series of voluntary movements that normally flow smoothly and in sequence into a succession of mechanical or puppet-like movements (decomposition of movement). It may also be manifest as movement abnormalities such as delayed starting and stopping of movements. Dysdiadochokinesia refers to an inability to perform rapid alternating movements, such as rapidly moving the tongue from one side of the mouth to the other and back several times. To be performed rapidly such movements require considerable co-ordination by the cerebellum and are, therefore, severely disturbed in patients with cerebellar disorders.

A decrease in the muscle tone (hypotonia) is usually evident in cerebellar disorders (as can be ascertained by palpation of the muscles) and muscles affected by cerebellar lesions tend to be weaker and tire more easily than normal muscles (asthenia). The reduction in muscle tone may result from reduced muscle spindle sensitivity, possibly because of an inadequate alpha-motor neurone discharge as a result of the cerebellar damage. Tremor is another feature of cerebellar disease. It usually takes the form of an intention tremor and is seen during movement but is absent at rest. Disturbances of posture and gait may be very pronounced, the patient possibly being unable to maintain an upright posture and walking in a staggering fashion with a broad base of support.

The presence of rebound phenomenon can be demonstrated by asking patients to flex their elbow against resistance offered by the observer when their hand is only a small distance from their face and then suddenly releasing the fore-arm. Normally the movement of the fore-arm in the direction of flexion (i.e. towards the face) is quickly arrested by contraction of the extensor muscle (triceps). Cerebellar damage, however, delays this contraction with the result that the patients may strike themselves in the face. Nystagmus may also be present, especially if the lesion encroaches upon the vermis.

Finally, as a result of dyssynergy and decomposition of the movement of the muscles of the speech mechanism during speech production, ataxic dysarthria may be present in association with some cerebellar lesions. The characteristics of ataxic dysarthria are discussed in detail below.

10.1.4 Diseases of the cerebellum associated with ataxic dysarthria

The cerebellum can be affected by a variety of different pathological conditions (see Table 10.4), all of which may be associated with the occurrence of ataxic dysarthria. In general the signs and symptoms of cerebellar dysfunction are the same, regardless of aetiology, with the exception that in those disorders in which the lesion is slowly progressive (e.g tumour cases), symptoms of cerebellar deficiency are much less severe than where the lesion develops acutely (e.g. cerebrovascular accidents and head trauma). Further, considerable recovery from the effects of an acute lesion can usually be expected.

10.1.5 Clinical characteristics of ataxic dysarthria

Ataxic dysarthria results from damage to the cerebellum and/or its connections. Although it may occur in association with severe, acute cerebellar lesions, being most evident in the early stages of such disorders and then slowly subsiding, ataxic dysarthria most frequently occurs as a late sign of more slowly developing bilateral or generalized cerebellar damage. If ataxic dysarthria is a prominent and early sign of cerebellar damage, it has been suggested that this indicates that the lesion involves the mid-portion of the vermis, as this region is believed by some researchers to be the primary locus for the co-ordination of motor speech (Lothman and Ferrendelli, 1980). Although the loci of cerebellar lesions associated with the occurrence of dysarthria are not entirely certain, the results of a study conducted by Lechtenberg and Gilman (1978) suggest that speech is localized to the left cerebellar hemisphere.

Ataxic dysarthria is a motor speech disorder in which a breakdown in the articulatory and prosodic aspects of speech are the predominant features. The imprecise articulation results in improper formation and separation of individual syllables leading to a reduction in intelligibility while the disturbance in prosody is associated with loss of texture, tone, stress and rhythm of individual syllables. The dysprosody results in slow, monotonous and improperly measured speech, often termed 'scanning speech'. Charcot (1877, p. 192) was probably the first to use the term 'scanning' to describe the speech of patients with cerebellar disorders. He stated that 'the words are as if measured or scanned: there is a pause after every syllable, and the syllables themselves are pronounced slowly'. It should be noted, however, that the term 'scanning speech' has been used by other authors to describe a different set of speech characteristics from those referred to by Charcot. Consequently, unless the term is fully explained, some authors (Darley, Aronson and Brown, 1975a) recommend that it not be used.

The most extensive description of the perceptual features of ataxic dysarthria reported to date is that provided by Brown, Darley and Aronson (1970). They divided the ten deviant speech dimensions they considered to be most characteristic of ataxic dysarthria into three clusters: articulatory inaccuracy; prosodic excess; and phonatory–prosodic insufficiency. Articulatory inaccuracy includes deviant speech dimensions such as imprecise consonants, irregular articulatory breakdown and distorted vowels. They believed that these articulatory problems were the product of ataxia of the respiratory and oral–buccal–lingual musculature. Prosodic excess, which includes deviant speech dimensions such as excess and equal stress, prolonged phonemes, prolonged intervals and slow rate were thought by these workers to result from slow movements, while they attributed the occurrence of phonatory–prosodic insufficiencies including monopitch, monoloudness and harsh voice to the presence of hypotonia. Subsequently, Darley et al. (1975a) indicated that inaccuracy of movement, slowness of movement and hypotonia are the prominent features of ataxia as it affects speech production.

Table 10.4 Diseases of the cerebellum associated with ataxic dysarthria

Diseases	Example	General features
Congenital anomalies	Cerebellar agenesis	Partial to almost total non-development of the cerebellum. May in some cases not be associated with any clinical evidence of cerebellar dysfunction. In other cases, however, a gait disturbance may be evident in addition to limb ataxia (especially involving the lower limbs) and dysarthria
Chromosomal disorders	Trisomy	Diffuse hypotrophy (under-development) of the cerebellum may be present which may be associated with either no clinical symptoms of cerebellar dysfunction through to marked limb ataxia
Trauma	Penetrating head wounds (e.g. bullet wounds)	May be associated with either mild slowly developing cerebellar dysfunction or rapid, severe cerebellar dysfunction
Vascular disease	Occlusion of anterior-inferior cerebellar artery	Hypotonia and ipsilateral limb ataxia
	Occlusion of posterior-inferior cerebellar artery	Dysarthria, nystagmus, ipsilateral limb ataxia and disordered gait and station
	Occlusion of superior cerebellar artery	Disordered gait and station. Ipsilateral hypotonia. Ipsilateral limb ataxia and intention tremor. Occasionally dysarthria
Infections	Cerebellar abscess	Most frequently caused by purulent bacteria but can also occur with fungi. Cerebellar abscesses most frequently arise by direct extension from adjacent infected areas such as the mastoid process or from otologic disease
Tumours	Medulloblastomas, astocytomas and ependynomas	Primary tumours of the cerebellum occur more frequently in children than adults. Medulloblastomas occur most

usually have a rapid course with a poor prognosis.

	Astrocytomas are more benign than medulloblastomas and generally occur in children of an older age group than medulloblastomas. Ependymomas are relatively slow growing and again are more common in children than adults
Toxic metabolic and endocrine disorders	
Exogenous toxins, e.g. industrial solvents, carbon tetrachloride, heavy metals, etc.	Signs of cerebellar involvement usually associated with symptoms of diffuse involvement of the central nervous system following these intoxications rather than appearing in isolation to other neurological deficits.
Enzyme deficiencies, e.g. pyruvate dehydrogenase deficiency	Ataxia most marked in the lower limbs
Hypothyroidism	Cretins show poor development of the cerebellum. Ataxia present in 20–30% of myxoedema cases
Hereditary ataxias	
Friedreich's ataxia	The most commonly encountered spinal form of hereditary ataxia. Pathological degeneration primarily involves the spinal cord with degeneration of neurones occurring in the spino-cerebellar tracts. Some degeneration of neurones in the dentate nucleus and brachium conjunctivum may also occur. The first clinical sign of the disease is usually clumsiness of gait. Later limb ataxia (especially involving the lower extremities) also occurs. A large percentage of cases also exhibit dysarthria and nystagmus and cognitive deficits may also be present.
Demyelinating disorders	
Multiple sclerosis	Usually associated with demyelination in a number of regions of the central nervous system including the cerebellum. Consequently the dysarthria, if present, usually takes the form of a mixed dysarthria rather than purely an ataxic dysarthria. Paroxysmal ataxic dysarthria may occur as an early sign of multiple sclerosis

Enderby (1986), using the Frenchay Dysarthria Assessmen (Enderby, 1983), identified the following ten features to be character istic of ataxic dysarthria: poor intonation and phonation; poor tongu movement in speech; poor alternating movement of the tongue i speech; reduced rate of speech; poor swallowing; reduced latera movement of the tongue; reduced elevation of the tongue; reducec intelligibility of conversation; poor alternating movement of the lips and poor lip movements in speech.

Hypernasality does not appear to be a prominent feature of ataxic dysarthria, suggesting that velo-pharyngeal function in patients with cerebellar lesions may be normal. Oral examination of these patients supports these perceptual findings, in that elevation of the soft palate during phonation appears normal in most cases.

In addition to the studies described above based on perceptual assessment of the speech of ataxic speakers, a number of researchers have used instrumental techniques to study the speech production of ataxic dysarthrics. In a case study that included spectrographic and cinefluorographic analysis of ataxic dysarthria, Kent and Netsell (1975) described articulatory (motoric)–acoustic correlates for many of the perceptually deviant speech dimensions reported by Brown, Darley and Aronson (1970). Among the motor abnormalities observed by Kent and Netsell (1975) were slow movements, prolonged articulatory settings for consonants and vowels, errors of direction and range of articulatory movements and reduction of overall articulatory mobility. Abnormal acoustic characteristics reported by these workers included monotone, lengthened vowel and consonant segments, lengthened formant transition durations, frictionalization of stop gaps and occasional erratic changes in fundamental frequency. These physiological and acoustic observations were consistent with the perceptual descriptions of ataxic dysarthria provided by Brown, Darley and Aronson (1970).

Kent, Netsell and Abbs (1979) conducted a spectrographic analysis of the speech of five patients with ataxic dysarthria. They reported that the most marked and consistent abnormalities observed in the spectrograms were alterations of the normal timing patterns and a tendency towards equalized syllable durations. The ataxic dysarthric patients, when compared to normal controls, all showed some degree of lengthening of word segments. The perceptually recognized 'scanning' element of the speech disorder was evidenced in the spectrographic data as a regularity of syllable production with distinct segregation of segments. Kent, Netsell and Abbs (1979) also noted two consistent features of the fundamental frequency contour. These were a flat lower contour with a top pattern showing a fall on each syllable. This pattern gives a monotonous character to the utterance and is consistent with the monotony of pitch reported in perceptual studies of ataxic dysarthria.

Using an X-ray micro-beam technique and electromyography, Hirose (1986) examined the movement patterns in the jaw and lower lip of ataxic dysarthrics and provided physiological evidence of inconsistency in their articulatory movements. In particular, he showed that the articulatory movements are characterized by inconsistency in both the range and velocity of movement. Further, Hirose (1986) found that the electromyographic patterns of two articulatory muscles (the anterior belly of the digastric and the mentalis) were irregular both in shape and timing and that there was a tendency towards a disturbance in the initiation of muscle activity in repetitive movement. The inconsistency in the articulatory movement observed by Hirose (1986) is compatible with the perceived irregular articulatory breakdown reported in perceptually-based studies of ataxic dysarthria.

A number of investigators have reported that ataxic dysarthrics demonstrate essentially normal patterns of succession for articulatory gestures (Kent and Netsell, 1975; Netsell and Kent, 1976; Hirose, 1986). In other words, the ordering of articulatory movements is not disturbed in ataxic dysarthria. At the same time, however, individual articulatory movements are abnormal as indicated above. The articulatory disturbance in ataxic dysarthria is not caused by a breakdown in the co-ordination of component articulatory movement, but rather represents a problem in controlling the execution of individual articulatory movements, the sequencing of which is essentially normal.

Kent and Netsell (1975) suggested that many of the prosodic disturbances seen in ataxic dysarthria could be the result of generalized hypotonia of the muscles of the speech mechanism. It is possible, for instance, that hypotonia could cause a delay in the generation of muscular force leading to prosodic disturbances such as prolongation of phonemes and prolonged intervals between phonemes. Further, hypotonia could conceivably reduce the rate of muscular contraction thereby producing slowness of movement. Although hypotonia may also be a possible explanation for some of the articulatory disturbances seen in ataxic dysarthria, other neuro-muscular explanations are required for other specific movement disturbances.

Based on a study of a case with paroxysmal ataxic dysarthria (an episodic form of ataxic dysarthria in which remissions and relapses occur at regular and frequent intervals), Netsell and Kent (1976) suggested that the role of the cerebellum in speech is to monitor the positions of the components of the speech mechanism and to generate the necessary motor instructions where appropriate to enable the articulatory targets intended by the cerebral cortex to be achieved. They argued that because, as indicated above, the co-ordination of articulatory movements is essentially normal in ataxic dysarthria, the successional pattern for speech movements is programmed by the cerebral cortex and that cerebellar dysfunction results in the delayed and inaccurate execution of the required component individual muscle movements. Thus the cerebellum may serve to supplement and revise as necessary the basic motor programme supplied by the motor cortex in line with the notions of cerebellar function in motor activities proposed by Eccles (1973) and Ito (1970). Hirose (1986) also concluded that the deterioration of speech in ataxic dysarthria is primarily caused by an impairment of the updating–revisory function of the cerebellum. Kornhuber (1977) also regarded the cerebellum as having a refining role in motor activities.

In summary, ataxic dysarthria is associated with some cerebellar lesions and is characterized by disturbances in the articulatory and prosodic aspects of speech. Control of individual articulatory movements is irregular due to factors such as a disturbance in the updating–revisory function of the cerebellum and possibly hypotonia in the muscles of the speech mechanism. This leads to articulatory problems such as inaccurate consonant production and distorted vowels. Rather than being consistent and regular, as in other types of dysarthria, the articulatory breakdown is irregular and transient. These articulatory characteristics lead to impairment of the overall intelligibility. Dysprosody is also an important feature of ataxic dysarthria. The prosodic disturbance includes features such as inappropriate stress on syllables, monopitch, monoloudness and a slow rate of speech. It has been suggested that these prosodic characteristics result from hypotonia of the speech muscles.

10.2 MIXED DYSARTHRIA

A number of nervous system disorders affect more than one level of the motor system. Consequently, in addition to the more 'pure' forms of dysarthria discussed previously, speech pathologists in clinical practice often encounter patients who exhibit what are called 'mixed dysarthrias'. Dysarthria of this latter type may occur in patients with cerebrovascular accidents, head trauma, brain tumours, inflammatory diseases and degenerative conditions. There are three neurological diseases that typically display symptoms of mixed dysarthria. These are amyotrophic lateral sclerosis, multiple sclerosis and Wilson's disease. The type of dysarthria associated with each of these conditions together with their epidemiology, prognosis and neurological symptoms is discussed below.

10.2.1 Amyotrophic lateral sclerosis

Amyotrophic lateral sclerosis is a form of motor neurone disease characterized by selective and progressive degeneration in the cortico-spinal and cortico-bulbar pathways and in the motor neurones associated with the cranial nerves and anterior horn cells of the spinal cord. The intellectual abilities of the patient remain intact. Because it involves the progressive degeneration of both upper and lower motor neurones, the condition is associated with a speech disorder with characteristics of both a pseudobulbar and bulbar palsy, i.e. a spastic–flaccid dysarthria. The incidence rate of amyotrophic lateral sclerosis is 1.4 per 100000 population (Rosenberg and Pettegrew, 1980). Despite numerous clinical investigations, little is known regarding the aetiology of amyotrophic lateral sclerosis. Bannister (1985) has suggested that the cause may be a variety of hereditary, traumatic, toxic or viral influences that bring about motor neurone degeneration in susceptible subjects. Hypothyroidism, lead poisoning and auto-immune disease have been reported to produce an amyotrophic lateral sclerosis-like condition, with the exception that the symptoms are reversible. Some authors have suggested that the presence of a persistent polio virus may be the cause of the disease (Schiffer *et al.*, 1987). Several different inherited forms of the disorder have been reported (Patten, 1987), including a sex-linked and autosomal dominant form.

Amyotrophic lateral sclerosis is a fatal disorder that develops primarily between the ages of 50 and 70 years with a median age of onset of 66 years (Mulder and Kurland, 1987). The median duration of the disease is about two years from onset to death, with longer durations occurring in the younger patients. The mode of onset is usually insidious and the relative prominence of symptoms depends upon which motor neurones are most affected.

Degeneration of both upper and lower motor neurones produces signs of both upper and lower motor neurone lesions, although the signs of one may predominate at any one time. Spastic paralysis is present unless degeneration of the lower motor neurone is far advanced. Upper motor neurone signs such as hyperactive muscle stretch reflexes, spasticity, a positive Babinski sign and a positive sucking reflex may be present in addition to signs of lower motor neurone involvement such as muscle atrophy and fasciculations. The most important feature of the disorder as far as speech is concerned is an all-pervading weakness of the bulbar musculature. Damage to the bulbar musculature inevitably produces problems in swallowing, aspiration, saliva control and speech (Hillel and Miller, 1987).

The presence of a severe speech disorder is a common finding in amyotrophic lateral sclerosis. Carrow *et al.* (1974) reported that 97% of their sample of 79 subjects exhibited one or more deviant speech symptoms. Because the disorder is progressive in nature, the characteristics of the speech disorder tend to change over time. The most deviant speech characteristics seen in association with this condition represent a mixture of those deviant speech dimensions which occur in bulbar (phonatory and resonatory incompetence) and pseudo-bulbar palsy (prosodic excess and insufficiency, articulatory and resonatory incompetence and phonatory stenosis) (Darley *et al.*, 1975a). The combination of spastic and flaccid dysarthria has a much more severe effect on speech intelligibility than either dysarthria alone. Overall, the pattern of speech deviations is more characteristic of spastic than flaccid dysarthria. Based on a perceptual analysis of the speech abilities of 30 patients with amyotrophic lateral sclerosis, Darley *et al.* (1975a) found that the ten most dominant speech deviations were imprecise consonants, hypernasality, harsh voice quality, slow rate, monopitch, short phrases, distorted vowels, low pitch, monoloudness and excess and equal stress. In fact characteristics such as distortion of vowels, slow rate, shortness of phrases and imprecision of consonants are more deviant in this condition than in any other neurological disorder. Further, they reported the presence of three prosodic alterations in amyotrophic lateral sclerosis that do not appear in either bulbar or pseudobulbar palsy: prolongation of intervals; prolongation of phonemes; and inappropriate silences.

In addition to the findings of studies based on perceptual analysis, further light has been shed on the nature of the speech disturbance in amyotrophic lateral sclerosis by instrumental investigations. In particular, tongue strength and mobility has been reported to be severely reduced in patients with this condition. Dworkin and Hartman (1979) used a combination of a pressure transducer and the measurement of lingual alternate motion rates to follow the progression of amyotrophic lateral sclerosis in a 49-year-old man. They found that tongue strength was 25% of that found in normal men and that the alternate motion rate of the tongue was nearly one-third of the rate usually found in normal adult men. A pressure gauge was also used by Dworkin (1980) to measure tongue strength in patients with this condition. He found that both the anterior and lateral tongue strength was reduced in amyotrophic lateral sclerosis. Further, Dworkin, Aronson and Mulder (1980) found that there was a high negative correlation between tongue force and severity of articulation defect and also between syllable rate and severity of articulation defect. The findings of Dworkin and co-workers has recently been supported by de Paul *et al.* (1988). Using custom-designed force transducers, they found that the tongue muscles are more severely affected in amyotrophic lateral sclerosis than the facial muscles and muscles of mastication.

Hirose, Kiritani and Sawashima (1982a) used a system called 'pellet tracking' for automatically tracking the moving articulators in patients with amyotrophic lateral sclerosis and pseudobulbar palsy. They found that in both of these groups of patients, articulatory movements were slow and the range of movement limited. The regularity of articulatory movements, on the other hand, was maintained. Further, the articulatory dynamics of both of these groups were found to be similar, leading these authors to suggest that the two groups be combined as 'paralytic dysarthria' to simplify the description of their articulatory characteristics.

On the basis that the results of the perceptual studies conducted by Carrow *et al.* (1974) and Darley *et al.* (1975a) suggested that the speech of amyotrophic lateral sclerosis patients is temporally disrupted, Caruso and Burton (1987) used broad-band sound spectrograms to investigate stop-gap duration, voice onset time and vowel duration in intelligible speakers with this disorder and normal speakers. Significant differences were found between the two groups for stop-gap and vowel duration but no significant differences were found in voice onset time. The impairment to the lingual musculature that these data indicate supports the findings of de Paul *et al.* (1988) that the hypoglossal nucleus is affected more early and more severely than the motor trigeminal and facial nuclei. The acoustic results obtained by Caruso and Burton (1987) also support the finding that tongue strength (Dworkin, Aronson and Mulder, 1980) and speed (Hirose, Kinitani and Sawashima, 1982b) are impaired in patients with amyotrophic lateral sclerosis.

10.2.2 Multiple sclerosis (disseminated sclerosis)

Multiple sclerosis is the most common primary demyelinating disease and is a major cause of neurological disability in young adults in most Western countries. In general, the disease occurs with a different incidence and prevalence in different geographic regions, being most common in the temperate areas of the world and rare in the tropics. In about two-thirds of cases the condition is characterized by periods of exacerbation and remission so that the course of the disease is variable. In the remaining one-third, the course of the disorder is progressive without remissions. Death may occur at 3–6 months after onset in some cases while other multiple sclerosis patients may still be active up to 40 years after onset. The period between exacerbations may be as short as a few weeks or as long as 20 to 30 years. The average number of fresh exacerbations is about 0.4 per patient per year. The prognosis in multiple sclerosis varies from case to case. Approximately 20% of patients have only minor symptoms. About 10–15% eventually become bedridden or wheelchair-bound. The majority of multiple sclerosis patients, however, have a significant disabling problem but never become bedridden or wheelchair-bound.

Multiple sclerosis principally attacks young adults, with the peak onset occurring between 20 to 40 years of age (Bannister, 1985). Most studies show the mean onset between 31 and 33 years of age. Although cases with onset at < 10 years and > 60 years of age have been documented, they are exceedingly rare. In addition, women are slightly more frequently affected than are men. The cause of the disease is unknown. About 10% of cases are familial and there is some evidence of metabolic, immunological, inflammatory and viral factors having an influence on the disease. The most popular aetiological theory centres around a 'slow virus' being activated by an unknown event. Precipitating factors have been found to include influenza, upper respiratory tract infections, fevers, pregnancy, etc.

The most important clinical and pathophysiological lesions in multiple sclerosis occur in the white matter of the central nervous system. Demyelination causes scarring (sclerosis) of the brain tissue which is evidenced at autopsy as irregular grey islands called 'plaques' in the white matter. The plaques may vary in size from microscopic to large lesions visible to the naked eye and represent areas of myelin loss with relative preservation of axons and neurone cell bodies. The significance of the plaques is that they interfere with nerve impulse conduction. As the plaques may affect a number of different parts of the central nervous system including the white matter of the cerebral hemispheres, brain stem, cerebellum, spinal cord and optic pathways, the neurological signs and symptoms of multiple sclerosis vary widely.

The signs and symptoms exhibited by patients with multiple sclerosis are dependent upon which part of the central nervous system is affected. Optic neuritis is often one of the first symptoms to be manifest and may in some cases precede other manifestations by years. Motor difficulties in the form of weakness, clumsiness and/or ataxia of gait are also commonly found and are often among the early signs of the disease. Visual complaints are also quite common and may involve acute or progressive loss of vision. Diplopia is also common and sensory symptoms including paresthesias of the extremities, dysaesthesias and loss or impairment of sensation are also frequent. Symptoms of autonomic dysfunction including evidence of incontinence and increased anal reflex may also occur. Bulbar signs are frequent and consequently dysarthria and dysphagia are common features and may progress to a level that makes speech unintelligible and swallowing food virtually impossible. Other less frequently seen symptoms include vertigo, tinnitus, hearing loss, seizures, trigeminal neuralgia and involuntary facial motor activity. Patients with multiple sclerosis may also exhibit symptoms related to mood, behaviour, memory or other higher cognitive functions. If present, symptoms related to memory defect and mild-to-moderate blunting of higher cortical functions usually occur late and in patients with rather extensive disease.

Reports vary as to the proportion of multiple sclerosis patients who present with dysarthria. Less than half of a series of 168 patients with multiple sclerosis examined by Darley, Brown and Goldstein (1972) exhibited significant speech deviations. Figures of between 20% and 41% have been suggested by other workers (Enderby and Phillip, 1986). The presence of aphasia in multiple sclerosis patients has also been reported (Friedman, Brem and Mayeux, 1983). Beukelman, Kraft and Freal (1985) found 23% of their sample of 656 patients with multiple sclerosis reported the presence of 'speech or other communication problems'; 4% of this sample indicated that strangers were unable to understand them and of this unintelligible group, nearly 30% used augmentative communication equipment.

The term 'scanning speech' has often been used to describe the dysarthria exhibited by multiple sclerosis patients (Charcot, 1877; Grinker and Sahs, 1966; West and Ansberry, 1968). Farmakides and Boone (1960) found that five deviant speech characteristics generally contributed to the dysarthria of their 82 patients with multiple sclerosis. These included: nasal voice quality; weak phonation and poor respiratory cycle; changes in pitch; slow rate; and intellectual deterioration and emotional lability. Fitzgerald, Murdoch and Chenery (1987) found deficits in all aspects of speech production, including respiration, phonation, prosody, articulation and resonance. The deviant speech dimension which contributed the most to variations in overall intelligibility of speech in their multiple sclerosis subjects was precision of consonant production. Impaired respiratory support for speech occurred in all 23 subjects examined by these workers and was found to be highly associated with deviations in vocal quality, volume control and articulation.

The most frequently occurring speech deviations observed by Darley, Brown and Goldstein (1972) were impaired loudness control and harshness. These latter authors also reported that articulation was defective to some degree in about half of their patients while impaired pitch control, impaired use of vocal variability for emphasis, hypernasality, inappropriate pitch level and breathiness were observed in 20–40% of the multiple sclerosis subjects. Based on these findings, these workers concluded that scanning speech is not a characteristic of multiple sclerosis, being surpassed in incidence by nine other deviant speech dimensions. These findings have also been supported by other workers (Fitzgerald, Murdoch and Chenery, 1987). Further, although in the past a number of authors have attributed the dysarthric symptoms seen in multiple sclerosis solely to cerebellar involvement (Hallpike, Adams and Tourtelotte, 1983), several investigations have indicated the presence of dysarthria in multiple sclerosis patients with involvement in parts of the central nervous system other than the cerebellum (Darley, Brown and Goldstein, 1972; Fitzgerald *et al.*, 1987).

10.2.3 Wilson's disease (hepato-lenticular degeneration)

Wilson's disease or hepato-lenticular degeneration is a rare, inborn metabolic disorder caused by the body's inability to process dietary copper. As a result, copper accumulates in the tissues of the body, especially in the brain, liver and cornea of the eye. The most severely affected parts of the brain are the basal ganglia, although lesser degrees of copper deposition may also occur in the cerebellum, brainstem and other parts of the cerebrum. Of the basal ganglia, the corpus striatum is most involved with damage usually more marked in the putamen than in the caudate nucleus. Cirrhosis of the liver is also present and the deposition of copper in the eye gives a greenish-brown colour to the cornea (Kayser–Fleisher rings).

The condition is inherited in an autosomal recessive manner (Bearn, 1960) and usually onsets between the ages of 10 and 25 years. Two separate neurological pictures may be manifest in Wilson's disease – an acute form characterized by dystonia and a chronic form characterized by tremor and rigidity. The initial symptoms of the acute form vary from mild slowing of voluntary movements, to dysarthria and mental changes. The latter include difficulty with memory and concentration as well as mild personality changes such as loss of interest and increased irritability. Dystonia appears later and serves to differentiate the acute form of the disorder from the more chronic form. The most common symptoms of the chronic form of the disorder are tremor and rigidity. Tremors may be of the intention type or may be alternating like the tremors involved in Parkinson's disease.

Commonly the tremors may be of the bizarre 'wing-beating' type. These latter tremors are confined to the upper extremities and represent involuntary movements of the shoulders. They are accentuated by extension of the upper limbs. Dystonic movements are not a prominent feature of the chronic form of the disorder.

The course of the disease is fluctuating and partial remissions and exacerbations commonly occur. The condition, however, is often fatal within ten years if untreated. Treatment is with a low copper diet and an agent to reduce the blood copper levels.

A mixed ataxic–hypokinetic–spastic dysarthria is a prominent feature of Wilson's disease. Berry et al. (1974) found positive neurological signs of ataxia, rigidity or spasticity in their 20 patients with Wilson's disease. They noted that the most prominent speech characteristics of the dysarthria exhibited by their subjects were similar to the most prominent deviant speech dimensions exhibited by patients with Parkinson's disease. Reduced stress, monopitch, monoloudness and imprecise consonants were the four most highly ranked features in the Wilson's disease patients. The same four deviant speech dimensions were also reported by Darley, Aronson and Brown (1969a,b) to be the most highly ranked speech abnormalities in Parkinson's patients. Further, subjects with Wilson's disease were reported by Berry et al. (1974) to share the deviant speech dimensions of irregular articulatory breakdown with cerebellar disorders, hypernasality with psueodbulbar palsy and inappropriate silences with Parkinson's disease. Overall, Berry et al. (1974) determined that the speech deviations seen in their patients with Wilson's disease fell into four clusters which included: prosodic insufficiency (monopitch, monoloudness and reduced stress) as seen in hypokinetic and spastic dysarthria; phonatory stenosis (low pitch, strained voice and harsh voice) as reported in spastic dysarthria; prosodic excess (slow rate, prolonged phonemes, prolonged intervals and excess and equal stress) as observed in ataxic dysarthria; and articulatory–resonatory inadequacy (hypernasality and imprecise consonants) as seen in spastic dysarthria and dysarthria associated with amyotrophic lateral sclerosis and chorea.

Pharmacological (D-penicillamine) and dietary treatments of Wilson's disease have been reported to improve the speech of patients with this condition provided that treatment is commenced prior to permanent neurologic or hepatic damage occurring (Berry et al., 1974; Darley et al., 1975a). Day and Parnell (1987) described a case of Wilson's disease in which a severe dysarthria persisted, however, despite drug and dietary control.

CHAPTER ELEVEN

Acquired aphasia in childhood

Childhood speech–language disorders can be divided into developmental disorders and acquired disorders (Ludlow, 1980). Developmental disorders of speech and language are those which onset prior to the emergence of language (i.e. between birth and one year of age). Consequently children with developmental language disorders have never developed language normally. Although it is usually presumed that primary developmental speech–language disorders are caused by dysfunctioning of the central nervous system, in most cases they have an idiopathic origin (i.e. the cause is unknown). Developmental speech–language disorders can, however, occur secondary to conditions such as peripheral hearing loss, mental retardation, cerebral palsy, child autism, birth trauma and environmental deprivation.

Acquired speech–language disorders, on the other hand, are disturbances in speech–language function that result from some form of cerebral insult after language acquisition has already commenced (Hecaen, 1976). The cerebral insult, in turn, can result from a variety of aetiologies, including head trauma, brain tumours, cerebrovascular accidents, infections, convulsive disorders (intractable epilepsy) and electroencephalographic abnormalities (Miller *et al.*, 1984). Typically these children have commenced learning language normally and were acquiring developmental milestones at an appropriate rate prior to injury.

Of the two types of childhood language disorder, the acquired variety most closely resembles the acquired adult communicative disorders discussed in earlier chapters. Unfortunately, many texts on the language disorders in the past have paid this important group of neurologically based speech–language disorders only scant attention. This chapter will review acquired childhood speech–language disorders in terms of their aetiology and clinical features.

11.1 ACQUIRED CHILDHOOD APHASIA

11.1.1 Clinical features of acquired childhood aphasia

Children with acquired language disorders are referred to as having acquired aphasia. The clinical features of acquired childhood aphasia are manifestly different in a number of ways to those of adult aphasia. In particular, there appear to be two major differences between acquired aphasia in children and aphasia in adults. First, the recovery process is described as being more rapid and complete in children (Lenneberg, 1967). Secondly, in the majority of cases, acquired childhood aphasia is predominantly non-fluent, its major features being mutism and lack of spontaneity of speech (Alajouanine and Lhermitte, 1965; Hecaen, 1976; Fletcher and Taylor, 1984). Further, with some rare exceptions, the acquired aphasia in children does not appear to fall into clear-cut syndromes evocative of the well-known aphasia sub-types described in adults (see Chapter 2).

Although there is some variation between reports in the literature, the symptoms most reported in the classical studies to be characteristic of acquired childhood aphasia include initial mutism (suppression of spontaneous speech) followed by: a period of reduced speech initiative; a non-fluent speech output; simplified syntax (telegraphic expression); impaired auditory comprehension abilities (particularly in the early stages post-onset); an impairment in naming; dysarthria; and disturbances in reading and writing (primarily in the acute stage post-onset). Most authors suggest that fluent aphasia and receptive disorders of oral speech such as literal and verbal paraphasic errors, logorrhoea, and perseverations are only rarely found in children with acquired aphasia. There is, however, evidence to suggest that the age of the child has a role to play in determining whether or not these symptoms occur in a particular case. Some authors are of the opinion that the primarily non-fluent pattern of aphasia is only prevalent in children who are less than ten years of age at the onset of the aphasia (Poetzl, 1926; Guttmann, 1942; Alajouanine and Lhermitte, 1965). For example, Alajouanine and Lhermitte (1965) found that the predominant features of the acquired aphasia demonstrated by children at < 10 years of age included decreased auditory comprehension, severe writing deficit, and no logorrhoea, paraphasias or perseveration. These same authors, however, reported that the acquired aphasia demonstrated by children > 10 years of age is a more fluent form of aphasia, with paraphasia present, less frequent articulatory and phonetic disintegration and disturbed written language. Other authors, however, are of the opinion that the non-fluent type of aphasia is the pattern present in older children as well as the younger ones (Basser, 1962; Benson, 1972; Assal and Campiche, 1973; Hecaen, 1976). In recent years the classical descriptions of acquired childhood aphasia have been questioned by the findings of studies that have shown that a fluent aphasia with paraphasias may be exhibited by children in the early stages post-onset. In this chapter, the features of acquired childhood aphasia as described in the classical studies will be outlined first and the findings of more recent studies discussed later.

Since the first reports of acquired childhood aphasia last century (Bernhardt, 1885; Freud, 1897) the condition has largely been regarded as being characterized by aphasia of the non-fluent type. Indeed, most contemporary authors describe non-fluency as the most prevailing clinical feature of acquired aphasia in children (Satz and Bullard-Bates, 1981; Carrow-Woolfolk and Lynch, 1982). Even those children with focal lesions involving the posterior language centre exhibit fluent aphasia less frequently than adults with similar lesions.

A reduction in all expressive activities including oral, written and gestural activities has been reported to occur in children with acquired aphasia (Alajouanine and Lhermitte, 1965). Certainly mutism is a frequently reported early symptom of acquired childhood aphasia. When present the period of suppressed spontaneous speech may last from a few days to months. Hecaen (1983) carried out a retrospective study involving 56 acquired aphasic children ranging from 3.5 to 15 years of age, with brain lesions resulting from a variety of aetiologies including trauma, tumour and haematoma. Hecaen found mutism to be the predominant clinical symptom in the initial stage post-onset, occurring in 47% of his child acquired aphasics. When analysed according to lesion type, however, Hecaen found that only 20% of cases with progressive brain lesions (e.g. tumours) exhibited an initial mutism while 85% of acute head trauma patients demonstrated this symptom in the early stages post-onset. Further, the findings of Hecaen's study showed that mutism was exhibited more frequently by children with fronto-rolandic lesions (63%) than children with temporal lesions (10%).

Following the period of mutism when speech returns, there is often a period during which the aphasic child is unwilling to speak (Guttmann, 1942; Alajouanine and Lhermitte, 1965). This period has also been described as representing a loss or reduction of 'speech initiative' (Hecaen, 1976, 1983). Increased incentives and encouragements are required in this period to get the child to produce even the few words they are capable of producing.

Alajouanine and Lhermitte (1965) offered a purely psychological reason for the presence of mutism in children with acquired aphasia. They proposed that the suppression of spontaneous speech might be the result of a psychological reaction experienced by children in response to their inability to communicate. Alajouanine and Lhermitte (1965) noted that children tend to isolation, refusal and silence in response to conflict or difficulties. Further, these authors also observed that the unwillingness of aphasic children to speak after speech returns is similar to the behaviour of normal children when faced with a difficult problem they cannot solve and wish to put aside.

Although a number of authors have reported that children with acquired aphasia exhibit telegraphic expression or simplified syntax (Bernhardt, 1885; Guttmann, 1942; Alajouanine and Lhermitte, 1965) few have described the spoken syntax of these children in detail. Aram, Ekelman and Whitaker (1986) examined the spontaneous spoken syntax of 16 children with acquired cerebral lesions to either the left or right hemisphere and compared it to that of a group of appropriately matched control subjects. These workers found that children with unilateral left hemisphere lesions performed less well on most measures of spoken syntax (including both simple and complex sentence structure) than the control subjects. Specifically, children with left cerebral lesions had a shorter mean length of utterance, lower developmental sentence scores, lower percentages of total sentences correct, a fewer number of main verbs and interrogative reversals, fewer sentences containing conjunctions or embedded clauses and produced a greater percentage of complex sentences in error compared to the children in the control group. On the other hand, the syntactic limitations of children with right cerebral lesions were limited primarily to errors on simple sentence measures and were far less pervasive than the syntactic errors exhibited by the left lesioned subjects. They had a shorter mean length of utterance, produced more simple sentences in error and produced develop-mentally earlier negatives than the control subjects.

Although there is general agreement that expressive language problems are common in acquired childhood aphasia, there has been considerable debate over the years concerning the presence of deficits in auditory comprehension in this group of children. By far the majority of early researchers in this field considered impairments in auditory comprehension to be a rare occurrence in children with acquired aphasia (Bernhardt, 1885; Guttmann, 1942). Alajouanine and Lhermitte (1965) reported the presence of auditory comprehension deficits in about one-third of their acquired childhood aphasia cases. It has been suggested by Hecaen (1976, 1983) that, when present, disturbances in auditory comprehension occur almost exclusively in the early stages post-onset of aphasia and disappear rapidly and virtually completely. Consequently, as different authors may have examined children with acquired aphasia at different stages post-onset, this may account for the variability relating to the presence of impaired auditory comprehension abilities in childhood acquired aphasics reported in the literature. Further confusion arises when patient age considerations are taken into account. In some studies, auditory comprehension problems have been reported to be rare in child acquired aphasics < 10 years of age (Guttmann, 1942; Basser, 1962; Assal and Campiche, 1973).

Diminished verbal stock or an impoverished lexicon is another commonly reported symptom of acquired childhood aphasia (Bernhardt, 1885; Alajouanine and Lhermitte, 1965; Collingnon, Hecaen and Angerlerques, 1968). This was noted by Hecaen (1983) to occur in the late stages. He reported that 44% of his sample of acquired childhood aphasics had naming problems (not of a paraphasic type) which tended to persist. In fact, Hecaen (1983) observed that naming problems are commonly present when the child returns to school and consequently is often mentioned explicitly in their school reports.

Hesitations have also been noted in the speech of children with acquired aphasia by various authors (e.g., Berhardt, 1885; Guttmann, 1942). It is possible that these are the result of either the dysarthria sometimes reported to be present in these children (Guttmann, 1942; Hecaen, 1976, 1983) or alternatively they may reflect the word-finding problems also reported to occur in this population. Hecaen (1983) listed dysarthria as a common finding in association with acquired childhood aphasia, 52% of his 56 subjects exhibiting a dysarthric disturbance. According to Hecaen, the factor having the greatest influence in determining the occurrence of dysarthria is the localization of the underlying lesion. Hecaen (1983) reported that 81% of children with anterior lesions exhibited a dysarthria as part of their aphasic disturbance, while only 20% of cases with temporal lesions exhibited a dysarthria. Alajouanine and Lhermitte (1965) found dysarthria with paralytic and dystonic features (see Chapter 9) in 22 out of their total of 32 cases.

A disorder vaguely referred to as 'reading problem' is often included by authors in the list of symptoms they regard as characteristic of acquired childhood aphasia. Alajouanine and Lhermitte (1965), for instance, reported the presence of a 'reading problem' in 18 out of the 32 children with acquired aphasia in their study. Unfortunately, however, in the majority of studies reported in the literature there has been no attempt made to define the nature of the observed reading disturbance in any further detail, other than to merely document its presence. Most authors, for instance, give no indication as to the relative effects of the condition on children's ability to read aloud versus their ability to read for comprehension. It is possible that the general lack of differentiation in most descriptions of the reading abilities of children with acquired aphasia may be due to the fact that reading disorders, although common in the acute stage post-onset, in most cases disappear rapidly and completely (Hecaen, 1976, 1983).

A writing deficit is also a commonly reported feature of acquired childhood aphasia (Branco-Lefevre, 1950; Alajouanine and Lhermitte, 1965; Hecaen, 1976, 1983). Alajouanine and Lhermitte (1965) noted that the written language of all 32 of their child subjects with acquired aphasia was disturbed, there being a severe disorder in spontaneous writing, writing to dictation and copying in over half of their subjects. In 25% of their cases, copied writing only was possible. Dysorthrographia (the misspellings often being based on phonetic disturbances) was also observed in the spontaneous writing and writing to dictation of a number (5 out of 32) of their acquired aphasic cases. Hecaen (1976) described the writing disorder observed in his subjects as one of the most frequent, most persistent and most variable of all the symptoms of acquired childhood aphasia.

Although, based on descriptions contained in 'landmark' studies such as those conducted by Alajouanine and Lhermitte (1965), Bernhardt (1885), Gloning and Hift (1970), Guttmann (1942) and Hecaen (1976), acquired childhood aphasia has, over past years, been generally regarded as being of a non-fluent type, a number of publications in more recent years have documented the occurrence of an initial fluent aphasia in children with neurological impairment (Visch-Brink and van de Sandt-Koenderman, 1984; van Dongen, Loonen and van Dongen, 1985; van Hout, Evrard and Lyon, 1985; van Hout and Lyon, 1986). In addition, although the majority of 'landmark' studies also stressed the absence or rarity of receptive speech disorders such as paraphasias, logorrhoea and perseveration, especially in children < 10 years of age, more recent reports in the literature have documented the occurrence of these features in the spontaneous speech and test responses of children with acquired aphasia. Consequently, although the rarity of paraphasic errors and logorrhoea was once thought to illustrate the unique character of acquired childhood aphasia compared to adult aphasia, the findings of recent studies suggest a need to re-appraise the traditional concept of the clinical features of acquired aphasia in children.

As pointed out by Visch-Brink and van de Sandt-Koenderman (1984), the term 'rarity' when used to describe the presence of paraphasias in cases of acquired childhood aphasia appears to be used somewhat loosely in the literature. Alajouanine and Lhermitte (1965), who stressed the 'rarity' of paraphasias, actually observed paraphasic errors in the spontaneous speech of 7 out of the 32 children with acquired aphasia in their study. Likewise, Gloning and Hift (1970) observed paraphasias in the spontaneous speech of 4 out of 8 of their acquired aphasic children. Included among Guttmann's (1942) subjects were 4 cases described extensively, 2 cases of whom produced paraphasias (1 child for at least a year post-onset).

Visch-Brink and van de Sandt-Koenderman (1984) found that, of the 14 children with acquired aphasia they studied, the spontaneous speech of 8 was marked by the occurrence of paraphasias: 4 of these children were reported in detail, with their aphasia classified according to type (fluent/non-fluent/mixed) and their paraphasic errors categorized as literal, verbal and neologism. They concluded that in no sense of the word could paraphasias be regarded as rare in their sample of children with acquired aphasia since over half of the children produced paraphasias in their spontaneous speech and a single child could produce many paraphasias. Moreover, paraphasias occurred in all forms of aphasia including fluent, non-fluent and mixed aphasias.

Van Hout, Evrard and Lyon (1985) examined from onset of aphasia or emergence from coma 16 children with acquired aphasia, 11 of whom (ranging in age from 4 to 10.8 years) presented with 'non-classic' symptoms. These authors also found paraphasias in the language output of the children once speech re-appeared after a period of initial mutism. Indeed, these authors suggested that paraphasias are the rule rather than the exception in acquired childhood aphasia. Furthermore, in half of the cases they studied, paraphasic errors were not limited only to the acute stage post-onset. They divided their subjects into three groups according to the evolution of paraphasias over time. In one group the paraphasic errors resolved within a few days, in the second group paraphasia resolved in a few months and in the third group paraphasia was still present at greater than one year post-onset.

A fluent aphasia was reported by van Dongen, Loonen and van Dongen (1985) in 3 of the 27 acquired aphasic children referred to their clinic over a four-year period. Their findings demonstrated that an adult-like fluent aphasia can occur in children < 10 years of age and consequently they challenged the view that acquired aphasia is always non-fluent and devoid of paraphasias. Although all three subjects with fluent aphasia had posterior lesions, these authors emphasized that posterior lesions do not invariably result in fluent aphasia in children.

A case of Wernicke's aphasia in a 10-year-old boy resulting from herpes simplex encephalitis was described by van Hout and Lyon (1986). The symptoms exhibited by this case were similar to those associated with Wernicke's aphasia in adults and included a severe comprehension deficit, jargon output, logorrhoea and anosognosia. The recovery pattern exhibited by this case resembled the pattern described for a sensory aphasia by Buckingham and Kertesz (1974). His rapid recovery of writing skills was atypical of that usually reported for cases of acquired childhood aphasia. Thus the case as described by van Hout and Lyon (1986) differs considerably from the usual case of acquired childhood aphasia where logorrhoea and anosognosia are rare and writing disorders are one of the more persistent language deficits. Van Hout and Lyon (1986) attributed the Wernicke's aphasia observed in their subject to the nature of the lesion itself and not to the age of the subject. In herpes simplex encephalitis, the lesions are profoundly destructive (Barringer, 1978) and consequently this condition causes more severe lesions in the temporal lobes than seen in other varieties of lesions which affect the brain in childhood.

Although it is not immediately clear why the clinical features of acquired childhood aphasia reported by earlier researchers (e.g. Alajouanine and Lhermitte, 1965) differ from those of more recent workers (e.g. van Dongen, Loonen and van Dongen, 1985), several explanations have been proposed. One possible reason for the disparity could be the difference in the time post-onset that the subjects were examined in the different studies. For instance, Visch-Brink and van de Sandt-Koenderman (1984) suggested that it is possible that the presence or absence of neologisms in the spontaneous speech output of children with acquired aphasia might be related to the time post-onset that the subjects were examined. In their study, the subjects were examined within a few days post-onset of aphasia and neologisms were recorded. On the other hand Alajouanine and Lhermitte (1965) made their observations a number of months post-onset and did not report the presence of neologisms in the spontaneous speech of their acquired aphasic children. It is possible, therefore, that by the time that Alajouanine and Lhermitte (1965) examined their subjects, a number of symptoms, including the presence of neologisms, may have disappeared. Van Hout, Evrard and Lyon (1985) also suggested that the time post-onset may be a factor underlying the disparity between the earlier and more recent studies in terms of whether so called positive signs such as paraphasias, perseveration, sterotypics, etc. were recorded or not. Since they observed that some positive signs persisted long after the acute stage in half of the cases they studied, these authors believed that the time post-onset of the language examination does not wholly account for the difference in clinical signs described in the earlier versus more recent studies of acquired childhood aphasia.

Another possible reason for the variation in clinical features of acquired childhood aphasia reported in earlier versus more recent studies lies in the nature of the criteria used to select the aphasic subjects. In many earlier studies, subjects were only included if they had a concomitant hemiparesis or hemiplegia, as this was taken as being indicative of the presence of brain damage and hence served to delineate acquired aphasia from developmental language disorders. As pointed out by Woods and Teuber (1978), however, such a selection criterion could result in a bias towards children with anterior lesions and hence a motor-type of aphasia. It is possible, therefore, that this could explain the lack of paraphasias, logorrhoea, etc. reported in many earlier studies of acquired childhood aphasia.

Van Dongen, Loonen and van Dongen (1985) suggested that differences in aetiology may provide another reason for the discrepancy between reports. They believed that when the aetiology is head trauma, recovery may be observed within a short time, so that the fluent characteristics of the aphasia may be either not recognized or not recorded.

Van Hout, Evrard and Lyon (1985) proposed that variations in methodology could best account for the differences between their findings and the descriptions of acquired childhood aphasia in earlier studies. As in the majority of other reports, their subjects were also reluctant to speak. If they had limited their study simply to a clinical examination of spontaneous speech, as many of the earlier researchers had done, they believed that positive signs such as paraphasias, logorrhoea, etc. would have been 'masked' by the children's lack of spontaneity or mutism. They suggested that it was the testing itself and the encouragement they provided that enabled the acquired aphasic children in their sample to overcome their unwillingness to speak and thereby produce the observed positive signs.

In summary, clinically the aphasia pattern observed in cases of acquired childhood aphasia appears to be predominantly that of a non-fluent aphasia. Often there is an initial mutism followed by a period of reduced speech initiative together with a diminished lexicon, simplified syntax, hesitations and dysarthria. Disturbances in reading and writing are also common. This pattern of language disturbance, however, is by no means invariant and fluent aphasia does occur in some cases. According to Satz and Bullard-Bates (1981) it is unknown whether the variations in the manifestations of acquired childhood aphasia are related to either age/maturation mechanisms or age-independent factors such as lesion site, lesion size, aetiology, type of lesion or time post-onset of assessment.

11.1.2 Recovery from acquired childhood aphasia

The consequences of cerebral lesions incurred in childhood are generally regarded as less serious than those incurred in adult life (Basser, 1962; Teuber, 1975). Consequently, it is generally agreed that the prognosis for recovery in acquired childhood aphasia is much better than that expected in adult aphasia (Guttmann, 1942; Basser, 1962; Alajouanine and Lhermitte, 1965; Lenneberg, 1967). The often described complete or near-complete recovery of language function following lesions of the left cerebral hemisphere in childhood is frequently cited as being indicative of the 'plasticity' of the immature brain, whereby the non-damaged areas of the brain are capable of assuming language function. Although the mechanisms underlying compensation are not fully understood, the good recovery from acquired childhood aphasia has been attributed to processes which include the transfer of language function to the undamaged portions of the left cerebral hemisphere and/or the intact right cerebral hemisphere. Some authors (e.g. Satz and Bullard-Bates, 1981), however, suggest that the speed of recovery sometimes witnessed in children with acquired aphasia is incompatible with a transfer of language to and a learning of language by the right hemisphere.

Another proposed explanation for the often good recovery exhibited by children with acquired aphasia is that both hemispheres contain mechanisms for language and that language therefore need not be re-learned in the minor hemisphere. Under normal circumstances, in the majority of children, the language mechanisms in the right hemisphere are inhibited by those in the left such that only the left hemisphere develops complex language function. According to this proposal, damage to the left hemisphere in children causes a 'release from inhibition' in the right hemisphere allowing it to assume a greater role in language function.

Although for some time it has been generally believed that the prognosis of acquired aphasia in children is good, the findings of several studies reported in the literature suggest that the recovery is not as complete as often stated. In fact, there are a variety of opinions expressed in the literature concerning the prognosis of acquired aphasia in children ranging from favourable declarations of complete recovery to more guarded predictions of only partial recovery. In describing recovery of language function in acquired childhood aphasia, many authors make reference only to 'clinical recovery'. A number of researchers, however, have emphasized that despite apparent 'clinical recovery', subtle but persistent language deficits may persist, even in those cases where the left hemisphere injury was acquired as early as during intra-uterine life (Alajouanine and Lhermitte, 1965; Woods and Carey, 1979; Rankin, Aram and Horwitz, 1981; Vargha-Khadem, Gorman and Watters, 1985).

Satz and Bullard-Bates (1981) reviewed the literature relating to the prognosis of acquired childhood aphasia and concluded that although spontaneous recovery occurs in the majority of children with this disorder, it by no means occurs in all cases. Of the cases included in the studies they reviewed, 25–50% remained unremitted by one year post-onset. Alajouanine and Lhermitte (1965) reported that 75% of their acquired aphasic subjects attained normal or near-normal language by one year post-onset. Of the 8 children who had an unfavourable course in their study, 6 had massive lesions, 1 showed mental deterioration and 1 died. One-third of the children with acquired aphasia studied by Hecaen (1976) attained complete recovery within a period of 6 weeks to 2 years post-onset. Carrow-Woolfolk and Lynch (1982) suggested that the recovery period in cases of acquired childhood aphasia may extend up to 5 years. Even in those cases where recovery from aphasia occurs, however, there are often serious cognitive and academic problems which remain (Satz and Bullard-Bates, 1981). For instance the majority of children with acquired aphasia appear to have difficulty following a normal progression through school (Hecaen, 1976; Chadwick, 1985). None of the 32 children with acquired aphasia studied by Alajouanine and Lhermitte (1965) showed normal progress in the long term. According to these authors, school subjects requiring the use of language (first language study, foreign languages, history, geography, etc.) are more difficult for these children than subjects such as mathematics.

A number of different factors which may influence the recovery of language in acquired childhood aphasia have been identified by various authors. These factors include: the site of lesion; the size and side of lesion; the aetiology; the associated neurological disturbances; the age at onset; the type and severity of the aphasia; and the presence of electroencephalographic abnormalities. As pointed out by van Dongen and Loonen (1977), however, the wide diversity of aetiologies, severities of aphasia, and length of follow-up reported in the various studies of acquired childhood aphasia, make it difficult to work out which of the factors are the most important in determining the final outcome of the aphasia. According to Satz and Bullard-Bates (1981) our current knowledge is inadequate for determining which factors assist and which factors impede recovery.

The prognostic factor that has perhaps received the greatest attention in the literature is the age at onset. Lenneberg (1967) stated that the prognosis of acquired aphasia in children is directly related to the age at onset of aphasia and that any aphasia incurred after the age of puberty does not remit entirely. Various authors agree that there are considerable differences in prognosis with age. Penfield (1965) claimed that in children < 10 years with acquired aphasia, there is a good chance of re-acquisition of lost verbal skills within one year, although such recovery may occur at the expense of other non-verbal skills. In support of this claim, Carrow-Woolfolk and Lynch (1982) suggested that children < 3 years of age with cerebral lesions follow normal language acquisition after an initial pause of all language development. Vargha-Kadem, Gorman and Watters (1985) found that as the age of aphasia onset increases (at least in children > 5 years old), the language impairment becomes progressively worse; 10 years of age is considered by many authors to be the upper limit for complete language recovery (Oelschlaeger and Scarborough, 1976), cerebral lesions incurred after this time causing a persistent language deficit. There is an unconfirmed premise that cerebral plasticity is lost by the age of 10 years as a result of development of cerebral dominance or lateralized specialization of language function.

Despite the support provided by the above studies that age at onset is an important prognostic determinant in acquired childhood aphasia, not all authors have been able to find a relationship between the age at aphasia onset and recovery. Further, a number of reports in the literature actually contradict the information provided by the more supportive studies outlined above. For instance, Hecaen (1976) described three cases of children with acquired aphasia with onset at 14 years of age, but who showed excellent recovery patterns. Although Alajouanine and Lhermitte (1965) reported a difference in the symptomatology between children with acquired aphasia at < 10 years and aphasic children over > 10 years of age, they found no significant difference in the speed of recovery between these two age groups. Likewise, van Hout, Evrard and Lyon (1985) found no direct relationship between the age at onset and the rate of disappearance of paraphasias in children with acquired aphasia. With such conflicting empirical data, it is obvious that as yet the relationship between age at onset and recovery in acquired childhood aphasia is uncertain.

With regard to the type of aphasia, Guttmann (1942) emphasized the good prognosis of a purely motor aphasia, especially in young children. Van Dongen and Loonen (1977) also found the type of aphasia exhibited in the acute stage post-onset to be of prognostic significance. They reported that 5 out of the 6 children with an initial amnestic aphasia recovered from aphasia, but only 1 of the 7 children with mixed aphasia showed recovery. The findings of van Dongen and Loonen (1977) lend support to an earlier claim by Assal and Campiche (1973) that mixed aphasia has a poor prognosis. Van Dongen and Loonen (1977) also found a significant relationship between the severity of the comprehension deficit at the onset of acquired childhood aphasia and a poor recovery from aphasia.

No systematic studies relating aetiology to recovery from acquired childhood aphasia have been reported in the literature. Guttmann (1942), however, reported that children with head trauma improve more than those with vascular disease. Likewise, van Dongen and Loonen (1977) also found that most children with traumatic aphasia recovered completely. Infection (e.g. encephalitis) was found to be more frequent in children with the most severe aphasia and persistent paraphasias by van Hout, Evrard and Lyon (1985). A number of authors, including Mantovani and Landau (1980) have suggested that the prognosis for recovery from acquired aphasia associated with convulsive disorder is much worse than for other acquired childhood aphasias.

The findings of a number of studies have suggested a link between the extent of the cerebral lesion and the persistence of aphasic symptoms in children with acquired aphasia (Alajouanine and Lhermitte, 1965; Hecaen, 1976, 1983). As mentioned previously, 6 of the 8 aphasic children with unfavourable language outcomes in the study conducted by Alajouanine and Lhermitte (1965) had extensive cerebral lesions suggesting that the larger the cerebral lesion the poorer the prognosis for recovery. Hecaen (1983) found bilateral lesions frequently linked with persistent aphasic symptoms. Van Hout, Evrard and Lyon (1985) also concluded that bilateral lesions may negatively influence the outcome of the language disturbance in acquired childhood aphasia based on their findings that the aphasic children with the most persistent paraphasias and who exhibited the most severe aphasic disorder in their sample, all had bilateral brain lesions. Other authors, including Collignon, Hecaen and Angerlerques (1968) and Gloning and Hift (1970) have also stressed that severe bilateral lesions are indicative of a poor prognosis for recovery in acquired childhood aphasia.

Based primarily on the findings of studies published prior to the 1940s, it was for many years considered that there was a higher incidence of aphasia following right cerebral lesions in children than in adults. This belief led to the formulation of the 'equipotentiality hypothesis' which states that at birth, each cerebral hemisphere has the same potential to develop language function and that the lateralization of language and the development of cerebral dominance occurs as the child matures (Lenneberg, 1967). More recent studies, however, have shown that the frequency of aphasia from right cerebral hemisphere lesions in right-handed children is similar to that in right-handed adults (Satz and Bullard-Bates, 1981; Carter, Hohenegger and Satz, 1982). Satz and Bullard-Bates (1981) stated that, after infancy, the risk of aphasia is significantly greater following left brain damage than right brain injury regardless of the age of the individual. Furthermore, 'crossed aphasia' (i.e. aphasia occurring after right hemisphere damage in right-handed individuals) is rare in both adults and children, particularly after 3 to 5 years of age and perhaps earlier. Woods and Teuber (1978) suggested that in the studies carried out earlier than the 1940s, prior to the widespread use of antibiotics, there may have been biased estimations relating to the handedness distribution of subjects with acquired aphasia in that, in the cases observed, there may have been undetected bilateral brain damage caused by infections.

In addition to the extent and side of lesion, it has also been suggested by several authors that the localization of the cerebral lesion influences recovery from acquired childhood aphasia. Van Hout, Evrard and Lyon (1985) suggested that the localization of the lesion is a more important prognostic variable than the extent of lesion. Alajouanine and Lhermitte (1965) also indicated that recovery from acquired aphasia in children is dependent upon the site as well as the extent and reversibility of the lesion. It should be noted, however, that although these latter authors suggested a link between lesion localization and recovery, they did not provide any specific examples of how the site of the cerebral damage influences the prognosis for recovery. Although Hecaen (1983) found that all language symptoms, including auditory and written comprehension, were more disturbed following anterior than following temporal lesions in children, he did not list the site of lesion as a factor involved in the determination of the outcome of the aphasic disturbance.

The relationship between changes in the electroencephalographic pattern and recovery in children with acquired aphasia is controversial. Some authors have reported that recovery correlates with a disappearance of abnormalities in the electroencephalographic trace while others have been unable to find any link between these two factors. Shoumaker et al. (1974) found that improved language abilities in children with acquired aphasia associated with convulsive disorder corresponded with an improvement in the electroencephalographic pattern. Likewise, van Dongen and Loonen (1977) found that recovery of language function in a group of children with acquired aphasia resulting from a variety of aetiologies (including convulsive disorder) was associated with a reduction in electroencephalographic abnormalities. Other authors, however, including Gascon *et al.* (1973) and McKinney and McGreal (1974) were unable to demonstrate a correlation between improved electroencephalographic patterns and improved language. Of the 24 children with acquired aphasia who were reported to have recovered in the study conducted by Alajouanine and Lhermitte (1965), 16 still suffered severe motor sequelae and electroencephalographic disturbances. Alajouanine and Lhermitte (1965) interpreted this finding as an indication that recovery does not result from the reversibility of the lesion.

Concomitant neurological disturbances represent another variable that has been implicated as a factor which influences recovery from acquired childhood aphasia. As in the case of the prognostic variables discussed above, however, the findings reported in the literature relating to the importance of associated neurological signs as a prognostic indicator in children with acquired aphasia tend to be contradictory. Lange-Cosack and Tepfner (1973) reported that there is minimal or no recovery in traumatic aphasic subjects who have suffered coma for more than seven days. On the other hand, Hecaen (1976) questioned the importance of coma as a prognostic indicator in cases of acquired childhood aphasia, he being unable to demonstrate a clear relationship between the occurrence and duration of coma and the severity and persistence of the language deficit.

In summary, although a number of different factors have been suggested as having prognostic significance in acquired childhood aphasia, currently there is an insufficient amount of information available to determine which of these factors are unequivocally favourable or unfavourable for recovery.

11.2 ACQUIRED CHILDHOOD APHASIA OF DIFFERENT AETIOLOGIES

The general clinical features of acquired childhood aphasia described previously are largely based on the findings of studies which have included aphasic children with a variety of underlying aetiologies including trauma, vascular lesions, tumours, infections and convulsive disorder. As well as influencing the prognosis for recovery of language function in cases of acquired childhood aphasia, there is some evidence that the aetiology also has an important influence on the type of aphasia that is exhibited. It is no longer possible to assume that the effects of slow-onset lesions (e.g. tumours) on language are the same as those of rapid-onset lesions (e.g. cerebrovascular accidents and head trauma). Further, children who have suffered traumatic head injuries characteristically exhibit expressive language deficits and typically show good recovery. In cases of acquired childhood aphasia following vascular lesions, the prognosis for recovery is poorer and the aphasic symptoms more variable and more persistent (Guttmann, 1942; van Dongen and Loonen, 1977). Consequently there is a need to examine the clinical features of the acquired childhood aphasia associated with each aetiology separately.

11.2.1 Acquired aphasia following vascular disorders

Few studies of acquired aphasia occurring secondary to vascular disorders in children have been reported in the literature. Those studies that have been published, however, suggest that the pattern of language symptoms is similar to that seen in cases of adult aphasia of vascular origin (Dennis, 1980; Aram et al., 1983).

Aram et al. (1983) documented the course of acquired aphasia following a vascular lesion in the putamen, anterior limb of the internal capsule and lateral portion of the head of the caudate nucleus in the left hemisphere of a right-handed 7-year-old girl. These workers reported that symptoms were similar to those listed for adult aphasics with equivalent sub-cortical lesions, the child exhibiting a right-sided hemiplegia, mutism, oral apraxia and a comprehension deficit but no dysarthria. At 6 months post-onset, the only persisting problems were a mild hemiparesis and minor spelling difficulties. Dennis (1980) noted a significant degree of both expressive and receptive aphasia in a 9-year-old girl following an infarct of vascular origin in the left temporo-parietal region.

In addition to the two single case studies mentioned above, a number of cases of acquired childhood aphasia resulting from vascular lesions have also been described as a part of some studies which have included children with acquired aphasia resulting from a variety of aetiologies (e.g. Hecaen, 1983). Hecaen (1983) noted that all 6 children in his sample with a vascular aetiology exhibited aphasia. He reported that the most frequent language symptom exhibited by these children was a writing disorder. The next most frequent symptoms were mutism, articulatory disorders and naming disorders. Auditory comprehension deficits were present in half of these cases while paraphasias and reading disorders were exhibited by only 2 out of the 6 children.

Overall, the lack of studies dealing specifically with acquired childhood aphasia following vascular disorder means that little is known about the prognosis for recovery from acquired aphasia in children with vascular origin. Both Guttmann (1942) and van Dongen and Loonen (1977), however, have reported that children with acquired aphasia of vascular origin recover language less well than those with aphasia resulting from traumatic head injuries.

11.2.2 Acquired aphasia following head trauma

Head injury is a major cause of acquired neurodevelopmental handicap in children (Chadwick, 1985). In Western countries, pedestrian road-traffic accidents account for the majority of severe childhood head injuries (Lange-Cosack and Tepfner, 1973; Moyes, 1980; Rutter *et al.*, 1980) and consequently, these are a common cause of acquired childhood aphasia. (The mechanisms of head injury are discussed in Chapter 4.)

The study conducted by Hecaen (1983) included an investigation of 12 children with acquired aphasia resulting from traumatic head injury. Based on his findings, Hecaen (1983) reported that the major aphasic symptoms exhibited by this group of children included, in order of decreasing frequency, writing disorders, mutism and articulation and reading disorders. Auditory comprehension and naming deficits occurred in half of the cases. Although none of Hecaen's head injured subjects was reported to exhibit paraphasias (language examination being carried out at approximately three months post-onset of aphasia in his study), such positive symptoms have been reported in studies where initial language examination was performed before one month post-onset (van Dongen, Loonen and van Dongen, 1985; van Hout, Evrard and Lyon, 1985). It is possible, therefore that paraphasic errors may be present in the speech output of head injured children in the acute stage post-onset.

Levin and Eisenberg (1979a) examined language function following closed head injury in 64 children and adolescents within six months of injury and reported that linguistic deficit was present in approximately one-third of these subjects. The most common language deficits identified were those of dysnomia for objects presented visually (13%) or tactually to the left hand (12%). Impaired auditory comprehension was identified in 11% of their subjects while verbal repetition was deficient in only 4%.

In that they observed a higher incidence of linguistic disturbance in head injured adults than head injured children, Levin and Eisenberg (1979b) concluded that recovery from traumatic aphasia is more rapid and complete in children than adults. Despite this finding however, long-term language deficits have been identified in children following traumatic head injury (Gaidolfi and Vignolo, 1980; Satz and Bullard-Bates, 1981; Jordan, Ozanne and Murdoch, 1988). Jordan, Ozanne and Murdoch (1988) assessed the language abilities of a group of 20 closed-head injured children at least 12 months post-injury and found them to be mildly impaired when compared to the language abilities of a group of age and sex matched controls. In particular, these workers identified a specific deficit in naming. The linguistic impairment exhibited by the closed head injury subjects studied by these authors however, did not conform to any recognized developmental language disorder. Rather, it was noted by these workers that the observed language disturbance was similar to that reported to occur following closed head injury in adults (see Chapter 4) in that their closed-head injured children also presented with a 'sub-clinical aphasia' characterized by dysnomia (Sarno, 1980). Jordan, Ozanne and Murdoch (1988) concluded that, in contrast to the traditional view that the immature brain makes a rapid and full recovery following traumatic injury (Basser, 1962), their findings indicated that closed head injury in children can produce long-term and persistent language deficits. Further, their findings lend support to the suggestion of Levin, Benton and Grossman (1982) that, although the immature brain may exhibit resilience to focal lesions, it may be more susceptible to the effects of the diffuse brain damage resulting from traumatic head injury.

11.2.3 Acquired aphasia associated with tumours

A number of authors have cited intra-cerebral tumours as a cause of acquired aphasia in children (Alajouanine and Lhermitte, 1965; Hecaen, 1976; Carrow-Woolfolk and Lynch, 1982; Brown, 1985; Rekate *et al.*, 1985; van Dongen, Loonen and van Dongen, 1985; Hudson, Murdoch and Ozanne, 1989). In addition, children with tumours have often been included as part of the subject group in studies aimed at investigating various clinical features of acquired childhood aphasia. Unfortunately, in the majority of studies, either the symptoms of the tumour cases have not been differentiated from those of other aetiologies or insufficient information regarding the language abilities of the children with tumours is provided to allow the language features attributable to the presence of the tumour to be identified. In fact, individually described cases of acquired childhood aphasia resulting from intra-cerebral tumours are rare. Hecaen (1976) included 2 cases of tumour in his sample of 26 children with acquired aphasia. He reported that a tumour in the left cerebral hemisphere resulted in muteness lasting two months as well as deficits in articulation, reading and writing from which there was no change. Another 2 children who had undergone surgery to remove astrocytomas were included in the group of 32 acquired aphasic children studied by Alajouanine and Lhermitte (1965). They noted that the most striking feature of the language disturbance exhibited by these 2 children was a reduction of expressive activities (including oral, written and gestural activities). In fact, these authors observed a similar reduction of expression in all 32 subjects regardless of aetiology.

Tumours involving structures in the posterior cranial fossa (i.e. the cerebellum, pons, fourth ventricle and cisterna magna) occur more commonly in children than supra-tentorial neoplasms (Matson, 1956; Gjerris, 1978; Segall et al., 1985). It is generally agreed that the most common posterior fossa tumours are astrocytomas, medulloblastomas and ependymomas (see Chapter 1). Although, considering the location of posterior fossa tumours, it would not be immediately expected that language disorders would be associated with these tumours, there are a number of secondary features of posterior fossa tumours that could lead to language problems. As these tumours in many cases either originate from or invade the fourth ventricle, hydrocephalus may occur as a secondary outcome due to the associated obstruction of the flow of cerebrospinal fluid. Subsequent compression of the cerebral cortex could lead to dysfunction in the central speech–language centres. In addition, radiotherapy administered after surgical removal of posterior fossa tumours, in order to prevent tumour spread of recurrence, has been reported to cause aphasia in some adults and intellectual deficits in some children (Broadbent, Barnes and Wheeler, 1981; Meadows et al., 1981; Duffner, Cohen and Thomas, 1983; Burns and Boyle, 1984). Further the negative effects of radiotherapy have been reported to appear as a delayed reaction, so that any associated language deficit may only appear in the long term (Hodges and Smith, 1983).

Hudson, Murdoch and Ozanne (1989) investigated the speech and language abilities of a group of children who had undergone surgery to remove a posterior fossa tumour. They reported that although dysarthria and language impairment were not the inevitable outcome of such surgery, in some cases speech and/or language deficits did occur. In 4 of the 6 cases they assessed there was a language impairment. In particular, they observed that those children who had undergone radiotherapy post-surgery appeared to have a greater chance of long-term impairment than those who did not. Although they recognized that radiotherapy may be vital for the long-term survival of children who had undergone surgery to remove brain tumours, these authors emphasized the need for the medical team (including speech–language pathologists) to be aware of the possible long-term effects that radiotherapy may have on language abilities.

Rekate *et al.* (1985) studied 6 children all of whom had experienced acute bilateral damage to large areas of both cerebellar hemispheres as a result of the presence of tumours. Their sample included 4 children with medulloblastoma, 1 with astrocytoma and 1 with ependymoma. All 6 children were described as being mute for 1–3 months post-surgery, the muteness in most cases resolving to a mild residual cerebellar (ataxic) dysarthria. One subject was described by Rekate *et al.* (1985) as having normal speech 6 months post-onset. Volcan, Cole and Johnston (1986) reported the case of an 8-year-old girl who had a right hemiparesis, truncal ataxia, signs of cerebellar dysfunction and muteness subsequent to removal of a medullo-blastoma from her fourth ventricle. Within 2 weeks post-surgery, she had regained monosyllabic speech but was described as having a monotone and dysarthria.

11.2.4 Acquired aphasia following infection

Infectious disorders of the nervous system are a well-recognized cause of acquired aphasia in children. As reported by van Dongen, Loonen and van Dongen (1985), of the 27 acquired aphasic children referred to their clinic over a four-year period, 15% had an aetiology of infectious disease. Likewise, of the 16 aphasic children examined by van Hout, Evrard and Lyon (1985), 38% had an aetiology of infectious disease. In the majority of cases reported in the literature the infectious disorder involved has been herpes simplex encephalitis.

As a result of the destructive nature of the lesions associated with infectious disorders, particularly herpes simplex encephalitis, it has been suggested that the language disorder associated with this disease is more severe than in acquired aphasia associated with other aetiologies. Van Hout and Lyon (1986) reported a case of Wernicke's aphasia in a 10-year-old boy subsequent to herpes simplex encephalitis. Their subject exhibited a number of features atypical of the usual descriptions of acquired childhood aphasia in that he exhibited symptoms such as a severe comprehension deficit, neo-logistic jargon, logorrhoea and anosognosia, the latter two features in particular usually being described as lacking in aphasic children. Van Hout and Lyon (1986) attributed the severe language disorder evidenced by their subject to the destructive bilateral damage to the temporal lobes caused by herpes simplex encephalitis.

11.2.5 Acquired aphasia associated with convulsive disorder

Landau and Kleffner (1957) reported the first cases of acquired child-hood aphasia with convulsive disorder. Landau–Kleffner syndrome, also known as 'acquired epileptic aphasia' or 'acquired verbal agnosia' appears to result from a heterogeneous group of conditions with variable aetiologies (Deonna et al., 1977). Onset has been reported to occur between the ages of 2 to 13 years with the majority of children experiencing their first loss of language function somewhere between 3 and 7 years of age. Most authors agree that males are affected twice as often as females (Cooper and Ferry, 1978; Msall et al., 1986).

Acquired epileptic aphasia is characterized by an initial deterioration of language comprehension followed by disruption of the child's expressive abilities. In some cases the onset of language deterioration is abrupt while in others the language disturbance develops gradually. Comprehension may be totally lost or reduced to understanding only short phrases and simple instructions (Worster-Drought, 1971). Often due to the reduced comprehension ability the presence of a hearing loss is suspected in the early stages of the disorder and many of the subjects are initially thought to be deaf. However, in the majority of cases their audiogram is within normal limits (Cooper and Ferry, 1978; van Harskamp, van Dongen and Loonen, 1978; van de Sandt-Koenderman et al., 1984). In association with the reduction in comprehension, the spontaneous speech of the child also changes. Expressively the child may become 'mute, use jargon or produce odd sounds, exhibit misarticulations, inappropriate substitution of words and anomia, or resort to gestures and grunts' (Cooper and Ferry, 1978, p. 177).

Preceding, co-occurring with, or following the language deterioration there may be a series of convulsive seizures (van de Sandt-Koenderman et al., 1984). Although seizures do occur often, they are not the defining feature of the syndrome. Miller et al. (1984) reported that of those cases that exhibit seizures, 43% experience the seizures before the language regression, 16% display co-occurrence of seizures and language regression and 41% experience seizures some time after the language regression. Regardless of whether or not there are clinically observable seizures, however, all patients with the syndrome exhibit epileptiform discharges in their electroencephalograms (Deonna, Fletcher and Voumard, 1982). The electroencephalographic abnormalities usually take the form of bilateral synchronous disturbances, frequently with a temporal predominance (Gascon et al., 1973; Deonna, Fletcher and Voumard, 1982). Other clinical measures, such as X-ray, arteriography, computed tomography and cerebrospinal fluid examination usually yield completely normal results.

In addition to the language disorder, a number of other associated problems may also occur in acquired epileptic aphasia. Emotional problems have been reported in a number of cases (Miller et al., 1984) and behavioural problems such as aggressiveness, temper outbursts, refusing to respond, inattention, withdrawal and hyperactivity occur frequently (Gascon *et al.*, 1973; Deonna *et al.*, 1977; Campbell and Heaton, 1978). One surprising feature of this syndrome is that the child's non-verbal intelligence usually remains unimpaired (Miller *et al.*, 1984).

As previously mentioned, the cause of acquired epileptic aphasia is unknown. Miller *et al.* (1984) pointed out that speculation regarding the neurological basis of this syndrome is likely to continue until a non-invasive assessment of cortical structure and function is better developed. Despite this, however, several hypotheses as to the pathogenesis of aphasia with convulsive disorder have been proposed. Landau and Kleffner (1957) together with Sato and Dreifuss (1973) postulated that the speech and language regression may be the result of functional ablation of the primary cortical language areas by persistent electrical discharges. Gascon *et al.* (1973), however, suggested that the electrical discharges displayed by these children occur secondary to a lower level sub-cortical de-afferenting process and that the discharges are not directly responsible for the aphasia.

As for the actual cause of the convulsive disorder, the data on which several hypotheses are based were obtained from patho-anatomical studies (Miller *et al.*, 1984). One hypothesis proposes that there exists a pathogenetic mechanism related in an unknown way to the convulsive disorder (Gascon *et al.*, 1973). For instance, there may be an unusual genetic pattern of cerebral organization that makes a child particularly sensitive to brain damage or seizure activity as far as language is concerned (Deonna *et al.*, 1977). Another hypothesis suggests that the convulsive disorder and language loss is caused by an active low-grade selective encephalitis that affects the temporal lobes (Worster-Drought, 1971; McKinney and McGreal, 1974). It has also been suggested that acquired epileptic aphasia may be caused by vascular disorders. A diminished vascular supply in the territory of the left middle cerebral artery was found in one subject with this disorder examined by Rapin *et al.* (1977).

The prognosis of aphasia with convulsive disorder is unclear. Van de Sandt-Koenderman *et al.* (1984) caution that many reports concentrate on the medical aspects of the syndrome, whilst the aphasia is poorly described. Consequently the many contradictory statements about the prognosis may be due to the variation in the particular aspect of the disorder being described as recovering. A medical examination of children may declare them 'completely recovered' when there may be still demonstrable aphasic characteristics evidenced if sufficiently sensitive testing is carried out. Often the language recovery in acquired epileptic aphasia is very limited. Miller *et al.* (1984) stated that > 80% of cases reported in the literature have receptive and expressive deficits that persist for longer than six months. In general it appears that the prognosis for recovery is poor if there has been no progress within one year post-onset. Some children with this disorder go through periods of exacerbations and remissions. Mantovani and Landau (1980) found that children who exhibit this latter type of course have a relatively good prognosis.

11.3 SUMMARY

Although for a long time it has been believed that acquired aphasia in children is primarily of the non-fluent type, in recent years this belief has been challenged by studies which have demonstrated that fluent aphasias can be observed in children if language examination is carried out in the early stages post-onset. In addition, there is now evidence to suggest that the type of aphasia resulting from brain damage in children varies according to the underlying aetiology. Clearly there is a need for more research to further elucidate the specific clinical features of each of the speech–language deficits associated with the various neurological disorders which may cause acquired childhood aphasia.

References

Abbs, J.H., Hunker, C.J. and Barlow, S.H. (1983) Differential speech motor subsystem impairments with suprabulbar lesions: neurophysiological framework and supporting data. In *Clinical Dysarthria* (ed. W.R. Berry), College Hill Press, San Diego.

Adamovich, B. and Henderson, J. (1983) Can we learn more from word fluency measures with aphasic, right brain injured and closed head trauma patients? In *Clinical Aphasiology Conference Proceedings* (ed. R.H. Brookshire), BRK, Minneapolis.

Adams, J.H., Mitchell, D.E., Graham, O.T. and Doyle, D. (1977) Diffuse brain damage of immediate impact type. *Brain*, **100**, 489–502.

Akbarova, N.A. (1972) Memory disorders in patients with closed cranio-cerebral trauma of mild and moderate severity. *Zhurnal Neuopatologii I Psikhiatri*, **11**, 1641–1646.

Alajouanine, T. and Lhermitte, F. (1965) Acquired aphasia in children. *Brain*, **88**, 653–662.

Albert, M.L. (1980) Language in normal and dementing elderly. In *Language and Communication in the Elderly* (eds L.K. Obler and M.L. Albert), Lexington Books, Toronto.

Alexander, M.P. and LoVerme, S.R. (1980) Aphasia after left hemispheric intra-cerebral hemorrhage. *Neurology*, **30**, 1193–1202.

Alexander, M.P., Naeser, M.A. and Palumbo, C.L. (1987) Correlations of subcortical CT lesion sites and aphasia profiles. *Brain*, **110**, 961–991.

Alzheimer, A. (1969) Alzheimer's Disease. In *Archives of Neurology*, (Trans. R.H. Wilkins and I.A. Brody), **21**, 109–110. (Reprinted from *Allgemeine Zeitschrift für Psychiatrisch und Psychischgerichtlich Medizin*, 1907, **64**, 146.)

American Psychiatric Association (1980) *Diagnostic and Statistical Manual of Mental Disorders*, 3rd edn. Washington, DC.

Andral, G. (1840) *Clinique Médicale*. Fortin, Massonet Cie, Paris.

Annegers, J.F., Grabow, J.D., Kurland, L.T. and Laws, E.R. (1980) The incidence, causes and secular trends of head trauma in Olmsted County, Minnesota. *Neurology*, **30**, 912–919.

Appell, J., Kertesz, A. and Fisman, M. (1982) A study of language functioning in Alzheimer's patients. *Brain and Language*, **17**, 73–91.

Aram, D.M., Ekelman, B.L. and Whitaker, H.A. (1986) Spoken syntax in children with acquired unilateral hemisphere lesions. *Brain and Language*, **27**, 75–100.

Aram, D.M., Rose, D.F., Rekate, H.L. and Whitaker, H.A. (1983) Acquired capsular/striatal aphasia in childhood. *Archives of Neurology*, **40**, 614–617.

Aronson, A.E. (1981) Motor speech signs of neurologic disease. In *Speech Evaluation in Medicine* (ed. F.L. Darley), Grune & Stratton, New York.

Aronson, A.E., Brown, J.R., Litin, E.M. and Pearson, J.S. (1968) Spastic dysphonia. II. Comparison with essential (voice) tremor and other neurologic and psychogenic dysphonias. *Journal of Speech and Hearing Disorders*, **33**, 219–231.

Arseni, C. (1958) Tumours of the basal ganglia. *A.M.A. Archives of Neurology and Psychiatry*, **80**, 18–26.

Assal, G. and Campiche, R. (1973) Aphasie et troubels du langage chez l'enfant après contusion cérébrale. *Neurochirurgie*, **19**, 399–406.

Assal, G., Buttet, J. and Jolivet, R. (1981) Disassociations in aphasia: A case report. *Brain and Language*, **13**, 223–240.

Australian Bureau of Statistics. (1982) *Australia's Age Population* (Catalogue No. 4109.0). Commonwealth Government Printing Office, Canberra.

Bannister, R. (1985) *Brain's Clinical Neurology*. Oxford University Press, Edinburgh.

Barat, M., Mazauz, J., Bioulac, B., Giroire, J., Vital, C. and Arne, L. (1981) Troubles du langage de type aphasique et lésions putamino caudées. *Revue Neurologique*, **137**, 343–356.

Barringer, J.R. (1978) Herpes simplex infections of the nervous system. In *Handbook of Clinical Neurology*, Vol. 34 (eds D. Vinken and G. Bruyn), North-Holland/Elsevier, Amsterdam.

Basser, L. (1962) Hemiplegia of early onset and the faculty of speech with special reference to the effects of hemispherectomy. *Brain*, **85**, 427–460.

Basso, A., Lecours, A.R., Morashini, S. and Vanier, M. (1985) Anatomo-clinical correlations of the aphasias as defined through computerized tomography: Exceptions. *Brain and Language*, **26**, 201–229.

Basso, A., Sala, S.D. and Farabola, M. (1987) Aphasia arising from purely deep lesions. *Cortex*, **23**, 29–44.

Basso, A., Taborelli, A. and Vignolo, L.A. (1978) Dissociated disorders of speaking and writing in aphasia. *Journal of Neurology, Neurosurgery and Psychiatry*, **41**, 556–563.

Bastian, H.C. (1898) *Aphasia and Other Defects*, H.K. Lewis, London.

Bates, E. (1976) *Language and context: The Acquisition of Pragmatics*. Academic Press, New York.

Bay, E. (1964) Principles of classification and their influence on our concepts of aphasia. In *Disorders of Language* (eds A.V.S. de Reuck and M. O'Connor), Little, Brown and Co., Boston.

Bayles, K.A. (1982) Language function in senile dementia. *Brain and Language*, **16**, 265–280.

Bayles, K.A. (1984) Language and dementia. In *Language Disorders in Adults: Recent Advances* (ed. A.L. Holland), College Hill Press, San Diego.

Bayles, K.A. (1985) Communication in dementia. In *The Aging Brain: Communication in the Elderly* (ed. H.K. Ulatowska), College Hill Press, London.

Bayles, K.A. and Boone, D.R. (1982) The potential of language tasks for identifying senile dementia. *Journal of Speech and Hearing Disorders*, **47**, 210–217.

Bayles, K.A. and Tomoeda, C.K. (1983) Confrontation naming impairment in dementia. *Brain and Language*, **19**, 98–114.

Bayles, K.A., Tomoeda, C.K. and Caffrey, J.T. (1982) Language and dementia producing diseases. *Communicative Disorders*, **7**, 131–146.

Bearn, A.G. (1960) A genetic analysis of thirty families with Wilson's disease (hepatolenticular degeneration). *Annals of Human Genetics*, **24**, 33–43.

Beaumont, J.G. (1974) Handedness and hemisphere function. In *Hemisphere Function in the Human Brain* (eds S.J. Diamond and J.G. Beaumont), Paul Elek, London.

Beauvois, M.F., Saillant, B., Meininger, V. and Lhermitte, F. (1978) Bilateral tactile aphasia: A tacto-verbal dysfunction. *Brain*, **101**, 381–402.

Behan, P.O. and Behan, W.M.H. (1979) Possible immunological factors in Alzheimer's disease. In *Alzheimer's Disease. Early Recognition of Potentially Reversible Deficits* (eds A.I.M. Glen and L.J. Whalley), Churchill-Livingstone, London.

Bell, D.S. (1968) Speech functions of the thalamus inferred from the effects of thalamotomy. *Brain*, **91**, 619–638.

Benowitz, L.I., Bear, D.M., Rosenthal, R., Mesulam, M.M., Zaidel, E. and Sperry, R.W. (1983) Hemispheric specialization in non-verbal communication. *Cortex*, **19**, 5–12.

Benson, D.F. (1967) Fluency in aphasia: Correlation with radioactive scan localization. *Cortex*, **3**, 373–394.

Benson, D.F. (1972) Language disturbances of childhood. *Clinical Proceedings Children's Hospital of Washington*, **28**, 93–100.

Benson, D.F. (1977) The third alexia. *Archives of Neurology*, **34**, 327–331.

Benson, D.F. (1979) *Aphasia, Alexia and Agraphia*. Churchill-Livingstone, New York.

Benson, D.F. (1986) Aphasia and the lateralization of language. *Cortex*, **22**, 71–86.

Benson, D.F. and Geschwind, N. (1969) The alexias. In *Handbook of*

Clinical Neurology, Vol. 4, (eds P.J. Vinken and G.W. Bruyn), North-Holland/Elsevier, Amsterdam.

Benton, A.L. (1977) Reflections on the Gerstmann syndrome. *Brain and Language*, **4**, 45–62.

Berg, L. (1985) Does Alzheimer's disease represent an exaggeration of normal aging? *Archives of Neurology*, **42**, 737–739.

Berlin, C.I. and McNeil, M.R. (1976) Dichotic listening. In *Contemporary Issues in Experimental Phonetics* (ed. N.J. Lass), Academic Press, New York.

Berlin, C.I., Lowe-Bell, S.S., Cullen, J.K., Thompson, C.L. and Loovis, C.F. (1973) Dichotic speech perception. An interpretation of right ear advantage and temporal offset effects. *Journal of the Acoustical Society of America*, **53**, 699–709.

Bernhardt, M. (1885) Ueber die spastiche cerebralparalyse im Kindersalter (Hemiplegia spastica infantalis), nebst einem excurse über Aphasie bei Kindern. *Archiv für Pathologische Anatomie und Physiologie und für Klinische Medecin*, **102**, 26–80.

Berry, W.R., Darley, F.L., Aronson, A.E. and Goldstein, N.P. (1974) Dysarthria in Wilson's disease. *Journal of Speech and Hearing Research*, **17**, 167–183.

Beukelman, D.R., Kraft, G.H. and Freal, J. (1985) Expressive communication disorders in persons with multiple sclerosis: a survey. *Archives of Physical Medicine and Rehabilitation*, **66**, 675–677.

Blackburn, I.M. and Tyrer, G.M.B. (1985) The value of Luria's neuropsychological investigation for the assessment of cognitive dysfunction in Alzheimer-type dementia. *British Journal of Clinical Psychology*, **24**, 171–179.

Boller, F., Mizutani, T., Roessmann, V. and Gambetti, P. (1980) Parkinson's disease, dementia and Alzheimer's disease: Clinicopathological correlations. *Annals of Neurology*, **7**, 329–335.

Bouillaud, M.J. (1825) Recherches cliniques propres à demontrer que la perte de la parole correspond à la lesion des lobules anterieurs du cerveau, et à confirmer l'opinion de M. Gall, sur le siège de l'orane du langage articulé. *Archives Générals de Médécine*, **8**, 22–45.

Bradshaw, J.L. and Nettleton, N.C. (1983) *Human Cerebral Asymmetry*, Prentice Hall, New Jersey.

Brain, L. and Walton, J.N. (1969) *Brain Diseases of the Nervous System*, 7th edn, Oxford University Press, New York.

Branco-LeFevre, A.F. (1950) Contribucao para o estudio da psicopatólogia da afasia en criancas. *Archivos Neuro-Psiquiatria (San Paulo)*, **8**, 345–393.

Broadbent, V.A., Barnes, N.D. and Wheeler, T.K. (1981) Medulloblastoma in childhood: Long-term results of treatment. *Cancer*, **48**, 26–30.

Broca, P. (1861) Portée de la parole. Ramollissement chronique et destruction partielle du lobe antérieur gauche du cerveau. *Paris Bulletin of Society of Anthropology*, **2**, 219.

Broca, P. (1865) Sur la faculté du langage articulé. *Paris Bulletin of Society of Anatomy*, **36**, 337–393.

Brooks, D.N., Aughton, M.E., Bond, M.R., Jones, P. and Rizvi, S. (1980) Cognitive sequelae in relationship to early indices of severity of brain damage after severe blunt head injury. *Journal of Neurology, Neurosurgery and Psychiatry*, **43**, 529–534.

Brookshire, R. (1973) *An Introduction to Aphasia*. BRK, Minneapolis.

Brown, J.K. (1985) Dysarthria in children: Neurologic perspective. In *Speech and Language Evaluation in Neurology: Childhood Disorders* (ed. J.K. Darby), Grune & Stratton, New York.

Brown, J.R. and Simonson, J. (1963) Organic voice tremor. *Neurology*, **13**, 520–525.

Brown, J.R., Darley, F.L. and Aronson, A.E. (1970) Ataxic dysarthria. *International Journal of Neurology*, **7**, 302–318.

Brown, J.W. (1974) *Aphasia, Apraxia and Agnosia: Clinical and Theoretical Aspects*. Charles C. Thomas, Illinois.

Brown, J.W. (1975) On the neural organization of language: Thalamic and cortical relationships. *Brain and Language*, **2**, 18–30.

Brunner, R.J., Kornhuber, H.H., Seemuller, E., Suger, G. and Wallesch, C.W. (1982) Basal ganglia participation in language pathology. *Brain and Language*, **16**, 281–299.

Buckingham, H. and Kertesz, A. (1974) A linguistic analysis of fluent aphasia. *Brain and Language*, **1**, 43–61.

Bugiani, D., Conforto, C. and Sacco, G. (1969) Aphasia in thalamic haemorrhage. *Lancet*, **i**, 1902.

Burns, M.S. and Boyle, M. (1984) Aphasia after successful radiation treatment: A report of two cases. *Journal of Speech and Hearing Disorders*, **19**, 107–111.

Burns, M.S. and Canter, G.J. (1977) Phonemic behaviour of aphasia patients with posterior cerebral lesions. *Brain and Language*, **4**, 492–507.

Caldas, A.C. and Botelho, M.A.S. (1980) Dichotic listening in the recovery of aphasia after stroke. *Brain and Language*, **10**, 145–151.

Caligiuri, M.P. and Murry, T. (1983) The use of visual feedback to enhance prosodic control in dysarthria. In *Clinical Dysarthria*, (ed. W.R. Berry), College-Hill Press, San Diego

Cambier, J., Elghozi, D., Signoret, J.L. and Henin, D. (1983) Contribution of the right hemisphere to language in aphasic patients. Disappearance of this language after a right-sided lesion. *Paris Review Neurologique*, **139**, 55–63.

Campbell, T.F. and Heaton, E.M. (1978) An expressive speech program for a child with acquired aphasia: A case study. *Human Communication* (summer), 89–102.

Canter, G. (1963) Speech characteristics of patients with Parkinson's disease. I. Intensity, pitch and duration. *Journal of Speech and Hearing Disorders*, **28**, 221–229.

Canter, G. (1965a) Speech characteristics of patients with Parkinson's disease. II. Physiological support for speech. *Journal of Speech and Hearing Disorders*, **30**, 44–49.

Canter, G. (1965b) Speech characteristics of patients with Parkinson's disease. III. Articulation, diadochokinesis and overall speech adequacy. *Journal of Speech and Hearing Disorders*, **30**, 217–224.

Cappa, S.F. and Vignolo, L.A. (1979) Transcortical features of aphasia following left thalamic hemorrhage. *Cortex*, **15**, 121–130.

Cappa, S.F., Cavallotti, G. and Vignolo, L. (1981) Phonemic and lexical errors in fluent aphasia: Correlation with lesion site. *Neuropsychologia*, **19**, 171–177.

Cappa, S.F., Cavallotti, G., Guidotti, M., Papagno, C. and Vignolo, L.A. (1983) Subcortical aphasia: Two clinical–CT scan correlation studies. *Cortex*, **19**, 227–241.

Cappa, S.F., Papagno, C., Vallar, G. and Vignolo, L.A. (1986) Aphasia does not always follow left thalamic hemorrhage: A study of five negative cases. *Cortex*, **22**, 639–647.

Carrow, E., Rivera, V., Maudlin, M. and Shamblin, L. (1974) Deviant speech characteristics in motor neurone disease. *Archives of Otolaryngology*, **100**, 212–218.

Carrow-Woolfolk, E. and Lynch, J. (1982) *An Integrative Approach to Language Disorders in Children.* Grune & Stratton, Orlando.

Carter, R.L., Hohenegger, M.K. and Satz, P. (1982) Aphasia and speech organization in children. *Science*, **218**, 797–799.

Caruso, A.J. and Burton, E.K. (1987) Temporal acoustic measures of dysarthria associated with amyotrophic lateral sclerosis. *Journal of Speech and Hearing Research*, **30**, 80–87.

Chadwick, O. (1985) Psychological sequelae of head injury in children. *Developmental Medicine and Child Neurology*, **27**, 72–75.

Charcot, J.M. (1877) *Lectures on the Diseases of the Nervous System.* New Sydenham Society, London.

Chedru, F. and Geschwind, N. (1972) Writing disturbances in acute confusional states. *Neuropsychologia*, **10**, 343–353.

Cheek, W.R. and Taveras, J. (1966) Thalamic tumours. *Journal of Neurosurgery*, **24**, 505–513.

Chenery, H.J. and Murdoch, B.E. (1986) A case of mixed transcortical aphasia following drug overdose. *British Journal of Disorders of Communication*, **21**, 381–392.

Chenery, H.J., Murdoch, B.E. and Ingram, J.C.L. (1988) Studies in Parkinson's disease. I. Perceptual speech analysis. *Australian Journal of Human Communication Disorders*, **16**, 17–29.

Chesson, A.L. (1983) Aphasia following a right thalamic hemorrhage. *Brain and Language*, **19**, 306–316.

Chui, H., Teng, E.L., Henderson, V.W. and May, A.C. (1985) Clinical subtypes of dementia of the Alzheimer's type. *Neurology*, **35**, 1544–1550.

Ciemens, V.A. (1970) Localized thalamic hemorrhage: A cause of aphasia. *Neurology*, **20**, 776–782.

Code, C. (1987) *Language, Aphasia and the Right Hemisphere*. John Wiley, Chichester.

Cohen, J.A., Gelfer, C.E. and Sweet, R.P. (1980) Thalamic infarction producing aphasia. *Mt Sinai Journal of Medicine*, **47**, 398–404.

Collingnon, R., Hecaen, H. and Angerlerques, G. (1968) A propos de 12 cas d'aphasie acquise chez l'enfant. *Acta Neurologica et Psychiatrica Belgica*, **68**, 245–277.

Collins, M. (1975) The minor hemisphere. In *Clinical Aphasiology Conference Proceedings* (ed. R. Brookshire), BRK, Minneapolis.

Constantinidis, J., Richards, J. and de Ajuriaguerra, J. (1978) Dementias with senile plaques and neurofibrillary changes. In *Studies in Geriatric Psychiatry* (eds A.D. Isaacs and F. Post), John Wiley, Brisbane.

Cooper, J.A. and Ferry, P.C. (1978) Acquired auditory verbal agnosia and seizures in childhood. *Journal of Speech and Hearing Disorders*, **43**, 176–184.

Corsellis, J.A.N. (1976) Ageing and the dementias. In *Greenfield's Neuropathology* (eds W. Blackwood and J.A.N. Corsellis), Year Book Medical Publishers, Chicago.

Coyle, J.T., Price, D.L. and de Long, M.R. (1983) Alzheimer's disease: A disorder of cortical cholinergic innervation. *Science*, **219**, 1194–1219.

Crane, G.E. (1968) Tardive dyskinesia in patients treated with major neuroleptics: A review of the literature. *American Journal of Psychiatry*, **24**, 40–48.

Crapper, D.R., Krishnan, S.S. and Quittkart, S. (1976) Aluminium neurofibrillary degeneration and Alzheimer's disease. *Brain*, **99**, 67–80.

Critchley, E.M.R. (1981) Speech disorders of Parkinsonism: a review. *Journal of Neurology, Neurosurgery and Psychiatry*, **44**, 751–758.

Critchley, M. (1949) Observations on essential (heredofamilial) tremor. *Brain*, **72**, 113–139.

Critchley, M. (1964) The neurology of psychotic speech. *British Journal of Psychiatry*, **110**, 353–364.

Critchley, M. (1966) The enigma of Gerstmann's syndrome. *Brain*, **89**, 183–198.

Critchley, M. (1970) *Aphasiology*, Edward Arnold, London.

Critchley, M. (1984) And all the daughters of musick shall be brought low: Language function in the elderly. *Archives of Neurology*, **41**, 1135–1139.

Crosson, B. (1985) Subcortical functions in language: A working model. *Brain and Language*, **25**, 257–292.

Cummings, J.L. (1982) Cortical dementias. In *Psychiatric Aspects of Neurologic Disease*, Vol. 2 (eds D.F. Benson and D. Blumer), Grune & Stratton, New York.

Cummings, J.L. and Benson, D.F. (1983) *Dementia: A Clinical Approach*. Butterworths, Boston.

Cummings, J.L. and Benson, D.F. (1984) Subcortical dementia. Review of an emerging concept. *Archives of Neurology*, **41**, 874–879.

Cummings, J.L. and Duchen, L.W. (1981) The Klüver–Bucy syndrome in Pick's disease. *Neurology*, **31**, 1415–1422.

Cummings, J.L., Benson, D.F. and Lo Verme, S. (1980) Reversible dementia. *Journal of the American Medical Association*, **243**, 2434–2439.

Cummings, J.L., Benson, D.F., Hill, M.A. and Read, S. (1985) Aphasia in dementia of the Alzheimer's type. *Neurology*, **35**, 394–397.

Cummings, J.L., Benson, D.F., Walsh, M.J. and Levine, H.L. (1979) Left-to-right transfer of language dominance. A case study. *Neurology*, **29**, 1547–1549.

Cutting, J. (1978) The relationship between Korsakoff's syndrome and alcoholic dementia. *British Journal of Psychiatry*, **132**, 240–251.

Damasio, A.R., Damasio, H.D., Rizzo, M., Varney, N. and Gersh, F. (1982) Aphasia with nonhemorrhagic lesions in the basal ganglia and internal capsule. *Archives of Neurology*, **39**, 15–20.

Damasio, H. (1981) Cerebral localization of the aphasias. In *Acquired Aphasia* (ed. M.T. Sarno), Academic Press, New York.

Damasio, H. and Damasio, A.R. (1980) The anatomical basis of conduction aphasia. *Brain*, **103**, 337–350.

Damasio, H., Damasio, A., Hamsher, K. and Varney, N. (1979) CT scan correlates of aphasia and allied disorders. *Neurology*, **29**, 572.

Darby, J.K. (1981) The interaction between speech and disease. In *Speech Evaluation in Medicine* (ed. J.K. Darby), Grune & Stratton, New York.

Darley, F.L. (1975) Treatment of acquired aphasia. In *Advances in Neurology*, Vol. 7 (ed. W.J. Friedlander), Raven Press, New York.

Darley, F.L. (1982) *Aphasia*. W.B. Saunders, Philadelphia.

Darley, F.L., Aronson, A.E. and Brown, J.R. (1969a) Differential diagnostic patterns of dysarthria. *Journal of Speech and Hearing Research,* **12**, 246–269.

Darley, F.L., Aronson, A.E. and Brown, J.R. (1969b) Clusters of deviant speech dimensions in the dysarthrias. *Journal of Speech and Hearing Research,* **12**, 462–496.

Darley, F.L., Aronson, A.E. and Brown, J.R. (1975a) *Motor Speech Disorders.* W.B. Saunders, Philadelphia.

Darley, F.L., Brown, J.R. and Goldstein, N.P. (1972) Dysarthria in multiple sclerosis. *Journal of Speech and Hearing Research,* **15**, 229–245.

Darley, F.L., Brown, J.R. and Swenson, W.M. (1975b) Language changes after neurosurgery for Parkinsonism. *Brain and Language,* **2**, 65–69.

Davies, P. (1983) An update on the neurochemistry of Alzheimer's disease. In *The Dementias,* (eds R. Mayeux and W.G. Rosen), Raven Press, New York.

Dax, M. (1836) Lésions de la moitié gauche de l'encéphale coincident avec trouble des signes de la pensie. Paper presented in Montepelier.

Day, L.S. and Parnell, M.M. (1987) Ten-year study of a Wilson's disease dysarthric. *Journal of Communication Disorders,* **20**, 207–218.

de Ajuriaguerra, J. and Tissot, R. (1975) Some aspects of language in various forms of senile dementia (comparisons with language in childhood). In *Foundations of Language Development* (eds E.H. Lennenberg and E. Lennenberg), Academic Press, New York.

de Boni, U. and Crapper, D.R. (1978) Paired helical filaments of the Alzheimer type in cultured neurons. *Nature,* **271**, 566–568.

Deal, J.C., (1974) Consistency and adaptation in apraxia of speech. *Journal of Communication Disorders,* **7**, 135–140.

Dejerine, J. (1891) Sur un cas de cécité verbal avec agraphie, suivi d'autopsie. *Compte Rendu des Séances de la Société de Biologie,* **3**, 197–201.

Dejerine, J. (1892) Contributions à l'étude anatomo-pathologique et clinique des differentes variétés de cécité. *Compte Rendu des Séances de la Société de Biologie,* **4**, 61–90.

Delis, D.C., Wapner, W., Gardner, H. and Moses, J.A. (1983) The contribution of the right hemisphere to the organization of paragraphs. *Cortex,* **19**, 43–50.

Dennis, M. (1980) Strokes in childhood. 1. Communicative intent, expression and comprehension after left hemisphere arteriopathy in a right-handed nine-year-old. In *Language Development and Aphasia in Children* (ed. R.W. Reiber), Academic Press, New York.

Dennis, M. and Whitaker, H.A. (1977) Hemispheric equipotentiality and language acquisition. In *Language Development and Neurological*

Theory, (eds S.J. Segalowitz and F. Gruber), Academic Press, New York.

Deonna, T., Beaumanoir, A., Gaillard, F. and Assal, G. (1977) Acquired aphasia in childhood with seizure disorder: A heterogeneous syndrome. *Neuropadiatrie*, **8**, 263–273.

Deonna, T., Fletcher, P. and Voumard, C. (1982) Temporary regression during language acquisition: A linguistic analysis of a two-and-a-half year old child with epileptic aphasia. *Developmental Medicine and Child Neurology*, **24**, 156–163.

de Paul, R., Abbs, J.H., Caligiuri, M., Gracco, V.L. and Brooks, B.R. (1988) Hypoglossal, trigeminal and facial motorneuron involvement in amyotrophic lateral sclerosis. *Neurology*, **38**, 281–283.

DeWitt, L.D., Grek, A.J., Buonanno, F.S., Levine, D.N. and Kistler, J.P. (1985) MRI and the study of aphasia. *Neurology*, **35**, 861–865.

Diggs, C.C. and Basili, A.G. (1987) Verbal expression of right CVA patients. Convergent and divergent language. *Brain and Language*, **30**, 130–147.

Dowson, J.H. (1982) Neuronal lipofuscin accumulation in ageing and Alzheimer dementia: A pathogenic mechanism? *British Journal of Psychiatry*, **140**, 142–148.

Duffner, P.K., Cohen, M.E. and Thomas, P.R.M. (1983) Late effects of treatment on the intelligence of children with posterior fossa tumours. *Cancer*, **51**, 233–237.

Dunlop, J. and Marquardt, T. (1977) Linguistic and articulatory aspects of single word production in apraxia of speech. *Cortex*, **13**, 17–29.

Dworkin, J.P. (1980) Tongue strength measurement in patients with amyotrophic lateral sclerosis: qualitative vs quantitative procedures. *Archives of Physical Medicine and Rehabilitation*, **61**, 422–424.

Dworkin, J.P. and Aronson, A.E. (1986) Tongue strength and alternate motion rates in normal and dysarthric subjects. *Journal of Communication Disorders*, **19**, 115–132.

Dworkin, J.P. and Hartman, D.E. (1979) Progressive speech deterioration and dysphagia in amyotrophic lateral sclerosis. *Archives of Physical Medicine and Rehabilitation*, **60**, 423–425.

Dworkin, J.P., Aronson, A.E. and Mulder, D.W. (1980) Tongue force in normals and in dysarthric patients with amyotrophic lateral sclerosis. *Journal of Speech and Hearing Research*, **23**, 828–837.

Eccles, J.A. (1973) A re-evaluation of cerebellar function in man. In *New Developments in Electromyography and Clinical Neurophysiology*, Vol. 3 (ed. J. Desmedt), Karger, Basel.

Edwards, M. (1984) *Disorders of Articulation: Aspects of Dysarthria and Verbal Dyspraxia*. Springer-Verlag, New York, Vienna.

Emmorey, K.D. (1987) The neurological substrates for prosodic aspects of speech. *Brain and Language*, **30**, 305–321.

Enderby, P. (1983) *Frenchay Dysarthria Assessment.* College Hill Press, California.

Enderby, P. (1986) Relationships between dysarthric groups. *British Journal of Disorders of Communication,* **21,** 189–197.

Enderby, P. and Phillip, R. (1986) Speech and language handicap: towards knowing the size of the problem. *British Journal of Disorders of Communication,* **21,** 151–165.

Farmakides, M.N. and Boone, D.R. (1960) Speech problems in patients with multiple sclerosis. *Journal of Speech and Hearing Disorders,* **25,** 385–390.

Fazio, C., Sacco, G. and Bugiani, O. (1973) The thalamic hemorrhage. *European Neurology,* **9,** 30–43.

Fedio, P. and Van Buren, J.M. (1975) Memory and perceptual deficits during electrical stimulation in the left and right thalamus and parietal subcortex. *Brain and Language,* **2,** 78–100.

Field, J.R., Corbin, K.B., Goldstein, N.P. and Klass, D.W. (1966) Gilles de la Tourette's syndrome. *Neurology,* **16,** 453–462.

Fingers, S. (1978) *Recovery from Brain Damage.* Plenum Press, New York.

Fisher, J., Kennedy, J., Caine, E. and Shoulson, I. (1983) Dementia in Huntington's disease: A cross-sectional analysis of intellectual decline. In *The Dementias* (eds R. Mayeux and W.G. Rosen), Raven Press, New York.

Fitzgerald, F.J., Murdoch, B.E. and Chenery, H.J. (1987) Multiple sclerosis: associated speech and language disorders. *Australian Journal of Human Communication Disorders,* **15,** 15–33.

Fletcher, J.M. and Taylor, H. (1984) Neuropsychological approaches to children: Towards a developmental neuropsychology. *Journal of Neuropsychology,* **6,** 39–57.

Flourens, P. (1824) *Recherches Expérimentales sur les Propriétés et les Fonctions du Systéme Nerveux.* Bailliére, Paris.

Foldi, N.S., Cicone, M. and Gardner, H. (1983) Pragmatic aspects of communication in brain-damaged patients. In *Language Functions in Brain Organization* (ed. S.J. Segalowitz), Academic Press, New York.

Freud, S. (1897) *Infantile Cerebral Paralysis* (Trans. L.A. Russin, 1968). University of Miami, Coral Gables.

Friedman, J.H., Brem, H. and Mayeux, R. (1983) Global aphasia in multiple sclerosis. *Annals of Neurology,* **13,** 222–223.

Gaidolfi, E. and Vignolo, L.A. (1980) Closed head injuries of school aged children: Neuropsychological sequelae in early adulthood. *Italian Journal of Neurological Sciences,* **1,** 65–73.

Gainotti, G., Caltagirone, C., Miceli, G. and Masullo, C. (1981) Selective semantic-lexical impairment of language comprehension in right brain damaged patients. *Brain and Language,* **13,** 201–211.

316 References

Galaburda, A.M., Le May, M., Kemper, T. and Geschwind, M. (1978) Right left asymmetries in the brain. *Science*, **199**, 852–856.

Gardner, H. (1975) *The Shattered Mind*. Knopf, New York.

Gardner, H., Brownell, H.H., Wapner, W. and Michelow, D. (1983) Missing the point: The role of the right hemisphere in the processing of complex linguistic materials. In *Cognitive Processing in the Right Hemisphere*, (ed. E. Perecman), Academic Press, New York.

Gardner, H., King, P., Flamm, L. and Silverman, J. (1975) Comprehension and appreciation of humorous material following brain damage. *Brain*, **98**, 399–412.

Gascon, G., Victor, D., Lombroso, C.T. and Goodglass, H. (1973) Language disorder, convulsive disorder and electroencephalographic abnormalities. *Archives of Neurology*, **28**, 156–162.

Gazzaniga, M.S. (1970) *The Bisected Brain*. Appleton-Century-Crofts, New York.

Gazzaniga, M.S. (1983) Right hemisphere following brain bisection. A 20 year perspective. *American Psychologist*, **38**, 525–537.

Gazzaniga, M.S. and Hillyard, S. (1971) Language and speech capacity of the right hemisphere. *Neuropsychologia*, **9**, 273–280.

Gazzaniga, M.S. and Sperry, R.W. (1967) Language after sectioning of the cerebral commissures. *Brain*, **90**, 131–148.

Gazzaniga, M.S., Charlotte, S., Smylie, S., Baynes, K., Hirst, W. and Smylie, C.S. (1984) Profiles of right hemisphere language and speech following brain bisection. *Brain and Language*, **22**, 206–220.

Gerstmann, J. (1931) Zur symptomatologie der hirnläsionen im übergangsgebiet der unteren parietal-hund mittleren occipital Windung. *Nervenarzt*, **3**, 691–695.

Geschwind, N. (1965a) Disconnection syndromes in animals and man. Part 1. *Brain*, **88**, 237–294.

Geschwind, N. (1965b) Disconnection syndromes in animals and man. Part 2. *Brain*, **88**, 585–644.

Geschwind, N. (1967) The varieties of naming errors. *Cortex*, **3**, 97–112.

Geschwind, N. (1969) Anatomical understanding of aphasias. In *Contributions to Clinical Neuropsychology* (ed. A.L. Benton), Aldine Publishing, Chicago.

Geschwind, N. (1971) Aphasia. *New England Journal of Medicine*, **284**, 654–656.

Geschwind, N. (1975) The apraxias: Neural mechanism of disorders of learned movements. *American Scientist*, **63**, 188–195.

Geschwind, N. and Levitsky, W. (1968) Human brain: Left–right asymmetries in temporal speech region. *Science*, **161**, 186–187.

Geschwind, N., Quadfasel, F.A. and Segarra, J.M. (1968) Isolation of the speech area. *Neuropsychologia*, **6**, 327–340.

Gilman, S. (1969) The mechanism of cerebellar hypotonia. *Brain*, **92**, 621–638.

Gjerris, F. (1978) Clinical aspects and long-term prognosis of infratentorial intracranial tumours in infancy and childhood. *Acta Neurologica Scandinavica*, **57**, 31–52.

Glaser, M.A. and Schafer, F.P. (1932) Skull and brain traumas: their sequelae: clinical review of 255 cases. *Journal of the American Medical Association*, **98**, 271–276.

Gloning, I., Gloning, K. and Hoff, H. (1963) Aphasia: A clinical syndrome. In *Problems of Dynamic Neurology*, (ed. L. Halpern), Hebrew University, Jerusalem.

Gloning, I., Gloning, K., Haub, C. and Quatember, R. (1969) Comparison of verbal behaviour in right-handed and non-right handed patients with anatomically verified lesion of one hemisphere. *Cortex*, **5**, 43–52.

Gloning, K. (1977) Handedness and aphasia. *Neuropsychologia*, **15**, 353–358.

Gloning, K. and Hift, E. (1970) Aphasie im Vorschulalter. *Zeitschrift für Nervenheilkunde*, **28**, 20–28.

Glosser, G., Kaplan, E. and LoVerme, S. (1982) Longitudinal neuropsychological report of aphasia following left-subcortical hemorrhage. *Brain and Language*, **15**, 95–116.

Golper, L.C., Nutt, J.G., Rau, M.T. and Coleman, R.O. (1983) Focal cranial dystonia. *Journal of Speech and Hearing Disorders*, **48**, 128–134.

Goodglass, H. and Kaplan, E. (1972, 1983) The Assessment of Aphasia and Related Disorders. Lea and Febiger, Philadelphia.

Goodglass, H. and Quadfasel, F.A. (1954) Language laterality in left handed aphasics. *Brain*, **77**, 510–548.

Gorelick, P.B., Hier, D.B., Benevento, L., Levitt, S. and Tan, W. (1984) Aphasia after left thalamic infarction. *Archives of Neurology*, **41**, 1296–1298.

Gottfries, C., Adolfsson, R., Aquilonius, S., Carlsson, A., Eckernas, S., Norberg, A., Oreland, L., Svennerholm, L., Wiberg, A. and Winblad, A. (1983) Biochemical changes in dementia disorders of Alzheimer type. *Neurobiology of Aging*, 4, 261–271.

Graff-Radford, N.R., Eslinger, P.J., Damasio, A.R. and Yamada, T. (1984) Nonhaemorrhagic infarction of the thalamus: Behaviour, anatomic and physiologoical correlates. *Neurology*, **34**, 14–23.

Greenfield, J.G. (1938) Some observations on cerebral injuries. *Proceedings of the Royal Society of Medicine*, **32**, 45.

Grinker, R.R. and Sahs, A.L. (1966) *Neurology*, Charles C. Thomas, Springfield.

Grober, E., Buschke, H., Kawas, C. and Fuld, P. (1985) Impaired ranking of semantic attributes in dementia. *Brain and Language*, **26**, 276–287.

Groher, M. (1977) Language and memory disorders following closed head trauma. *Journal of Speech and Hearing Research*, **20**, 212–223.

Groher, M. (1983) Communication disorders. In *Rehabilitation of the Head Injured Adult* (eds M. Rosenthal, E.R. Griffith, M.R. Bond and J.D. Miller), F.A. Davis, Philadelphia.

Gustafson, L., Hagberg, B. and Ingvar, D. (1978) Speech disturbances in presenile dementia related to local cerebral blood flow abnormalities in the dominant hemisphere. *Brain and Language*, **5**, 103–118.

Guttmann, E. (1942) Aphasia in children. *Brain*, **65**, 205–219.

Haase, G. (1977) Diseases Presenting as Dementia. In *Dementia: Contemporary Neurology Series* (ed. C.E. Wells), E.A Davis, Philadelphia.

Hachinski, V.C., Iliff, L.D., Zilhka, E., du Boulay, G.H., McAllister, V.L., Marshall, J., Russell, R.W.R. and Symon, L. (1975) Cerebral blood flow in dementia. *Archives of Neurology*, **32**, 632–637.

Hagen, C. (1984) Language disorders in head trauma. In *Language Disorders in Adults: Recent Advances* (ed. A.L. Holland), College Hill Press, San Diego.

Hagen, C., Malkmus, D. and Burditt, G. (1979) *Intervention Strategies for Language Disorders Secondary to Head Trauma: Short Course*. American Speech-Language and Hearing Association, Atlanta.

Haller, R.G. (1980) Alcoholism and neurologic disorders. In *Neurology*, Vol. 5 (ed. R.N. Rosenberg), Grune & Stratton, New York.

Hallpike, J.F., Adams, C.W.M. and Tourtelotte, W.N. (1983) *Multiple Sclerosis*. Chapman and Hall, London.

Halpern, H., Darley, F.L. and Brown, J.R. (1973) Differential language and neurologic characteristics in cerebral involvement. *Journal of Speech and Hearing Disorders*, **38**, 162–173.

Hanson, W.R. and Metter, E.J. (1980) DAF as instrumental treatment for dysarthria in progressive supernuclear palsy: a case report. *Journal of Speech and Hearing Disorders*, **45**, 268–275.

Hardy, J.E. (1964) Lung function of athetoid and spastic quadriplegia children. *Developmental Medicine and Child Neurology*, **6**, 378–388.

Hayward, R.W., Naeser, M.A. and Zatz, L.M. (1977) Cranial computed tomography in aphasia. *Radiology*, **123**, 653–660.

Head, H. (1926) *Aphasia and Kindred Disorders*. Cambridge University, London.

Hecaen, H. (1976) Acquired aphasia in children and the ontogenesis of hemispheric functional specialization. *Brain and Language*, **3**, 114–134.

Hecaen, H. (1983) Acquired aphasia in children: revisited. *Neuropsychologia*, **21**, 581–587.

Hecaen, H. and Albert, M.L. (1978) *Human Neuropsychology*. John Wiley, New York.

Hecaen, H. and Sauguet, J. (1971) Cerebral dominance in left handed subjects. *Cortex*, **7**, 19–48.

Heilman, K.M. and Valenstein, C. (1979) *Clinical neuropsychology*. Oxford University Press, New York.

Heilman, K.M., Safran, A. and Geschwind, N. (1971) Closed head trauma and aphasia. *Journal of Neurology, Neurosurgery and Psychiatry*, **34**, 265–269.

Hermann, K., Turner, J.W., Gillingham, F.J. and Gaze, R.M. (1966) The effects of destructive lesions and stimulation of the basal ganglia on speech mechanisms. *Confina Neurologica*, **27**, 197–207.

Heston, L.L., Mastrik, A.R., Anderson, E. and White, J. (1981) Dementia of the Alzheimer type. *Archives of General Psychiatry*, **38**, 1085–1090.

Hier, D.B. and Kaplan, J. (1980) Verbal comprehension deficits after right hemisphere damage. *Applied Psycholinguistics*, **1**, 279–294.

Hier, D.B. and Mohr, J.P. (1977) Incongruous oral and written naming. *Brain and Language*, **4**, 115–126.

Hier, D.B., Hagenlocker, K. and Shindler, A.G. (1985) Language disintegration in dementia: Effects of etiology and severity. *Brain and Language*, **25**, 117–133.

Hier, D.B., Mogil, S.I., Rubin, N.P. and Komros, G.R. (1980) Semantic aphasia: A neglected entity. *Brain and Language*, **10**, 120–131.

Hillbom, E. (1959) Delayed effects of traumatic brain injuries: neurological remarks. *Acta Psychiatrica Scandinavica*, **137**, 7.

Hillel, A.D. and Miller, R.N. (1987) Management of bulbar symptoms in amyotrophic lateral sclerosis. In *Advances in Experimental Medicine and Biology: Amyotrophic Lateral Sclerosis, Therapeutic, Psychological and Research Aspects* Vol. 209 (eds V. Cosi, A.C. Kato, P. Parlette and M. Poloni), Plenum Press, New York.

Hirose, H. (1986) Pathophysiology of motor speech disorders (dysarthria). *Folia Phoniatrica*, **38**, 61–68.

Hirose, H., Kiritani, S., Ushijima, T., Yoshioka, H. and Sawashima, M. (1981) Patterns of dysarthric movements in patients with Parkinsonism. *Folia Phoniatrica*, **33**, 204–215.

Hirose, H., Kiritani, S. and Sawashima, M. (1982a) Patterns of dysarthric movement in patients with amyotrophic lateral sclerosis and pseudo-bulbar palsy. *Folia Phoniatrica*, **34**, 106–112.

Hirose, H., Kiritani, S. and Sawashima, M. (1982b) Velocity of articulatory movements in normal and dysarthric subjects. *Folia Phoniatrica*, **34**, 210–215.

Hodges, S. and Smith, R.W. (1983) Intracranial calcification and child-hood medulloblastoma. *Archives of Disease in Childhood*, **58**, 663–664.

Holland, A. (1977) Some practical considerations in aphasic rehabilitation. In *Rationale for adult aphasia therapy*, (eds M. Sullivan and M.S. Kommers), University of Nebraska Press, Nebraska.

Holland, A.L. (1980) *Communicative Abilities in Daily Living*, University Park Press, Baltimore.

Holland, A.L. (1982) When is aphasia aphasia? The problem of closed head injury. In *Clinical Aphasiology Conference Proceedings* (ed. R.H. Brookshire), BRK, Minneapolis.

Holland, A.L., McBurney, D.H., Moossy, J. and Reinmuth, O.M. (1985) The dissolution of language in Pick's disease with neurofibrillary tangles: A case study. *Brain and Language*, **24**, 36–38.

Horenstein, S., Chung, H. and Brenner, S. (1978) Aphasia in two verified cases of left thalamic hemorrhage. *Transactions of the American Neurological Association*, **103**, 193–198.

Horowitz, N. and Rizzoli, H.V. (1966) Complications following the surgical treatment of head injuries: clinical neurosurgery. *Proceedings of the Congress of Neurological Surgeons*, 277–287.

Hudson, L.J., Murdoch, B.E. and Ozanne, A.E. (1989) Posterior fossa tumours in childhood: Associated speech and language disorders post-surgery. *Aphasiology*, **3**, 1–18.

Huff, F.J., Corkin, S. and Growdon, J.H. (1986) Semantic impairment and anomia in Alzheimer's disease. *Brain and Language*, **28**, 235–249.

Hunker, C.J. and Abbs, J.H. (1984) Physiological analyses of Parkinsonian tremors in the orofacial system. In *The Dysarthrias: Physiology, Acoustics, Perception, Management* (eds M.R. McNeil, J.C. Rosenbek and A.E. Aronson), College Hill Press, San Diego.

Hunker, C.J., Abbs, J.H. and Barlow, S.M. (1982) The relationship between Parkinsonian rigidity and hypokinesia in the orofacial system: a quantitative analysis. *Neurology*, **32**, 749–754.

Irigaray, L. (1967) Approche psycho-linguistique du langage des dements. *Neuropsychologia*, **5**, 25–52.

Irigaray, L. (1973) *Le language de Dements*. Mouton, The Hague.

Ito, M. (1970) Neurophysiological aspects of cerebellar motor control system. *International Journal of Neurology*, **7**, 162–176.

Jackson, J.H. (1932) *Selected Writing*. Hodder and Stoughton, London.

Jamison, D.L. and Kaye, H.H. (1974) Accidental head injury in children. *Archives of Diseases of Childhood*, **49**, 376–381.

Jennet, B. (1983) Scope and scale of the problem. In *Rehabilitation of the Head Injured Adult* (ed. J.D. Miller), F.A. Davies, Philadelphia.

Jennet, B. and Teasdale, G. (1981) *Management of Head Injuries.* F.A. Davis, Philadelphia.

Jervis, G.A. (1937) Alzheimer's disease. *Psychiatry Quarterly,* **11**, 5–18.

Jervis, G.A. (1971) Alzheimer's disease. In *Pathology of the Nervous System,* Vol. 2, (ed. J. Minkler), McGraw-Hill, New York.

Johns, D.F. and Darley, F.L. (1970) Phonemic variability in apraxia of speech. *Journal of Speech and Hearing Research,* **13**, 556–583.

Johnson, J., Sommers, R. and Weidner, W. (1977) Dichotic ear preference in aphasia. *Journal of Speech and Hearing Research,* **20**, 116–129.

Jonas, S. (1982) The thalamus and aphasia, including transcortical aphasia: a review. *Journal of Communication Disorders,* **15**, 31–41.

Jordan, F.M., Ozanne, A.C. and Murdoch, B.E. (1988) Long-term speech and language disorders subsequent to closed head injury in children. *Brain Injury,* **2**, 179–185.

Joynt, R.J. (1984) The language of dementia. In *Advances in Neurology,* Vol. 42, (ed. F.C. Rose), Raven Press, New York.

Kalsbeek, W.D., McLawin, R.L., Harris, B.S. and Miller, J.D. (1980) Spinal cord injury: major findings. *Journal of Neurosurgery,* **53**, 519–524.

Kaplan, E. and Goodglass, H. (1981) Aphasia-related disorders. In *Acquired Aphasia,* (ed. M.T. Sarno), Academic Press, New York.

Karis, R. and Horenstein, S. (1976) Localization of speech parameters by brain scan. *Neurology,* **26**, 226–231.

Kemp, J.M. and Powell, T.P.S. (1971) The connections of the striatum and globus pallidus: Synthesis and speculation. *Philosophical Transactions of the Royal Society, London,* **262**, 441–457.

Kemper, T. (1984) Neuroanatomical and neuropathological changes in normal aging and in dementia. In *Clinical Neurology of Aging* (ed. M.L. Albert), Oxford University Press, New York.

Kent, R. and Netsell, R. (1975) A case study of an ataxic dysarthric: cineradiographic and spectrographic observations. *Journal of Speech and Hearing Disorders,* **40**, 115–134.

Kent, R.D. and Netsell, R. (1978) Articulatory abnormalities in athetoid cerebral palsy. *Journal of Speech and Hearing Disorders,* **43**, 353–374.

Kent, R.D. and Rosenbek, J.C. (1982) Prosodic disturbance and neurologic lesion. *Brain and Language,* **15**, 259–291.

Kent, R., Netsell, R. and Abbs, J.H. (1979) Acoustic characteristics of dysarthria associated with cerebellar disease. *Journal of Speech and Hearing Research,* **22**, 627–648.

Kerschensteiner, M., Poeck, K. and Lehmkuhl, G. (1975) Die Apraxien. *Aktuelle Neurologie,* **2**, 171–178.

Kertesz, A. (1979) *Aphasia and Associated Disorders: Taxonomy, Localization and Recovery*, Grune & Stratton, New York.

Kertesz, A. (1980) *Western Aphasia Battery*, University of Western Ontario, London, Canada.

Kertesz, A. (1982) *The Western Aphasia Battery*, Grune & Stratton, New York.

Kertesz, A., Harlock, W. and Coates, R. (1979) Computer tomographic localization of lesion size and prognosis in aphasia and nonverbal impairment. *Brain and Language*, **8**, 34–50.

Kertesz, A., Lesk, D. and McCabe, P. (1977) Isotope localization of infarcts in aphasia. *Archives of Neurology*, **34**, 590–601.

Kertesz. A., Sheppard, A. and Mackenzie, R. (1982) Localization in transcortical sensory aphasia. *Archives of Neurology*, **39**, 475–478.

Kidd, M. (1964) Alzheimer's disease: An electron microscopical study. *Brain*, **87**, 307–320.

Kinsbourne, M. (1971) The minor cerebral hemisphere as a source of aphasic speech. *Archives of Neurology*, **25**, 303–306.

Kinsbourne, M. and Cook, J. (1971) Generalized and lateralized effects on concurrent verbalization on a unimanual skill. *Quarterly Journal of Experimental Psychology*, **23**, 341–345.

Kinsbourne, M. and Hicks, R.E. (1978) Functional cerebral space: A model for overflow transfer and interference effects in human performance. In *Attention and Performance VII* (ed. J. Requin), Erlbaum, Hillsdale.

Kinsella, G. (1986) Rehabilitation of prosodic impairment following right hemisphere lesions. In *Proceedings of the Tenth Annual Brain Impairment Conference*, (eds V. Anderson, J. Ponsford and P. Snow), Australian Society for the Study of Brain Impairment, Melbourne.

Kirshner, H.S. and Kistler, K.H. (1982) Aphasia after right thalamic hemorrhage. *Archives of Neurology*, **39**, 667–669.

Kirshner, H.S., Webb, W.G. and Kelly, M.P. (1984) The naming disorder of dementia. *Neuropsychologia*, **22**, 23–30.

Kitselman, K. (1981) Language impairment in aphasia, delirium, dementia and schizophrenia. In *Speech Evaluation in Medicine*, (ed. J.E. Darby), Grune & Stratton, New York.

Kleist, K. (1922) In *Handbuch der argblichen Erfahrungen*, (ed. O. Schjernings), Banth, Leipzig.

Kleist, K. (1962) *Sensory Aphasia and Amusia*, Pergamon Press, London.

Klingman, K.C. and Sussman, H.M. (1983) Hemisphericity in aphasic language recovery. *Journal of Speech and Hearing Research*, **26**, 248–256.

Konorski, J. (1967) *Integrative Activity of the Brain*. University of Chicago Press, Chicago.

Kornhuber, H.H. (1974) Cerebral cortex, cerebellum and basal ganglia: An introduction to their motor functions. In *The Neurosciences: Third Study Program*, (eds F.O. Schmitt and F.G. Warden), MIT Press, Cambridge.

Kornhuber, H.H. (1977) A reconsideration of the cortical and subcortical mechanisms involved in speech and aphasia. In *Language and Hemispheric Specialization in Man: Cerebral ERPs*, (ed. J.E. Desmedt), Karger, Basel.

Kornhuber, H.H. (1980) Physiologie und Pathophysiologie der cortikalen und subcortikalen Bewegungssteuerung. In *Pathologische Erregbarkeit des Nervensystems und ihre Behandlung* (eds H.G. Mertens and H. Przuntek), Springer-Verlag, Berlin.

Krauthamer, G.M. (1979) Sensory functions of the neostriatum. In *The Neostriatum* (eds I. Divac and R.G.E.H. Oberg), Pergamon Press, Oxford.

Kreindler, A. and Fradis, A. (1968) *Performances in Aphasia*, Gauthier-Villars, Paris.

Kreul, E.J. (1972) Neuromuscular control examination (NMC) for Parkinsonism: Vowel prolongations and diadochokinetic and reading rates. *Journal of Speech and Hearing Disorders*, **15**, 72–83.

Krigman, M.R., Feldman, R.G. and Bensch, K. (1965) Alzheimer's presenile dementia: A histochemical and electron microscopic study. *Laboratory Investigations*, **14**, 381–396.

Laine, T. and Matilla, R.J. (1981) Pure agraphia: A case study. *Neuropsychologia*, **19**, 311–316.

Landau, W.M. and Kleffner, F.R. (1957) Syndrome of acquired aphasia with convulsive disorder in children. *Neurology*, **10**, 915–921.

Lange-Cosack, H. and Tepfner, G. (1973) *Das Hirntrauma im Kinder und Jugendalter*. Springer-Verlag, Berlin.

LaPointe, L.L. and Johns, D.F. (1975) Some phonemic characteristics in apraxia of speech. *Journal of Communication Disorders*, **8**. 259–269.

Larrson, T., Sjörgren, T. and Jacobson, G. (1963) Senile dementia. *Acta Psychiatrica Scandinavica*, Suppl. 167, 1–259.

Lawson, J.S. and Barker, M.G. (1968) The assessment of nominal dysphasia in dementia – the use of reaction time measures. *British Journal of Medical Psychology*, **41**, 411–414.

Leader, B.J. (1983) Aphasia following left putaminal hemorrhage: A longitudinal case report. *Topics in Language Disorders* (September), 77–86.

Lechtenberg, R. and Gilman, S. (1978) Speech disorders in cerebellar disease. *Annals of Neurology*, **3**, 285–290.

Le May, M and Culebras, A. (1972) Human brain. Morphological differences in the hemispheres demonstratable by carotid arteriography. *New England Journal of Medicine*, **287**, 168–170.

Lenneberg, E. (1967) *Biological Foundations of Language*, John Wiley, New York.

Lesser, R. (1974) Verbal comprehension in aphasia: An English version of three Italian tests. *Cortex*, **10**, 247–263.

Levin, H.S. and Eisenberg, G.M. (1979a) Neuropsychological impairment after closed head injury in children and adolescents. *Journal of Pediatric Psychology*, **4**, 389–402.

Levin, H.S. and Eisenberg, G.M. (1979b) Neuropsychological outcome of closed head injury in children and adolescents. *Child's Brain*, **5**, 281–292.

Levin, H.S., Benton, A.L. and Grossman, R.G. (1982) *Neurobehavioural Consequences of Closed Head Injury*, Oxford University Press, New York.

Levin, H.S., Ewing-Cobbs, L. and Benton, A.L. (1984) Age and recovery from brain damage: a review of clinical studies. In *Age and Recovery of Function in the Central Nervous System*, (ed. S.W. Scheff), Plenum, New York.

Levin, H.S., Grossman, R.G. and Kelly, P.J. (1976) Aphasic disorders in patients with closed head injuries. *Journal of Neurology, Neurosurgery and Psychiatry*, **39**, 1062–1070.

Levin, H.S., Grossman, R.G., Sarwar, M. and Meyers, C.A. (1981) Linguistic recovery after closed head injury. *Brain and Language*, **12**, 360–374.

Levine, D.N. and Calvanio, R. (1982) Conduction aphasia. In *The Neurology of Aphasia: Neurolinguistics*, (eds H.S. Kirshner and F.R. Freeman), Swets and Zeitlinger, Amsterdam.

Levine, D.N. and Mohr, J.P. (1979) Language after bilateral cerebral infarctions: Role of the minor hemisphere in speech. *Neurology*, **29**, 927–938.

Levy, J. (1970) Information Processing and Higher Psychological Functions in the Disconnected Hemispheres of Commissurotomy Patients. Doctoral Dissertation: California Institute of Technology.

Levy, J. (1974) Cerebral asymmetries as manifested in split brain man. In *Hemispheric Disconnection and Cerebral Function*, (eds M. Kinsbourne and W.D. Smith), Charles C. Thomas, Springfield.

Levy, J. (1983) Language, cognition and the right hemisphere. A response to Gazzaniga. *American Psychologist*, **38**, 538–541.

Levy, J., Trevarthen, C. and Sperry, R.W. (1972) Perception of bilateral chimeric figures following hemispheric disconnection. *Brain*, **95**, 61–78.

Lhermitte, F. (1984) Language disorders and their relationship to thalamic lesions. In *Advances in Neurology*, (ed. F.C. Rose), Raven Press, New York.

Lichtheim, L. (1885) On aphasia. *Brain*, **7**, 433–484.

Lieberman, A., Dziatolowski, M., Kupersmith, M., Serby, M., Goodgold, A., Korein, J. and Goldstein, M. (1979) Dementia in Parkinson's disease. *Annals of Neurology*, **6**, 355–359.

Liepmann, H. (1908) *Drei Aufsätze aus dem Apraxiegebiet*. Karger, Berlin.

Linebaugh, C.W. (1978) Dichotic ear preference in aphasia: Another view. *Journal of Speech and Hearing Research*, **21**, 598–600.

Linebaugh, C.W. and Wolfe, V.E. (1984) Relationships between articulation rate, intelligibility and naturalness in spastic and ataxic speakers. In *The Dysarthrias: Physiology, Acoustics, Perception, Management*, (eds M.R. McNeil, J.C. Rosenbek and A.E. Aronson), College Hill Press, San Diego.

Lishman, W.A. (1981) Cerebral disorder in alcoholism: Syndromes of impairment. *Brain*, **104**, 1–20.

Logemann, J.A. and Fisher, H.B. (1981) Vocal tract control in Parkinson's disease: phonetic feature analysis of misarticulations. *Journal of Speech and Hearing Disorders*, **46**, 348–352.

Logemann, J.A., Fisher, H.B., Boshes, B. and Blonsky, E.R. (1978) Frequency and co-occurrence of vocal tract dysfunctions in the speech of a large sample of Parkinson's patients. *Journal of Speech and Hearing Disorders*, **43**, 47–57.

Lothman, E.W. and Ferrendelli, J.A. (1980) Disorders and diseases of the cerebellum. In *Neurology*, Vol. 5, (ed. R.N. Rosenberg), Grune & Stratton, New York.

Lowe, S.S., Cullen, J.K., Berlin, C.I., Thompson, C.L. and Willet, M.E. (1970) Perception of simultaneous dichotic and monotic monosyllables. *Journal of Speech and Hearing Research*, **13**, 812–822.

Ludlow, C. (1980) Children's language disorders. Recent research advances. *Annals of Neurology*, **7**, 497–507.

Ludlow, C.L. and Bassich, C.J. (1984) Relationships between perceptual ratings and acoustic measures of hypokinetic speech. In *The Dysarthrias: Physiology, Acoustics, Perception, Management*, (eds M.R. McNeil, J.C. Rosenbek, A.E. Aronson), College Hill Press, San Diego.

Ludlow, C.L., Connor, N.P. and Bassich, C.J. (1987) Speech timing in Parkinson's and Huntington's disease. *Brain and Language*, **32**, 185–214.

Lundgren, K., Moya, K. and Benowitz, L. (1983) Perception of non-verbal cues after right brain damage. In *Clinical Aphasiology Conference Proceedings*, (ed. R.H. Brookshire), BRK Publishers, Minneapolis.

Luria, A.R. (1963) *Restoration of Function after Brain Injury*. Pergamon Press, London.

Luria, A.R. (1970) *Traumatic Aphasia*, Mouton, The Hague.

Madden, D.J. and Nebes, R.D. (1980) Visual perception and memory. In *The Brain and Psychology*, (ed. M.C. Wittrock), Academic Press, New York.

Mantovani, J.F. and Landau, W.M. (1980) Acquired aphasia with convulsive disorder: Course and prognosis. *Neurology*, **30**, 524–529.

Marie, P. (1906) The third frontal convolution plays no special role in the function of language. *Semaine Médecine*, **26**, 241–247.

Marsden, C.D. (1982) The mysterious motor function of the basal ganglia: The Robert Wartenberg lecture. *Neurology*, **32**, 514–539.

Marshall, J.C. and Newcombe, F. (1966) Syntactic and semantic errors in paralexia. *Neuropsychologia*, **4**, 169–176.

Marshall, J.C. and Newcombe, F. (1973) Patterns of paralexia: A psycholinguistic approach. *Journal of Psycholinguistic Research*, **2**, 175–199.

Martin, A. and Fedio, P. (1983) Word production and comprehension in Alzheimer's Disease: The breakdown of semantic knowledge. *Brain and Language*, **19**, 124–141.

Martin, A., Brouwers, P., Cox, C. and Fedio, P. (1985a) On the nature of the verbal memory deficit in Alzheimer's Disease. *Brain and Language*, **25**, 323–341.

Martin, A., Cox, C., Brouwers, P. and Fedio, P. (1985b) A note on different patterns of impaired and preserved cognitive abilities and their relation to episodic memory deficits in Alzheimer's patients. *Brain and Language*, **26**, 181–185.

Mateer, C.A. and Ojemann, G.A. (1983) Thalamic mechanisms in language and memory. In *Language Functions and Brain Organization*, (ed. S.J. Segalowitz), Academic Press, New York.

Matson, D.D. (1956) Cerebellar astrocytoma in childhood. *Pediatrics*, **18**, 150–158.

Maxwell, S., Massengill, R. and Nashold, B. (1970) Tardive dyskinesia, *Journal of Speech and Hearing Disorder*, **35**, 33–36.

Mazzocchi, F. and Vignolo, L.A. (1979) Localization of lesions in aphasia: Clinical-CT scan correlations in stroke patients. *Cortex*, **15**, 627–654.

McCarthy, R. and Warrington, E.K. (1984) A two route model of speech production: Evidence from aphasia. *Brain*, **107**, 463–486.

McDuff, T. and Sumi, S.M. (1985) Subcortical degeneration in Alzheimer's disease. *Neurology*, **35**, 123–126.

McFarling, D., Rothi, L.J. and Heilman, K.M. (1982) Transcortical aphasia from ischaemic infarcts of the thalamus: A report of two cases. *Journal of Neurology, Neurosurgery and Psychiatry*, **45**, 107–112.

McKinney, W. and McGreal, D.A. (1974) An aphasic syndrome in children. *Canadian Medical Association Journal*, **110**, 637–639.

McMenemy, W.H. (1940) Alzheimer's disease. *Journal of Neurology and Psychiatry*, **3**, 211–240.

Meadows, A.T., Massari, D.J., Fergusson, J., Gordon, J., Littman, P. and Moss, K. (1981) Decline in IQ scores and cognitive dysfunctions in children with acute lymphocytic leukaemia treated with cranial irradiation. *Lancet*, **ii**, 5–1018.

Mendez, M.F. and Benson, D.F. (1985) Atypical conduction aphasia: A disconnection syndrome. *Archives of Neurology*, **42**, 886–891.

Metter, E.J. (1987) Neuroanatomy and physiology of aphasia: Evidence from positron emission tomography. *Aphasiology*, **1**, 3–33.

Metter, E.J. and Hanson, W.R. (1986) Clinical and acoustic variability in hypokinetic dysarthria. *Journal of Communication Disorders*, **19**, 347–366.

Metter, E.J., Riege, W.H., Hanson, W.R., Kuhl, E.E., Phelps, M.E., Squire, L.R., Wasterlain, C.G. and Benson, D.F. (1983) Comparisons of metabolic rates, language and memory in subcortical aphasias. *Brain and Language*, **19**, 33–47.

Meyer, J.A., Sakai, F., Yamaguchi, F., Yamamoto, M. and Shaw, T. (1980) Regional changes in cerebral blood flow during standard behavioural activation in patients with disorders of speech and mentation compared to normal volunteers. *Brain and Language*, **9**, 61–77.

Miceli, G., Silveri, C. and Caramazza, A. (1985) Cognitive analysis of a case of pure dysarthria. *Brain and Language*, **25**, 187–212.

Millar, J.M. and Whitaker, H.A. (1983) The right hemisphere's contribution to language: A review of the evidence from brain-damaged subjects. In *Language Functions and Brain Organization*, (ed. S.J. Segalowitz), Academic Press, New York.

Miller, B.F. and Keane, C.B. (1978) *Encyclopedia and Dictionary of Medicine, Nursing and Allied Health*, W.B. Saunders, Philadelphia.

Miller, J.F., Campbell, T.F., Chapman, R.S. and Weismer, S.E. (1984) Language behaviour in acquired aphasia. In *Language Disorders in Children*, (ed. A. Holland), College Hill Press, Baltimore.

Miller, N. (1986) *Dyspraxia and its Management*, Croom Helm, London.

Milton, S.B., Prutting, C.A. and Binder G.M. (1984) Appraisal of communicative competence in head injured adults. In *Clinical Aphasi-*

ology Conference Proceedings, (ed. R.H. Brookshire), BRK Publishers, Minneapolis.

Mitler, A.E., Neighbor, A., Katzman, R., Aronson, M. and Lipkowitz, R. (1981) Immunological studies in senile dementia of the Alzheimer type: Evidence for enhanced suppressor cell activity. *Annals of Neurology*, **10**, 506–510.

Mohr, J.P. (1976) Broca's area and Broca's aphasia. In *Studies in Neurolinguistics*, (Vol. 1), (eds H. Whitaker and H.A. Whitaker), Academic Press, New York.

Mohr, J.P., Funkenstein, H.H., Finkelstein, S., Pessin, M., Duncan, G.W. and Davis, K. (1975) Broca's area infarction versus Broca's aphasia. *Neurology*, **25**, 349.

Mohr, J.P., Pessin, M., Finkelstein, S., Funkenstein, H., Duncan, G. and Davis, K. (1978) Broca aphasia: Pathologic and clinical. *Neurology*, **28**, 311–324.

Mohr, J.P., Watters, W.C. and Duncan, G.W. (1975) Thalamic hemorrhage and aphasia. *Brain and Language*, **2**, 3–17.

Moore, W. and Weidner, W. (1974) Bilateral tachistoscopic word perception in aphasic and normal subjects. *Perceptual and Motor Skills*, **39**, 1003–1011.

Morris, J.C., Cole, M., Banker, B.O. and Wright, D. (1984) Hereditary dysphasic dementias and the Pick-Alzheimer spectrum. *Annals of Neurology*, **16**, 455–466.

Moyes, C.D. (1980) Epidemiology of serious head injuries in childhood. *Child: Care, Health and Development*, **6**, 1–9.

Msall, M., Shapiro, B., Balfour, P.B., Niedermeyer, E. and Capute, A.J. (1986) Acquired epileptic aphasia: Diagnostic aspects of progressive language loss in preschool children. *Neurology*, **25**, 248–251.

Mueller, P.B. (1971) Parkinson's disease: motor-speech behaviour in a selected group of patients. *Folia Phoniatrica*, **23**, 333–346.

Mulder, D.W. and Kurland, L.T. (1987) Motor neuron disease: epidemiologic studies. In *Advances in Experimental Medicine and Biology, Amyotrophic Lateral Sclerosis, Therapeutic, Psychological and Research Aspects*, Vol. 209, (eds V. Cosi, A.C. Kato, P. Parlette and M. Poloni), Plenum Press, New York.

Murdoch, B.E. (1987) Aphasia following right thalamic hemorrhage in a dextral. *Journal of Communication Disorders*, **20**, 459–468.

Murdoch, B.E., Afford, R.J., Ling, A.R. and Ganguley, B. (1986a) Acute computerized tomographic scans: Their value in the localization of lesions and as prognostic indicators in aphasia. *Journal of Communication Disorders*, **19**, 311–345.

Murdoch, B.E., Chenery, H.J., Boyle, R. and Wilks, V. (1988) Functional communicative abilities in dementia of the Alzheimer type. *Australian Journal of Human Communication Disorders*, **16**(1), 11–21.

Murdoch, B.E., Chenery, H.J., Wilks, V. and Boyle, R. (1987) Language disorders in dementia of the Alzheimer type. *Brain and Language*, **31**, 122–137.

Murdoch, B.E., Thompson, D., Fraser, S., and Harrison, L. (1986b) Aphasia following non-haemorrhagic lesions in the left striato-capsular region. *Australian Journal of Human Communication Disorders*, **14**, (2), 5–21.

Murry, T. (1983) The production of stress in three types of dysarthric speech. In *Clinical Dysarthria* (ed. W.R. Berry), College Hill Press, San Diego.

Myers, P.S. (1979) Profiles of communication deficits in patients with right cerebral hemisphere damage: Implications for diagnosis and treatment. In *Clinical Aphasiology Conference Proceedings*, (ed. R.H. Brookshire), BRK Publishers, Minneapolis.

Myers, P.S. (1984) Right hemisphere impairment. In *Language Disorders in Adults*, (ed. A. Holland), College Hill Press, San Diego.

Myers, P.S. (1986) Right hemisphere communication impairment. In *Language Intervention Strategies in Adult Aphasia*, (ed. R. Chapey), Williams & Wilkins, Baltimore.

Myers, P.S. and Linebaugh, C.W. (1981) Comprehension of idiomatic expressions by right hemisphere damaged adults. In *Clinical Aphasiology Conference Proceedings*, (ed. R.H. Brookshire), BRK Publishers, Minneapolis.

Naeser, M.A. and Hayward, R.W. (1978) Lesion localization in aphasia with cranial computed tomography and the B.D.A.E. *Neurology*, **28**, 545–551.

Naeser, M.A., Alexander, M.P., Helm-Eastabrooks, M., Levine, H.L., Laughlin, M.A. and Geschwind, N. (1982) Aphasia with predominantly subcortical lesion sites. *Archives of Neurology*, **39**, 2–12.

Najenson, T., Sazbon, L., Fiselzon, J., Becker, E. and Schechter, I. (1978) Recovery of communicative functions after prolonged traumatic coma. *Scandianvian Journal of Rehabilitation Medicine*, **10**, 15–21.

Nebes, R.D. and Sperry, R.W. (1971) Hemisphere disconnection syndrome with cerebral birth injury in the dominant arm area. *Neuropsychologia*, **9**, 247–259.

Netsell, R. and Kent, R. (1976) Paroxysmal ataxic dysarthria. *Journal of Speech and Hearing Disorders*, **41**, 93–109.

Nicolosi, L., Harryman, E. and Krescheck, J. (1983) *Terminology of Communication Disorders*, 2nd edn, Williams & Wilkins, Baltimore.

Nielsen, J.M. (1936) *Agnosia, Apraxia and Aphasia: Their Value in Cerebral Localization.* Hafner, New York.

Nielsen, J.M. (1962) *Agnosia, Apraxia, Aphasia.* Hafner, New York.

Obler, L.K. and Albert, M.L. (1981) Language in the elderly aphasic and in the dementing patient. In *Acquired Aphasia*, (ed. M.T. Sarno), Academic Press, New York.

Obler, L.K., Albert, M.L., Goodglass, H. and Benson, D.F. (1978) Aging and aphasia types. *Brain and Language*, **6**, 318–322.

Oelschlaeger, M.L. and Scarborough, J. (1976) Traumatic aphasia in children: A case study. *Journal of Communication Disorders*, **9**, 281–288.

Ojemann, G.A. (1975) Language and the thalamus: Object naming and recall during and after thalamic stimulation. *Brain and Language*, **2**, 101–120.

Ojemann, G.A. (1976) Subcortical language mechanisms. In *Studies in Neurolinguistics*, (eds H. Whitaker and H.A. Whitaker), Academic Press, New York.

Ojemann, G.A. (1977) Asymmetric function of the thalamus in man. *Annals of the New York Academy of Science*, **299**, 380–396.

Ojemann, G.A. and Whitaker, H.A. (1978) Language localization and variability. *Brain and Language*, **6**, 239–260.

Olsen, T.S., Bruhn, P. and Oberg, R.G.E. (1984) Cause of aphasia in stroke patients with subcortical lesions. *Acta Neurologica Scandinavica* (Suppl.), **69**, 311–312.

Olsen, T.S., Bruhn, P. and Oberg, R.G.E. (1986) Cortical hypoperfusion as a possible cause of subcortical aphasia. *Brain*, **109**, 393–410.

Olsen, T.S., Larsen, B., Herning, M., Skriver, E.B. and Lassen, N.A. (1983) Blood flow and vascular reactivity in collaterally perfused brain tissue. *Stroke*, **14**, 332–341.

Osgood, C. and Miron, M. (1963) *Approaches to the Study of Aphasia.* University of Illinois Press, Chicago.

Oxtoby, M. (1982) *Parkinson's Disease patients and their Social needs.* Parkinson's Disease Society, London.

Palmer, M.R. (1952) Speech therapy in cerebral palsy. *Journal of Pediatrics*, **40**, 514–524.

Panchal, V.G., Parikh, V.R. and Karapurkar, A.P. (1974) Thalamic abscesses. *Neurology (India)*, **22**, 106–110.

Patten, B.M. (1987) The syndromic nature of amyotrophic lateral sclerosis. In *Advances in Experimental Medicine and Biology: Amyotrophic Lateral Sclerosis, Therapeutic, Psychological and Research Aspects*, Vol. 209, (eds V. Cosi, A.C. Kato, P. Parlette and M. Poloni), Plenum Press, New York.

Peach, R.K. and Tonkovich, J.D. (1983) Subcortical aphasia: A report of three cases. In *Clinical Aphasiology Conference Proceedings*, (ed. R.H. Brookshire), BRK, Minneapolis.

Penfield, W. (1965) Conditioning the uncommitted language cortex for language learning. *Brain*, **88**, 787–798.

Penfield, W. and Roberts, L. (1959) *Speech and Brain Mechanisms*, Princeton University Press, Princeton.

Penfield, W. and Roberts, L. (1966) *Speech and Brain Mechanisms*, Atheneum, New York.

Penn, A.S. (1980) Myasthenia gravis. In *Neurology* (Vol. 5), (ed. R.W. Rosenberg), Grune & Stratton, New York.

Perani, D., Gerundini, P., di Piero, V., Savi, A., Carenzi, M., Vanzulli, A., Del Mashio, A., Lenzi, G.L. and Fazio, F. (1985) Remote cortical effects in patients with ischemic lesions of subcortical structures studied with [123]HIPDM and SPECJ. *Journal of Nuclear Medicine and Allied Sciences*, **29**, 129–130.

Pettit, J.M. and Noll, J.D. (1979) Cerebral dominance in aphasia recovery. *Brain and Language*, **7**, 191–200.

Phillips, W.A. and Singer, W. (1974) Function and interaction of on and off transients in vision. I. Psychophysics. *Experimental Brain Research*, **19**, 493–506.

Pick, A. (1892) On the relation between aphasia and senile atrophy of the brain. In *Neurological Classics in Modern Translation*, (eds D.A. Rottenberg, and F.H. Hochberg), (Trans. W.C. Schoene), Hafner Press, New York.

Pick, A. (1931) *Aphasia*, Charles C. Thomas, Springfield.

Platt, L.J., Andrews, G. and Howie, P.M. (1980) Dysarthria of adult cerebral palsy: II. Phonemic analysis of articulation errors. *Journal of Speech and Hearing Research*, **23**, 41–55.

Platt, L.J., Andrews, G., Young, M. and Quinn, P.T. (1980) Dysarthria of adult cerebral palsy: I. Intelligibility and articulatory impairment. *Journal of Speech and Hearing Research*, **23**, 28–40.

Plum, F. (1979) Dementia: An approaching epidemic. *Nature*, **279**, 372–373.

Poeck, K. and Lehmkuhl, G. (1980) Ideatory apraxia in a left-handed patient with right sided brain lesion. *Cortex*, **16**, 273–284.

Poeck, K., Kerschensteiner, M. and Hartje, W. (1972) A qualitative study on language understanding in fluent and non-fluent aphasia. *Cortex*, **8**, 299–304.

Poetzl, T. (1926) Ueber sensorische aphasia in Kindersalter. *Hals-Nasen-Ohrenklin.*, **14**, 109–118.

Porch, B.E. (1967) *Porch index of communicative ability*. Consulting Psychologists Press, Palo Alto.

Portnoy, R.A. (1979) Hyperkinetic dysarthria as an early indicator of impending tardive dyskinesia. *Journal of Speech and Hearing Disorders*, **44**, 214–219.

Portnoy, R.A. and Aronson, A.E. (1982) Diadochokinetic syllable rate and regularity in normal and in spastic and ataxic dysarthric subjects. *Journal of Speech and Hearing Disorders*, **47**, 324–328.

Pratt, R.T.C. and Warrington, E.K. (1972) The assessment of cerebral dominance in unilateral ECT. *British Journal of Psychiatry*, **121**, 327–328.

Prutting, C. and Kirchner, D. (1983) Applied pragmatics. In *Pragmatic Assessment and Intervention Issues in Language*, (eds T. Gallagher and C. Prutting), College Hill Press, San Diego.

Prutting, C. and Kirchner, D. (1987) A clinical appraisal of the pragmatic aspects of language. *Journal of Speech and Hearing Disorders*, **52**, 105–119.

Rankin, J.M., Aram, D.M. and Horwitz, S.J. (1981) Language ability in right and left hemiplegic children. *Brain and Language*, **14**, 292–306.

Rapin, I., Mattis, S., Rowan, J.A. and Golden, G.G. (1977) Verbal auditory agnosia in children. *Developmental Medicine and Child Neurology*, **119**, 192–207.

Ratcliff, G., Dila, C., Taylor, L. and Milner, B. (1980) The morphological asymmetry of the hemispheres and cerebral dominance for speech: A possible relationship. *Brain and Language*, **11**, 87–98.

Reifler, B.V., Larson, E. and Hanley, R. (1982) Co-existence of cognitive impairment and depression in geriatric outpatients. *American Journal of Psychiatry*, **139**, 623–629.

Reisberg, B. (1981) *Brain Failure*. The Free Press, New York.

Rekate, H.L., Grubb, R.L., Aram, D.M., Hahn, J.F. and Ratcheson, R.A. (1985) Muteness of cerebellar origin. *Archives of Neurology*, **42**, 697–698.

Reynolds, A.F., Harris, A.B., Ojemann, G.A. and Turner, P.T. (1978) Aphasia and left thalamic hemorrhage. *Journal of Neurosurgery*, **48**, 570–574.

Reynolds, A.F., Turner, P.T., Harris, A.B., Ojemann, G.A. and Davis, L.E. (1979) Left thalamic hemorrhage with dysphasia: A report of five cases. *Brain and Language*, **7**, 62–73.

Riklan, M. and Cooper, I.S. (1975) Psychometric studies of verbal functions following thalamic lesions in humans. *Brain and Language*, **2**, 45–64.

Riklan, M. and Levita, E. (1965) Laterality of subcortical involvement and psychological functions. *Psychological Bulletin*, **64**, 217–224.

Riklan, M. and Levita, E. (1970) Psychological studies of thalamic lesions in humans. *Journal of Nervous and Mental Disorders*, **150**, 251–265.

Rivers, D.L. and Love, F.J. (1980) Language performance on visual processing tasks in right hemisphere lesion cases. *Brain and Language*, **10**, 348–366.

Roberts, C., Kinsella, G. and Wales, R. (1982) Disturbances in processing prosodic features of language following right hemisphere lesions. In *Proceedings of the 1982 Brain Impairment Workshop*, (eds G.V. Stanley and K.W. Walsh), Australian Society for the Study of Brain Impairment, Melbourne.

Robin, D.A. and Schienberg, S. (1983) Isolated thalamic lesion and aphasia: A case study. In *Clinical Aphasiology Conference Proceedings*, (ed. R.H. Brookshire), BRK, Minneapolis.

Rochford, G. (1971) A study of naming errors in dysphasic and in demented patients. *Neuropsychologia*, **9**, 437–443.

Roland, P.E., Meyer, E., Shibasaki, T., Yamamota, Y.L. and Thompson, C.J. (1982) Regional cerebral blood flow changes in cortex and basal ganglia during voluntary movements in normal human volunteers. *Journal of Neurophysiology*, **48**, 467–480.

Rolls, E.T., Perrett, D.I., Caan, A.W. and Wilson, F.A.W. (1982) Neural responses related to visual recognition. *Brain*, **105**, 611–646.

Romanul, F.C.A. and Abramowitz, A. (1961) Changes in brain and pial vessels in arterial borderzones. *Archives of Neurology*, **11**, 40–49.

Ron, M.A. (1977) Brain damage in chronic alcoholism: A neuropathological, neuroradiological and psychological review. *Psychological Medicine*, **7**, 103–112.

Rosati, G. and de Bastiani, P. (1979) Pure agraphia: A discrete form of aphasia. *Journal of Neurology, Neurosurgery and Psychiatry*, **42**, 266–269.

Rosenbek, J.C., Kent, R.D. and LaPointe, L.L. (1984) Apraxia of speech and some perspectives. In *Apraxia of Speech: Physiology, Acoustics, Linguistics, Management*, (eds J.C. Rosenbek, M.R. McNeil and A.R. Aronson), College Hill Press, California.

Rosenbek, J.C., Wertz, R.T. and Darley, F.L. (1973) Oral sensation and perception in apraxia of speech and aphasia. *Journal of Speech and Hearing Research*, **16**, 22–36.

Rosenberg, R.N. and Pettegrew, J.W. (1980) Genetic diseases of the nervous system. In *Neurology*, Vol. 5 (ed. R.N. Rosenberg), Grune & Stratton, New York.

Rosenthal, M. (1983) Behavioural sequelae. In *Rehabilitation of the Head Injured Adult*, (eds M. Rosenthal, E.R. Griffith, M.R. Bond and J.D. Miller), F.A. Davis, Philadelphia.

Ross, E.D. (1980a) Disorders of higher cortical functions: Diagnosis and treatment. In *Neurology*, Vol. 5, (ed. R.N. Rosenberg), Grune & Stratton, New York.

Ross, E.D. (1980b) Left medial parietal lobe and receptive language functions: Mixed transcortical aphasia and left anterior cerebral artery infarction. *Neurology*, **30**, 144–151.

Ross, E.D. (1981) The aprosodias: Functional anatomic organization of the affective components of language in the right hemisphere. *Archives of Neurology*, **38**, 561–569.

Ross, E.D. (1985) Modulation of affect and nonverbal communication by the right hemisphere. In *Principles of Behavioural Neurology*, (ed. M.M. Mesulam), F.A. Davis, Philadelphia.

Ross, E.D. and Mesulam, M.M. (1979) Dominant language functions of the right hemisphere: Prosody and emotional gesturing. *Archives of Neurology*, **36**, 144–148.

Rossi, G. and Rosandini, G. (1967) Experimental analysis of cerebral dominance in man. In *Brain Mechanisms Underlying Speech and Language*, (eds J. Millikan and F. Darley), Grune & Stratton, New York.

Rubens, A.B. (1976) Transcortical motor aphasia. In *Studies in Neurolinguistics*, (eds H. Whitaker and H.A. Whitaker), Academic Press, New York.

Rubens, A.B., Mahowald, M.W. and Hutton, J.T. (1976) Asymmetry of the lateral (Sylvian) fissures in man. *Neurology*, **26**, 620–624.

Russell, W.R. (1932) Cerebral involvement in head injury. *Brain*, **55**, 549–603.

Russell, W.R. and Espir, M.L.E. (1961) *Traumatic Aphasia: A Study of Aphasia in War Wounds of the Brain*, Oxford University Press, Oxford.

Russell, W.R. and Smith, A. (1961) Post-traumatic amnesia in closed head injury. *Archives of Neurology*, **5**, 4–17.

Rutherford, B.R. (1944) A comparative study of loudness, pitch, rate, rhythms and quality of the speech of children handicapped by cerebral palsy. *Journal of Speech Disorders*, **9**, 263–271.

Rutter, M., Chadwick, O., Shaffer, D. and Brown, G. (1980) A prospective study of children with head injuries: I. Design and methods. *Padiatrie*, **10**, 105–127.

Sadock, J. (1974) *Towards a Linguistic Theory of Speech Acts*, Academic Press, New York.

Sandson, J., Obler, L.K. and Albert, M.L. (1987) Language changes in healthy aging and dementia. In *Advances in Applied Psycholinguistics*, (ed. S. Rosenberg), Cambridge University Press, Cambridge, Mass.

Sarno, M.T. (1980) The nature of verbal impairment after closed head injury. *Journal of Nervous and Mental Disease*, **168**, 685–692.

Sarno, M.T. (1984) Verbal impairment after closed head injury: report of a replication study. *Journal of Nervous and Mental Disease*, **172**, 475–479.

Sarno, M.T. and Levin, H.S. (1985) Speech and language disorders after closed head injury. In *Speech and Language Evaluation in Neurology: Adult Disorders*, (ed. J.K. Darby), Grune & Stratton, New York.

Sato, S. and Dreifuss, F.E. (1973) Electroencephalographic findings in a patient with developmental expressive aphasia. *Neurology*, **23**, 181–185.

Satz, P. and Bullard-Bates, C. (1981) Acquired aphasia in children. In *Acquired Aphasia*, ed. M.T. Sarno), Academic Press, New York.

Schiffer, E., Brignolio, F., Chio, A., Leone, M. and Rosso, M.G. (1987) A study of prognostic factors in motor neuron disease. In *Advances in Experimental Medicine and Biology: Amyotrophic Lateral Sclerosis, Therapeutic, Psychological and Research Aspects*, Vol. 209, (eds V. Cosi, A.C. Kato, P. Parlette and M. Poloni), Plenum Press, New York.

Schuell, H., Jenkins, J. and Jiminez-Pabon, E. (1964) *Aphasia in Adults*, Harper and Row, New York.

Schulhoff, C. and Goodglass, H. (1969) Dichotic listening, side of brain injury and cerebral dominance. *Neuropsychologia*, **7**, 148–160.

Schwartz, M.F., Marin, O.S.M. and Saffran, E.M. (1979) Dissociations of language function in dementia: A case study. *Brain and Language*, **7**, 277–306.

Searle, J. (1969) *Speech Acts*. Cambridge University Press, Cambridge.

Searleman, A. (1977) A review of right hemisphere linguistic capabilities. *Psychological Bulletin*, **84**, 503–528.

Sears, E.S. and Franklin, G.M. (1980) Diseases of the cranial nerves. In *Neurology*, Vol. 5, (ed. R.N. Rosenberg), Grune & Stratton, New York.

Segall, H.D., Batnitzky, S., Zee, S., Ahmadi, J., Bird, C.R. and Cohen, M.E. (1985) Computed tomography in the diagnosis of intracarotid neoplasms in children. *Cancer*, **56**, 1748–1755.

Selby, G. (1968) Parkinson's disease. In *Handbook of Clinical Neurology*, Vol. 6, (eds P.J. Vinken and G.W. Bruyn), North-Holland, Amsterdam.

Seltzer, B. and Sherwin, I. (1983) A comparison of clinical features in early and later onset primary degenerative dementia. *Archives of Neurology*, **40**, 143–146.

Shore, D., Overman, C.A. and Wyatt, R.J. (1983) Improving accuracy in the diagnosis of Alzheimer's Disease. *Journal of Clinical Psychiatry*, **44**, 207–212.

Shoumaker, R., Bennett, D., Bray, P. and Curless, R. (1974) Clinical and EEG manifestations of an unusual aphasic syndrome in children. *Neurology*, **24**, 10–16.

Sinico, S. (1926) Neoplasia della seconda circonvoluzione frontale sinistra: Agrafia pura. *Gazetta degli Ospedali e delle Cliniche (Milano)*, **47**, 627–631.

Slaby, A.E. and Wyatt, R.J. (1974) *Dementia in the Presenium*. Charles C. Thomas, Springfield.

Smith, A. (1966) Speech and other functions after left dominant hemispherectomy. *Journal of Neurology, Neurosurgery and Psychiatry*, **29**, 467–471.

Smith, A. and Burklund, C.W. (1966) Dominant hemispherectomy. *Science*, 153–1280.

Smith, A. and Sugar, O. (1975) Development of above normal language intelligence 21 years after left hemispherectomy. *Neurology*, **25**, 813–818.

Smith, S.R., Murdoch, B.E. and Chenery, H.J. (1989) Semantic abilities in dementia of the Alzheimer type: 1. Lexical semantics. *Brain and Language*, **36**, 314–324.

Smyth, G.E. and Stern, K. (1938) Tumours of the thalamus: A clinico-pathological study. *Brain*, **61**, 339–360.

Snell, R.S. (1980) *Clinical Neuroanatomy for Medical Students*, Little, Brown Co., Boston,

Sparks, R., Goodglass, H. and Nickel, B. (1970) Ipsilateral versus contralateral extinction in dichotic listening resulting from hemispheric lesions. *Cortex*, **6**, 249–260.

Sperry, R.W. and Gazzaniga, M.S. (1967) Language following surgical disconnection of the hemispheres. In *Brain Mechanisms Underlying Speech and Language*, (eds G.H. Millikan and F. Darley), Grune & Stratton, New York.

Spreen, O. and Benton, A.L. (1969) *Neurosensory Centre Comprehensive Examination for Aphasia: Manual of Directions*, University of Victoria, Victoria, B.C.

Steele, J.C., Richardson, J.C. and Olszewski, J. (1964) Progressive supranuclear palsy. *Archives of Neurology*, **10**, 333.

Sterzi, R. and Vallar, G. (1978) Frontal lobe syndrome as a disconnection syndrome: Report of a case. *Acta Neurologica*, **33**, 419–425.

Stone, J.L., Lopes, J.R. and Moody, R.A. (1978) Fluent aphasia after closed head injury. *Surgical Neurology*, **9**, 27–29.

Strauss, E., Wada, J. and Kosaka, B. (1983) Writing hand posture and cerebral dominance for speech. *Cortex*, **20**, 143–147.

Strich, S.J. (1956) Diffuse degeneration of the cerebral white matter in severe dementia following head injury. *Journal of Neurology, Neurosurgery and Psychiatry*, **19**, 163–185.

Strich, S.J. (1969) The pathology of brain damage due to blunt injuries. In *The Late Effects of Head Injury*, (eds A.E. Walker, W.F. Caveness and M. Critchley), Charles C. Thomas, Springfield.

Svennilson, E., Torvik, A., Lowe, R. and Leksell, L. (1960) Treatment of Parkinsonism by stereotactic thermolesions in the pallidal region. *Acta Psychiatrica et Neurologica Scandinavica*, **35**, 358–377.

Sweet, R.D., Solomon, G.E., Wayne, H., Shapiro, E. and Shapiro, A.K. (1973) Neurological features of Gilles de la Tourette's syndrome. *Journal of Neurology, Neurosurgery and Psychiatry*, **36**, 1–9.

Tabaddor, K., Mattis, S. and Zazula, T. (1984) Cognitive sequelae and recovery after moderate and severe head injury. *Neurology*, **14**, 701–708.

Teasdale, G. and Jennet, B. (1974) Assessment of coma and impaired consciousness: a practical scale. *Lancet*, **ii**, 81–84.

Terry, R.D. and Katzman, R. (1983) Senile dementia of the Alzheimer type. *Annals of Neurology*, **14**, 497–506.

Terry, R.D., Peck, A., de Teresa, R., Schechter, R. and Horoupian, D.S. (1981) Some morphometric aspects of the brain in senile dementia of the Alzheimer type. *Annals of Neurology*, **10**, 184–192.

Teuber, H.L. (1975) Recovery of function after brain injury in man. In *Outcome of Severe Damage of the Central Nervous System*, (*CIBA Foundation Symposium No. 34*) (eds R. Porter, and D.W. Fitzsimons), Elsevier/Excerpta Medica, Amsterdam.

Thomsen, I.V. (1975) Evaluation and outcome of aphasia in patients with severe closed head trauma. *Journal of Neurology, Neurosurgery and Psychiatry*, **38**, 713.

Tikofsky, R.S., Kooi, K.A. and Thomas, M.H. (1960) Electroencephalographic findings and recovery from aphasia. *Neurology*, **10**, 154–156.

Tomlinson, B.E. (1964) Pathology. In *Acute Injuries of the Head, 4th edn*, (ed. G.F. Rowbotham), Livingstone, Edinburgh.

Tomlinson, B.E. (1977) The pathology of dementia. In *Dementia: Contemporary Neurology Series*, (ed. C.E. Wells), E.A. Davis, Philadelphia.

Tomlinson, B.E., Blessed, G. and Roth, M. (1970) Observations on the brains of demented old people. *Journal of the Neurological Sciences*, **11**, 205–242.

Trost, J.E. and Canter, G.J. (1974) Apraxia of speech in patients with Broca's aphasia: A study of phoneme production accuracy and error patterns. *Brain and Language*, **1**, 63–79.

Tsunoda, T. (1975) Functional differences between right and left cerebral hemispheres detected by the key tapping method. *Brain and Language*, **2**, 152–170.

Tucker, D.M., Watson, R.T. and Heilman, K.M. (1977) Discrimination and evocation of affectively intoned speech in patients with righ parietal disease. *Neurology*, **27**, 947–950.

Uziel, A., Bohe, M., Cadilhac, J. and Passouant, P. (1975) Les trouble de la voix et de la parole dans les syndromes Parkinsoniens. *Foli Phoniatrica*, **27**, 166–176.

Van Buren, J.M. (1975) The question of thalamic participation i speech mechanisms. *Brain and Language*, **2**, 31–44.

Van de Sandt-Koenderman, W.M.E., Smit, I.A.C., van Dongen, H.R. and van Hest, J.B.C. (1984) A case of acquired aphasia and convulsiv disorder: Some linguistic aspects of recovery and breakdown. *Brai and Language*, **21**, 174–183.

van Dongen, H.R. and Loonen, M.C.B. (1977) Factors related t prognosis of acquired aphasia in children. *Cortex*, **13**, 131–136.

van Dongen, H.R., Loonen, M.C.B. and van Dongen, K.J. (1985 Anatomical basis of acquired fluent aphasia in children. *Annals o Neurology*, **17**, 306–309.

van Harskamp, F., van Dongen, H.R. and Loonen, M.C.B. (1978 Acquired aphasia with convulsive disorders in children: A cas study with a seven-year follow up. *Brain and Language*, **6**, 141–148.

van Hout, A. and Lyon, G. (1986) Wernicke's aphasia in a 10-year-ol boy. *Brain and Language*, **29**, 268–285.

van Hout, A., Evrard, P. and Lyon, G. (1985) On the positive semiolog of acquired aphasia in children. *Developmental Medicine and Chil Neurology*, **27**, 231–241.

Vargha-Khadem, F., Gorman, A.M. and Watters, G.V. (1985) Aphasi and handedness in relation to hemispheric side, age and injury an severity of cerebral lesion during childhood. *Brain*, **108**, 677–696.

Vignolo, L.A. (1983) Modality-specific disorders of written language. I *Localization in Neuropsychology*, (ed. A. Kertesz), Academic Press New York.

Vignolo, L.A., Boccardi, E. and Caverni, L. (1986) Unexpected CT-sca findings in global aphasia. *Cortex*, **22**, 55–69.

Visch-Brink, E.G. and van de Sandt-Koenderman, M. (1984) Th occurrence of paraphasias in the spontaneous speech of childre with an acquired aphasia. *Brain and Language*, **23**, 258–271.

Volcan, I., Cole, G.P. and Johnson, K. (1986) A case of muteness o cerebellar origin. *Archives of Neurology*, **43**, 313–314.

Wada, J. and Rasmussen, T. (1960) Intracartoid injection of sodiun amytal for the lateralization of cerebral speech dominance: Experi mental and clinical observations. *Journal of Neurosurgery*, **17**, 266–252

Wada, J.A., Clarke, R. and Hamm, A. (1975) Cerebral hemispheri asymmetry in humans. *Archives of Neurology*, **32**, 239–246.

Wagenaar, W., Snow, C. and Prins, R. (1975) Spontaneous speech o aphasic patients: a psycholinguistic analysis. *Brain and Language*, **2**, 281–303.

Wallesch, C.W. (1985) Two syndromes of aphasia occurring with ischemic lesions involving the left basal ganglia. *Brain and Language*, **25**, 357–361.

Wallesch, C.W., Kornhuber, H.H., Brunner, R.J., Kunz, T., Hollerbach, B. and Suger, G. (1983) Lesions of the basal ganglia, thalamus and deep white matter: Differential effects on language functions. *Brain and Language*, **20**, 286–304.

Waltz, J.M., Riklan, M., Stellar, S. and Cooper, I.S. (1966) Crytothalamectomy for Parkinson's disease. *Neurology*, **16**, 994–1002.

Wapner, W., Hamby, S. and Gardner, H. (1981) The role of the right hemisphere in the apprehension of complex linguistic materials. *Brain and Language*, **14**, 15–33.

Warrington, E.K. (1975) The selective impairment of semantic memory. *Quarterly Journal of Experimental Psychology*, **27**, 635–657.

Warrington, W.K. and Pratt, R.T.C. (1973) Language laterality in left handers assessed by unilateral ECT. *Neuropsychologia*, **11**, 423–428.

Wechsler, A.F., Verity, M., Rosenschein, S., Fried, I. and Scheibel, A.B. (1982) Pick's disease. *Archives of Neurology*, **39**, 287–290.

Wechsler, D. (1945) *Wechsler Memory Scale Form 1*. Psychological Corporation, New York.

Weinstein, E.A. and Kahn, R.L. (1952) Non-aphasic misnaming (paraphasia) in organic brain disease. *Archives of Neurology and Psychiatry*, **107**, 72–79.

Weisenburg, T.S. and McBride, K.L. (1935) *Aphasia*, Hafner, New York.

Weismer, G. (1984) Articulatory characteristics of Parkinsonian dysarthria: segmental and phrase-level timing, spirantization and glottal-supraglottal co-ordination. In *The Dysarthrias: Physiology, Acoustics, Perception, Management*, (eds M.R. McNeil, J.C. Rosenbek and A.E. Aronson), College Hill Press, San Diego.

Wells, C.E. (1981) A deluge of dementia. *Psychosomatics*, **22**, 837–838.

Wepman, J.M. (1951) *Recovery from Aphasia*, Ronald, New York.

Wernicke, C. (1874) *Der Aphasische Symptomencomplex*, Cohn and Weigert, Breslau.

Wertz, R. (1978) Neuropathologies of speech and language: an introduction to patient management. In *Clinical Management of Neurogenic Communicative Disorders*, (ed. D.F. Johns), Little, Brown and Co. Boston.

Wertz, R.T., LaPointe, L.L. and Rosenbek, J.C. (1984) Characteristics of apraxia of speech. In *Apraxia of Speech in Adults: The Disorder and its Management*, (eds R.T. Wertz, L.L. LaPointe and J.C. Rosenbek), Grune & Stratton, New York.

West, R.W. and Ansberry, M. (1968) *The Rehabilitation of Speech*, Harper and Row, New York.

Whitaker, H. (1976) A case of isolation of language functions. In *Studies in Neurolinguistics*, Vol. 2), (eds H. Whitaker and H.A. Whitaker), Academic Press, New York.

Whitaker, H.A. and Ojemann, G.A. (1977) Lateralization of higher cortical functions: A critique. *Annals of the New York Academy of Sciences*, **229**, 459–473.

Whitehouse, P.J., Price, D.L., Clark, A.W., Coyle, J.T. and de Long, M.R. (1981) Alzheimer disease: Evidence for selective loss of cholinergic neurons in the nucleus basalis. *Annals of Neurology*, **10**, 122–126.

Wilson, S.K.A. (1912) Progressive lenticular degeneration: familial nervous disease associated with cirrhosis of the liver. *Brain*, **34**, 295.

Wisniewski, H., Terry, R.D. and Hirano, A. (1970) Neurofibrillary pathology. *Journal of Neuropathology and Experimental Neurology*, **29**, 163–176.

Witelson, S.F. and Pallie, W. (1973) Left hemisphere specialization for language in the newborn. Neuroanatomical evidence of asymmetry. *Brain*, **96**, 641–646.

Woods, B.T. and Carey, S. (1979) Language deficits after apparent clinical recovery from childhood aphasia. *Annals of Neurology*, **6**, 405–409.

Woods, B.T. and Teuber, H.L. (1978) Changing patterns of childhood aphasia. *Annals of Neurology*, **3**, 273–280.

Worster-Drought, C. (1971) An unusual form of acquired aphasia in children. *Developmental Medicine and Child Neurology*, **13**, 563–571.

Yarnell, P., Monroe, M.A. and Sobel, M.A. (1976) Aphasia outcome in stroke: A clinical neuroradiological correlation. *Stroke*, **7**, 516–522.

Zaidel, E. (1973) Linguistic competence and related functions in the right hemisphere of man following cerebral commissurotomy and hemispherectomy. Doctoral dissertation, California Institute of Technology.

Zaidel, E. (1976) Auditory vocabulary of the right hemisphere following brain bisection or hemidecortication. *Cortex*, **12**, 191–211.

Zaidel, E. (1978a) Concepts of cerebral dominance in the split brain. In *Cerebral Correlates of Conscious Experience*, (eds P.A. Buser and A. Rougeul-Buser), Elsevier North-Holland, Biomedical Press, Amsterdam.

Zaidel, E. (1978b) Auditory language comprehension in the right hemisphere following cerebral commissurotomy and hemispherectomy: A comparison with child language and aphasia. In *Language, Acquisition and Language Breakdown: Parallels and*

Divergencies, (eds A. Caramazza and E.B. Zurif), Johns Hopkins University Press, Baltimore.

aidel, E. (1983) A response to Gazzaniga. Language in the right hemisphere, convergent perspectives. *American Psychologist*, **38**, 542–546.

aidel, E. (1985) Right hemisphere language. In *The Dual Brain: Hemispheric Specialization in Humans*, (eds D.F. Benson and E. Zaidel), The Guilford Press, York.

aidel, E. and Peters, A.M. (1981) Phonological encoding and ideographic reading by the disconnected right hemisphere: Two case studies. *Brain and Language*, **14**, 205–234.

angwill, O.L. (1960) *Cerebral dominance and its relation to psychological function*, Charles C. Thomas, Springfield.

angwill, O.L. (1964) Intelligence in aphasia. In *Disorders of Language*, (eds A.V.S. De Ruek and M. O'Connor), Little, Brown and Co., Boston.

angwill, O.L. (1967) Speech and the minor hemisphere. *Acta Neuropsychologica*, **67**, 1013–1020.

iegler, W. and von Cramon, D. (1986) Spastic dysarthria after acquired brain injury: an acoustic study. *British Journal of Disorders of Communication*, **21**, 173–187.

urif, E.B. and Bryden, M.P. (1969) Familial handedness and left-right differences in auditory and visual perception. *Neuropsychologia*, **7**, 179–187.

yski, B.J. and Weisiger, B.E. (1987) Identification of dysarthria types based on perceptual analysis. *Journal of Communication Disorders*, **20**, 367–378.

Index

Accessory nerve, *see* Cranial
    nerves
Acoustic neuroma, *see*
    Neurofibroma
Acquired childhood aphasia
    282–304
    aetiology 296–304
    clinical features 283–91
    convulsive disorder 301–4
    definition 282
    head trauma 298–9
    infections 301
    recovery 291–6
    tumours 299–301
    vascular disorders 297–8
Agnosia 201–3
    auditory 202–3
    definition 201
    finger 203
    special forms 203
    tactile 203
    visual 201–2
Agraphia 195, 197–8, 200–1
    aphasic 200–1
    apraxic 200
    constructional 200
    definition 195
    pure 200
    with alexia 197–8
Akinesia, *see* Hypokinesia
Alexia 194–200
    agnosic, *see* Alexia, without
        agraphia
    deep 199–200
    frontal 198–9
    occipital, *see* Alexia, without
        agraphia

parietal, *see* Alexia, with
    agraphia
phonemic, *see* Alexia, deep
pure, *see* Alexia, without
    agraphia
with agraphia 197–8
without agraphia 195–7
Alzheimer's disease, *see* Dement
Amygdaloid nucleus 98–9
Amyotrophic lateral sclerosis
    275–7
Aneurysm 41, 53, 213, 215
Anton's disease 203
Aphasia 49–50, 60–95, 97–119,
    180–3
    acoustic, *see* Aphasia, sensory
    acoustic-mnestic 89
    afferent motor 88
    anomic 84–5
    apraxic, *see* Aphasia, afferent
        motor
    auditory, *see* Pure word
        deafness
    Broca's 71–3
    central 70
    classification 67–71
        Bostonian system 69–71,
            71–85
        Lurian system 71, 85–90
    conduction 76–8
    definition 49–50
    efferent motor 86
    expressive, *see* Aphasia, motor
    fluent 68
    frontal 87
    global 78–9
    history 60–7

isolation 82–4
lesion localization
   methods 90–6
mixed transcortical, see
   Aphasia, isolation
motor 67–8
non-fluent 68
pericentral 70
premotor 87
receptive, see Aphasia, sensory
relationship to language of
   dementia 180–3
semantic 89–90
sensory 67–8, 88–9
subcortical 85, 97–119
   striato-capsular 110–13
   thalamic 106–10
transcortical motor 80–1
transcortical sensory 81–2
Wernicke's 73–6
phasic agraphia, see Agraphia
phemia 62
praxia 49–50, 184–94
bucco-facial, see Apraxia, oral
constructional 189–90
definition 184
dressing 190
ideational 188–9
ideomotor 185–8
kinetic, see Apraxia,
   limb-kinetic
limb-kinetic 189
oral 185
verbal, see Apraxia of speech
praxia of speech 49–50, 191–4
clinical characteristics 191–3
definition 49–50, 191
relationship to aphasia 193–4
relationship to dysarthria 194
praxic agraphia, see Agraphia
queduct of Sylvius, see Cerebral
   aqueduct
rachnoid, see Meninges

Archicerebellum 257
Arteries 40–4, 105–6
anterior cerebral 40–2
anterior choroidal 40–1, 105–6
anterior communicating 40–1
anterior inferior cerebellar 43
anterior spinal 43
basilar 40–4
internal carotid 40–2
lateral striate 105–6
medial striate 105–6
middle cerebral 40–4, 105–6
ophthalmic 40
posterior cerebral 40–4, 105–6
posterior choroidal 106
posterior communicating 40–1
posterior inferior
   cerebellar 40, 43
superior cerebellar 40, 43
thalamogeniculate 106
thalamoperforating 106
vertebral 40–3
Ataxic dysarthria, see Dysarthria
Athetosis 250–1
Auditory agnosia, see Agnosia
Autonomic nervous system 32–4

Ballism 249
Basal ganglia 11, 98–101
Binswanger's disease 165, 170
Blood–brain barrier 45
Blood supply to brain 40–5
arterial supply 40–4
   see also Arteries
venous supply 44–5
   see also Veins
Brachium conjunctivum 258
Brachium pontis 257–8
Brain 7–27
Brainstem 18–26
Broca's aphasia, see Aphasia
Broca's area 16, 46–7
Bulbar palsy 215

Caudate nucleus 98–9, 101
Central lobe, see Insula
Central nervous system 2–29
Centrum semiovale 227
Cerebellar nuclei, see Deep nuclei
Cerebellar peduncles 257–8
Cerebellum 26–7, 255–68
 afferent nerve supply 259, 260
 damage 266–8
 diseases 268, 270–1
 efferent nerve supply 259, 261,
  262
 neuroanatomy 255–62
 role in voluntary motor
  activities 263–6
Cerebral angiography 90, 91
Cerebral aqueduct 21, 35, 36
Cerebral cortex 9–11
Cerebral embolism, see
  Cerebrovascular disorders
Cerebral peduncles 22
Cerebral thrombosis, see
  Cerebrovascular disorders
Cerebrospinal fluid 38–40
Cerebrovascular disorders 51–4
 cerebral embolism 51–2
 cerebral thrombosis 51–2
 haemorrhagic stroke 51, 53–4
 intracerebral haemorrhage 51
 intracranial aneurysms 53
 ischaemic stroke 51–2
 transient ischaemic attacks
  52–3
Cerebrum 8–18
Chorda tympani 211
Chorea 246–9
 Huntington's 247–8
 Sydenham's 248
Circle of Willis 41
Claustrum 98, 99, 100
Commissurotomy 153–6
Computed tomography 91–2
Concussion 123

Conduction aphasia, see Aphasia
Constructional agraphia, see
  Agraphia
Constructional apraxia, see
  Apraxia
Contra-coup effect 123
Contusions 123–4
Corpora quadrigemina 22
Corpus callosum 12
Corpus striatum 100
Cranial nerves 30, 31, 128
 abducens 30, 209
 accessory 30, 214
 facial 30, 211–12
 glossopharyngeal 30, 212
 hypoglossal 30, 214–15
 occulomotor 30, 209
 olfactory 30, 209
 optic 30, 209
 trigeminal 30, 209–10
  mandibular branch 210
  maxillary branch 210
  ophthalmic branch 210
 trochlear 30, 209
 vagus 30, 212–14
 vestibulo-cochlear 30, 209

Deep alexia, see Alexia
Deep nuclei 257
Degenerative disorders 56–7
Dementia 163–83
 Alzheimer's disease 165,
  166–9, 173–9
  clinical characteristics
   166–9
  language disorder 173–9
 cortical 165, 166–70, 173–80
 Korsakoff's syndrome 172–3
  clinical characteristics
   172–3
 multi-infarct 165, 170–1
  clinical characteristics
   170–1

Pick's disease 165, 169–70,
    179–80
  clinical characteristics
    169–70
  language disorder 179–80
  subcortical (extrapyramidal)
    165, 171–2
  clinical characteristics
    171–2
  types 164–6
Demyelinating disorders 57–8,
    277–80
Dentate nucleus 257
  see also Deep nuclei
Dichotic listening test 144,
    159–60
Diencephalon 18–21
Disseminated sclerosis, see
    Multiple sclerosis
Dressing apraxia, see Apraxia
Dura mater, see Meninges
Dynamic aphasia, see Aphasia
Dysarthria 49, 205–81
  ataxic 255–74
  clinical characteristics
    268–74
  diseases 268, 270–1
  defined 49, 205
  flaccid 207–24
  facial nerve lesions 222–3
  hypoglossal nerve lesions
    222–3
  multiple cranial nerve lesion
    223–4
  phrenic and intercostal
    nerve lesions 219–20
  trigeminal nerve lesions
    222–3
  vagus nerve lesions 220–2
  hyperkinetic 244–54
  quick 245–9
  slow 249–53
  hypokinetic 235–44

  clinical characteristics
    238–44
  neurological disorders
    235–8
  lower motor neurone, see
    Dysarthria, flaccid
  mixed 274–81
  spastic 224–33
  clinical characteristics
    230–8
  neurological disorders
    229–30
  upper motor neurone, see
    Dysarthria, spastic
Dysdiadochokinesia 267
Dyskinesia 251–2
  levo-dopa-induced 251–2
  lingual-buccal-facial 251–2
  tardive 251–2
Dyslexia, see Alexia
Dysmetria 265, 267
Dysphasia, see Aphasia
Dyssynergia 267
Dystonia 252–3

Electroencephalography 90, 159,
    161, 302
Emboliform nucleus 257
  see also Deep nuclei
Encephalitis 58–9, 301
  bacterial 58–9
  viral 58, 301
Ependymoma, see Neoplasm
Essential tremor 254
Extradural haematoma 125–6,
    127
Extrapyramidal system 225–6

Facial nerve, see Cranial nerves
Falx cerebelli 36
Falx cerebri 36
Fastigial nucleus 257
  see also Deep nuclei

Flaccid dysarthria, *see* Dysarthria
Flocculonodular lobe 256
    see also Cerebellum
Foramen of Luschka 36
Foramen of Magendie 36
Foramen of Monro 34
Freidrich's ataxia 271
Frontal alexia, *see* Alexia
Frontal eye field 15, 16
Frontal lobe 13, 14–16

Gerstmann syndrome 203–4
Gilles de la Tourette's syndrome
    246
Glioma, *see* Neoplasm
    astrocytoma 55, 300
    microglioma 55
    oligodendrocytoma 55
Global aphasia, *see* Aphasia
Globose nucleus 257
    see also Deep nuclei
Globus pallidus 98–9
Glossopharyngeal nerve, *see*
    Cranial nerves

Haemorrhagic stroke, *see*
    Cerebrovascular disorders
Head injury 56, 120–41, 298–9
    complications 122–9
    language disturbance 130–41
    mechanisms of injury 129–30
    neurological status 135–6
    neuropsychological sequelae
        141
    prognostic indicators 136–7
    recovery mechanisms
        138–41
    speech disturbance 137–8
    types 120–2
        closed 120–2
        open 120–2
Hemiballism, *see* Ballism
Hemispherectomy 152–3

Hepato-lenticular degeneration,
    see Wilson's disease
Huntington's chorea, *see* Chorea
Hydrocephalus 39
Hyperkinesia 234, 244
Hyperkinetic dysarthria, *see*
    Dysarthria
Hypoglossal nerve, *see* Cranial
    nerves
Hypokinesia 234
Hypokinetic dysarthria, *see*
    Dysarthria
Hypothalamus 18, 20–1

Ideational apraxia, *see* Apraxia
Ideomotor apraxia, *see* Apraxia
Infectious disorders 58–9
Insula 12, 13
Internal capsule 100–2
Intracarotid amobarbital test 144,
    159, 160
Intracerebral abscess 59
Intracerebral haemorrhage, *see*
    Cerebrovascular disorders
Intracranial tumours, *see*
    Neoplasm
Ischaemic stroke, *see*
    Cerebrovascular disorders

Jakob-Creutzfeldt disease 164,
    165

Kinetic aphasia, *see* Aphasia,
    efferent motor
Korsakoff's syndrome, *see*
    Dementia

Laceration 124
Landau-Kleffner syndrome, *see*
    Acquired childhood
    aphasia, convulsive
    disorder
Language lateralization 142–3

Language of dementia,
relationship to aphasia
180–3
Lateral medullary syndrome 217,
220–1
Lenticular nucleus 98, 99–100
Lentiform nucleus, see Lenticular
nucleus
Leptomeninges, see Meninges
Levo-dopa-induced dyskinesia,
see Dyskinesia
Limbic lobe 12, 13–14, 18
Limb-kinetic apraxia, see Apraxia
Lingual-buccal-facial dyskinesia,
see Dyskinesia
Lower motor neurone lesions
215–19, 225

Magnetic resonance imaging
93–5
Medial medullary syndrome 217,
221, 223
Medulla oblongata 24–6
Medulloblastoma, see Neoplasm
Meninges 36–8
arachnoid 37
dura mater 36–7
leptomeninges 38
pia mater 37–8
Meningioma, see Neoplasm
Meningitis 58
bacterial 58
pyogenic 58
tuberculous 58
viral 58
Mid-brain 21–3
Mixed dysarthria, see Dysarthria
Mobius syndrome 212, 218, 223
Models of language 60–7
Motor end plate 6
Motor homunculus 14, 15
Multi-infarct dementia, see
Dementia

Multiple sclerosis 277–80
Muscular dystrophy 219
Myasthenia gravis 224
Myoclonus 245–6

Neocerebellum 256
Neoplasm 54–6, 299–301
extra-cerebral tumours 54–5
adenomas 55
meningiomas 55
neurofibromas 55
osteomas 55
intra-cerebral tumours 54–5
ependymomas 55, 300
gliomas 55
medulloblastomas 55, 300
Nerves 29–32, 208–15
cranial, see Cranial nerves
pharyngeal 213
recurrent laryngeal 213
spinal 31–2
superior laryngeal 213
Nerve supply to speech
mechanism 208–15
Nervus intermedius 211
Neuroeffector junction 5–6
Neurofibroma, see Neoplasm
Neurone 2, 4–5
Nucleus ambiguus 25, 212, 214
Nucleus basalis of Meynert 168

Occipital alexia, see Alexia,
without agraphia
Occipital lobe 12, 13, 18
Organic voice tremor, see
Essential tremor
Otorrhea 128

Parasympathetic nervous
system 33
Parietal alexia, see Alexia, with
agraphia
Parietal lobe 12, 13, 16–17

Parkinson's disease 235-7
  see also Hypokinetic dysarthria
Peripheral nervous system 29-34
Pharyngeal nerve, see Nerves
Phonemic alexia, see Alexia, deep
Pia mater, see Meninges
Pick's disease, see Dementia
Pons 23-4
Positron emission tomography
  95-6
Post-traumatic epilepsy 128-9
Post-traumatic vertigo 128-9
Pre-motor area 15
Primary motor area 14-15
Progressive supra-nuclear palsy
  237-8
Prosopagnosia 202
Pseudobulbar palsy 229-30
Pure agraphia, see Agraphia
Pure alexia, see Alexia, without
  agraphia
Pure word deafness 76
Putamen 98, 99
Pyramidal system 225-9
  cortico-bulbar tracts 227-9
  cortico-mesencephalic tracts
  227
  cortico-spinal tracts 226-7
Pyramids 25, 227

Radioisotope brain scanning 90
Rebound phenomenon 267, 268
Recurrent laryngeal nerve, see
  Nerves
Red nucleus 22, 23
Regional cerebral blood flow 90,
  160-1
Restiform body 258
  see also Deep nuclei
Reticular formation 26
Rhinorrhea 128
Right cerebral
  hemisphere 142-62

  anatomical differences to left
    hemisphere 156-7
  lesions 144-52
    extra-linguistic deficits
    148-52
    linguistic deficits 145-8
    neuropsychological sequelae
    157-8
    role in recovery from aphasia
    158-62

Scanning speech 269, 279
Secondary motor area 15, 16
Secondary speech area 16
Sensory homunculus 16-17
Skull fractures 122-3
Spastic dysarthria, see Dysarthria
Speech and language centres
  45-8
Spinal cord 27-9
Spinal nerves, see Nerves
Striato-capsular region 98-102,
  105-6, 113-19
  blood supply 105-6
  language role 113-19
  neuroanatomy 98-102
Subarachnoid haemorrhage 127
Subcortical dementia, see
  Dementia
Subcortical structures 97-106,
  113-19
  language role 113-19
  neuroanatomy 97-106
Subdural haematoma 126, 127
Substantia nigra 22, 236
Superior laryngeal nerve, see
  Nerves
Supplementary motor area, see
  Secondary speech area
Supranuclear bulbar palsy 229
Sydenham's chorea, see Chorea
Sympathetic nervous system
  33-4

Synapse 5

Tachistoscopy 144, 156
Tactile agnosia, *see* Agnosia
Tardive dyskinesia, *see*
    Dyskinesia
Temporal lobe 12, 17–18
Tentorial hiatus 125
Tentorium cerebelli 36
Thalamus 18–20, 102–10, 113–19
    blood supply 105–6
    language role 113–19
    neuroanatomy 102–4
Tics 246
Tonsilar herniation 125
Toxic disorders 57
Transcortical motor aphasia, *see*
    Aphasia
Transcortical sensory aphasia, *see*
    Aphasia
Transient ischaemic attacks, *see*
    Cerebrovascular disorders
Traumatic head injury, *see* Head
    injury
Trigeminal nerve, *see* Cranial
    nerves
Tumours, *see* Neoplasm

Upper motor neurone lesions
    224–30
    neurological disorders 229–30

Vagus nerve, *see* Cranial nerves

Veins 44–5
    cerebral 44–5
    emissary 45
    jugular 45
    vein of Galen 45
Venous sinuses 44–5
    cavernous 44
    occipital 44
    petrosal 44
        inferior 44
        superior 44
    sagittal 44
        inferior 44
        superior 44
    straight 44–5
    transverse 44–5
Ventricular system 34–6
    fourth ventricle 35, 36
    lateral ventricles 34, 36
    third ventricle 34–6
Verbal apraxia, *see* Apraxia of
    speech
Visual agnosia, *see* Agnosia

Wada test, *see* Intracarotid
    amobarbital test
Wernicke–Korsakoff syndrome
    165, 172–3
Wernicke–Lichtheim model 63,
    64
Wernicke's aphasia, *see* Aphasia
Wernicke's area 47
Wilson's disease 280–1